Rock Mechanics

Felsmechanik

Mécanique des Roches

Supplementum 7

Geologische Vorerkundung. Tunnelbau – Bergbau – Gebirgssicherung – Kraftwerksbau

Vorträge des 26. Geomechanik-Kolloquiums
der Österreichischen Gesellschaft für Geomechanik

Geological Reconnaissance. Tunneling – Mining – Rock Support – Power Plant Construction

Contributions to the 26th Geomechanical Colloquium
of the Austrian Society for Geomechanics

Salzburg, 13. und 14. Oktober 1977

Herausgegeben für / Edited for
Österreichische Gesellschaft für Geomechanik
von / by L. Müller, Salzburg

1978 Springer-Verlag Wien New York

Mit 192 Abbildungen

ISBN-13:978-3-211-81498-7 e-ISBN-13:978-3-7091-8529-2
DOI: 10.1007/978-3-7091-8529-2

Inhaltsverzeichnis — Index — Table des matières

Rock Mechanics, Suppl. 7, 1—2 (1978)

Rock Mechanics
Felsmechanik
Mécanique des Roches

Geleitwort

Die Tätigkeit der in der Praxis stehenden Ingenieure und Geologen wird u. a. von drei wesentlichen Aspekten bestimmt: Verantwortung, Wissen und Interesse an der Arbeit.

Was das erste betrifft, heißt es beispielsweise im Österreichischen Verwaltungsverfahrensgesetz in bezug auf die Sachverständigentätigkeit trocken: „Wer die Wissenschaft, die Kunst oder das Gewerbe öffentlich als Erwerb ausübt oder zu deren Ausübung öffentlich ermächtigt ist, hat einer Bestellung (von seiten der Behörde) als Sachverständiger Folge zu leisten. Er ist für seine Aussage verantwortlich". Verantwortlichkeit kann aber für den Betroffenen zu schwerer psychischer Belastung führen, wofür es leider genügend Beispiele gibt.

Zum zweiten: Dort, wo der Ingenieur bei der Beurteilung ihm gestellter Aufgaben anhand des zur Verfügung stehenden Unterlagenmaterials mit logischem Denken nicht mehr weiterkommt, kann er Zusammenhänge nicht mehr beweisen, sondern solche nur intuitiv erfassen. Er kann dabei unsicher werden, wenn ihm nicht sehr große Erfahrung zur Verfügung steht.

Schließlich: Das Streben, Zusammenhänge physikalischer Kräfte und deren Auswirkungen zu ergründen, ist seit je die Triebfeder des Fortschrittes der Naturwissenschaften, in zunehmendem Maße auch der Geowissenschaften.

Auch die Tätigkeit des Felsmechanikers wird vom Spannungsfeld der genannten Aspekte bestimmt. Es wird von ihm gefordert, felsmechanische Prognosen zu erstellen, wobei es oft unmöglich ist, die tatsächlichen geomechanischen Verhältnisse quantitativ festzustellen. Dieses Dilemma haben vor etwa zweieinhalb Jahrzehnten die Gründer des Salzburger Kreises erkannt; damals, als sich nur wenige Enthusiasten in Dr. Müllers Wohnung versammelten. An der Unmöglichkeit der genauen Abgrenzung des Fachgebietes Felsmechanik, aber auch an der Vielschichtigkeit der Materie lag es, daß dieser Kreis vorerst um Anerkennung in Fachkreisen ringen mußte.

Welche Schlußfolgerungen sind nun aus der Tatsache zu ziehen, daß während der vergangenen Jahre die Teilnehmerzahl an den jährlichen Salzburger Kolloquien von anfänglich etwa 20 auf das ungefähr Vierzigfache angestiegen ist? Liegt das allein an der Zunahme des bauwirtschaftlichen Potentials oder sind seither die felsmechanischen Probleme größer und schwieriger geworden?

In der Einführung zum XXVI. Kolloquium, im Herbst 1977, stellte Professor Müller fest und bedauerte, daß heute der informierte, in der Praxis stehende, von theoretischem Wissen getragene Felsmechaniker zugeben muß, daß Ingenieurgeologie, Felsmechanik, Fels- und speziell das Teilgebiet Tunnelbau in gefährlicher Weise auseinanderklaffen und sich sogar auseinanderentwickeln. Dabei bezog er sich auf konkrete Fälle — auf Mißerfolge, die dadurch entstanden waren, daß ingenieurgeologische Daten falsch oder gar nicht interpretiert worden waren, oder weil neueste Ergebnisse der Felsmechanik außer acht gelassen worden waren.

Hier Abhilfe zu versuchen, ist der Grund der Veranstaltungen der Österreichischen Gesellschaft für Geomechanik.

Zum XXVI. Kolloquium vereinigten sich nun Fachleute, um über das Ausmaß und die Verwertung von Erkundungen im Gelände und vor Ort, über die Ausführung und Wirkung von Verankerungen — jeweils unter Berücksichtigung der Wechselwirkung zwischen Ausbaumaßnahmen und Gebirge —, um Probleme der möglichen Prognose, der unmittelbaren Anpassung des Bauvorganges an die Gebirgsverhältnisse sowie Fragen der Kostenschätzung im Tunnel- und Bergbau zu diskutieren.

Die Zeit für die Diskussionen, die sich an die einzelnen Vortragsgruppen anschlossen, war knapp; die Diskussionen müssen daher weitergeführt werden. Deshalb darf auch in diesem Rahmen darauf hingewiesen werden, daß beim XXVII. Kolloquium im Herbst 1978 vor allem Fortschritte in der theoretischen Behandlung und praktischen Anwendung der Neuen Österreichischen Tunnelbauweise, baugeologische Auswertungen von Erkundungsstollen und Stabilitätsprobleme hoher Felsböschungen behandelt werden sollen.

Das XXVI. Kolloquium ist ein Steinchen in der Entwicklung der Felsmechanik. Die Bewegung dieses Steinchens ist bereits eine Folge jenes zunächst unscheinbaren Steinchens, das seinerzeit die idealistischen Kräfte Prof. Müllers in Bewegung setzten und das in Salzburg unter Mitnahme weiterer Steine und Gesteinsblöcke von Jahr zu Jahr zur Freude der Theoretiker und Praktiker, der Ingenieure, Geologen und Bauwirtschafter und damit letztlich zum Wohle aller weiterrollt.

A. Niel

Rock Mechanics, Suppl. 7, 3—11 (1978)

Rock Mechanics
Felsmechanik
Mécanique des Roches

Geologische Vorerkundung und baugeologische Erfahrungen — ein kritischer Vergleich am Beispiel des Mitterbergtunnels

Von

W. Fürlinger

Mit 5 Abbildungen

Zusammenfassung — Summary

Geologische Vorerkundung und baugeologische Erfahrungen — ein kritischer Vergleich am Beispiel des Mitterbergtunnels. Beim Mitterbergtunnel haben gewisse quantitative und qualitative Abweichungen der tatsächlich angetroffenen Gebirgsverhältnisse von den in der Ausschreibung prognostizierten zu Schwierigkeiten geführt.

Im folgenden wird versucht, aus baugeologischer Sicht einige dieser Schwierigkeiten zu beleuchten und daraus Anregungen für Verbesserungen in künftigen Vorerkundungen und Ausschreibungen abzuleiten.

Geological Reconnaissance and Engineering-Geological Experiences — a Critical Comparison With Regard to the Mitterberg Tunnel. The highway from Graz to Klagenfurt (Südautobahn) requires three tunnels to pass through the Pack mountain (province of Styria), the middle of which is the Mitterbergtunnel. The complex consists of two separate slightly curved tubes, each 1100 m in length and with a distance of 35 to 40 m separating them. The maximum length of overburden is 185 m. The area of the tunnels cross section ranges between 65 and 76 sqm. The tunnel lies in flat to moderately steep South-dipping micaschists and gneisses of the Koralm crystalline complex.

Exploratory studies had been made by the contractor (Amt der Steiermärkischen Landesregierung) while planning the tunnels. Armed with a foreknowledge of regional geology seven core holes were drilled into the potential tunnel site followed by the excavation of an exploratory gallery to confirm the rock qualities which would be encountered in the final construction. The results of this exploration were used to establish the geological criteria of the work order.

After the first 200 meters of construction it developed that excavation methods were not meeting the geological requirements that were prior determined. This discrepancy did not allow for the desired smooth profile and further led to a partial collapse of the structure.

Consequently the construction consortium had made a restudy of the engineering geological conditions, along with a more detailed investigation carried out on the basis of a tunnel documentation. The continuous survey of the tunnel face revealed a degree of weathering and mechanical fracturing of the rock mass which exceeded exspectations. In addition to this the presence of expandable clay minerals in mylonites and joint fillings turned out to be of negative influence on the stability of the rock mass, making it hard to obtain the desired smoothness of the tunnel walls.

Der Mitterbergtunnel ist das mittlere von drei Tunnelbauwerken, mit denen die Südautobahn (Graz – Klagenfurt) auf steirischer Seite den Rücken der Pack überquert. Die Fahrbahnen verlaufen in zwei leicht gekrümmten, je 1100 m langen Röhren im Abstand von 35—40 m. Die maximale Überlagerung liegt bei 185 m. Der Ausbruchsquerschnitt beträgt zwischen 65 und 76 m^2, je nach GGKL.

Der Tunnel liegt in flach bis mittelsteil gegen S fallenden Granatglimmerschiefern und Augengneisen des Koralm-Kristallins.

Für Projektierung und Bau des Mitterbergtunnels wurden vom Bauherrn (Amt der Steierm. Landesregierung) geologische Vorerkundungen ausgeführt. Es handelt sich dabei um eine Verarbeitung regionalgeologischer Kenntnisse aus umfangreichen Kartierungen im Kristallingebiet der Koralpe. Außerdem wurden Aufschlußbohrungen im Bereich der Tunneltrasse gemacht und ein Sondierstollen, der im Zentrum der südlichen Röhre des MT verläuft, gebaut. Die Ergebnisse dieser Vorerkundung wurden in Form eines geologischen Gutachtens mit schematischen Planbeilagen und einer Beschreibung der Gebirgsgüteklassen der Ausschreibung zugrundegelegt.

Die Gebirgsgüteklassifizierung des Sondierstollens lieferte erste Grundlagen für die Prognose der Gebirgsverhältnisse beim Auffahren der beiden Röhren.

Eine einfache summarische Gegenüberstellung der im Vollausbruch angetroffenen GGKL zeigt gegenüber dem Sondierstollen insgesamt eine deutliche Verschiebung zum Schlechteren.

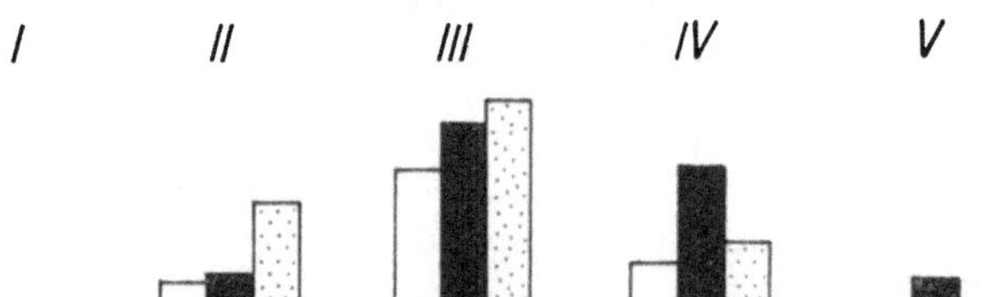

Verteilung der einzelnen Gebirgsgüteklassen — Prozentzahlen. Leeres Feld: GGKL lt. Sondierstollen; volles Feld: Tatsächlich angetroffene GGKL; punktiert: GGKL lt. Leistungsverzeichnis der Ausschreibung (Prognose)

Die Mengenanteile der Klassen I, II und III blieben merkbar hinter den in der Ausschreibung prognostizierten Werten zurück. Die Anteile der schlechten Klassen IV und V haben sich beträchtlich erhöht.

Diese relativ schlechte Bilanz ist wohl zum Teil rein aus der Tatsache des ungünstigeren Gebirgsverhaltens beim Übergang vom kleinen Sondierstollenquerschnitt auf den vollen Querschnitt der Tunnelröhre zu verstehen und war zu erwarten.

Nach den Erfahrungen von Prof. E. H. Weiss macht diese zu erwartende, oder „normale" Verschlechterung 1 bis 1 1/2 Klassen aus.

Es gibt aber auch zahlreiche Fälle, bei denen der Sprung größer als 1 Klasse ist. Die Ursachen der Verschlechterung des Gebirgsverhaltens sind dann nicht nur aus der Vergrößerung des Querschnittes bzw. der freien Stützweite zu erklären, sondern es wurden zusätzliche unvorhergesehene Einflüsse berücksichtigt.

In solchen Fällen sprechen wir von einer „effektiven" Verschlechterung der Gebirgsgüte.

Wenn man einen solchen Vergleich am MT anstellt, so sieht man, daß sich auf mehr als 1/3 der ausgebrochenen Tunnelstrecke das Gebirge um mindestens 2 Klassen, also „effektiv" schlechter erwiesen hat, als aufgrund der Voruntersuchungen angenommen wurde.

Eine tabellarische Bilanz zeigt auch die großen Unterschiede zwischen S- und N-Röhre bezogen auf den Sondierstollen. Während die effektive Verschlechterung bei der Südröhre (in deren Zentrum ja der Sondierstollen lag) im Mittel 26% beträgt, liegt dieser Wert für die etwa 40 m entfernte Nordröhre bei 45%.

Bilanz der GGKL-Relation zwischen dem Sondierstollen und den beiden Tunnelröhren

Südröhre	Ost	West
Gleichbleibend und besser	39,09%	24,94%
Verschlechterung um 1 Klasse	43,67%	39,16%
Verschlechterung um mehr als 1 Klasse	17,24%	35,90%
Nordröhre		
Gleichbleibend und besser	41,50%	14,40%
Verschlechterung um 1 Klasse	28,65%	25,66%
Verschlechterung um mehr als 1 Klasse	29,85%	59,94%

An den Anfang der Suche nach den Ursachen dieser effektiven Verschlechterung wollen wir zunächst eine Beschreibung des Gebirges stellen:

Das Gebirge, in dem der Mitterbergtunnel liegt, wird von zwei verschiedenen Gesteinsarten aufgebaut. Es sind dies einmal dunkelfarbige Granatglimmerschiefer, die eine gewisse lithologische Variationsbreite von Biotitschiefern bis zu sogenannten Disthenflasergneisen haben können. Zum anderen sind hell- bis mittelgraue augige bis linsig-lagige Gneise von pegmatoidem Charakter am Aufbau des Gebirges beteiligt.

Beide Gesteinsarten treten zu etwa gleichen Anteilen im Tunnel auf. Sie gehören der Serie der „venitisch durchtränkten Granatglimmerschiefer" (Homann, 1962; Beck-Mannagetta, 1975) des Kristallins der Koralpe an.

Das zunächst auffälligste Gefügemerkmal ist eine streng geregelte Schieferung, die dem Gebirge einen Habitus verleiht, der von grobbankig bis dünnschiefrig reicht. Alle denkbaren Kombinationen von hell/dunkel wurden im Laufe des Vortriebes beobachtet: Volle Querschnitte in ausschließlich dunklen Schiefern bzw. hellen Gneisen; vorwiegend heller Gneis mit dunklen

Bänken und umgekehrt. Verzahnungsbereiche der beiden Gesteinstypen boten die ästhetisch schönsten Aufschlußbilder und lassen Schlüsse auf die Verformungs- und Entstehungsgeschichte des Gesteins zu.

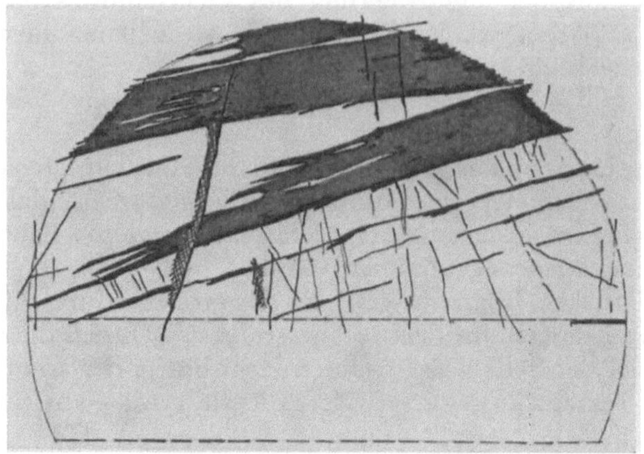

Abb. 1. Verzahnung heller pegmatoider Gneise mit dunklen Granatglimmerschiefern
Light pegmatite-type gneisses interfingering dark colored garnet-bearing mica-schists

Abb. 2. Verfaltung von hellen Gneisen und dunklen Granatglimmerschiefern. Störungen
Light colored gneisses folded together with dark colored mica-schists. Faults

Diese Bilder charakterisieren also das Gesteinsmaterial mit dem wir es beim Mitterbergtunnel zu tun haben: Eine Serie von wechselnd mächtigen, z. T. seitlich auskeilenden und ausdünnenden hellen und dunklen kristallinen Schiefern bzw. Gneisen, die im plastischen Zustand innig ineinander verwalzt, verschuppt und zu wilden Faltenstrukturen verformt wurden.

Eine jüngere, brechend verformende Gebirgsbildungsphase hat diesem Material dann jene Strukturen auf- bzw. eingeprägt, die die Eigenschaften des Gebirges im Sinne der Baugeologie und der Felsmechanik ausmachen. Mit ihren negativen Einflüssen auf das Festigkeitsverhalten sieht sich der Tunnelbauer von Abschlag zu Abschlag aufs Neue konfrontiert: Störungen, Zerrüttungszonen und Mylonite ...

Je nach Überwiegen der einen oder anderen Gesteinsart ändert sich auch das jeweils charakteristische mechanische Verhalten des Gebirges in einer Weise, daß sich aus der Fülle der Aufschlußbilder, die der Mitterbergtunnel bisher bot, einige strukturelle Grundtypen herausschälen lassen.

Die überwiegend grobkristallinen, augig-linsigen hellen pegmatoiden Gneise, in denen Quarz und Feldspat gegenüber schuppenförmigen Glimmermineralien vorherrschen, zeigen unter bruchtektonischer Beanspruchung sprödes Materialverhalten und neigen zur Ausbildung eng gescharter bankrechter Klüfte und Störungen.

Als erster Strukturtyp läßt sich also charakterisieren: Grobbankig, flaches bis mäßig steiles Fallen der Schieferungsflächen; steilstehende eng gescharte Klüftung, vereinzelte Harnischflächen an Klüften und Schieferungsfugen, z. T. mit dünnen mylonitischen Bestegen.

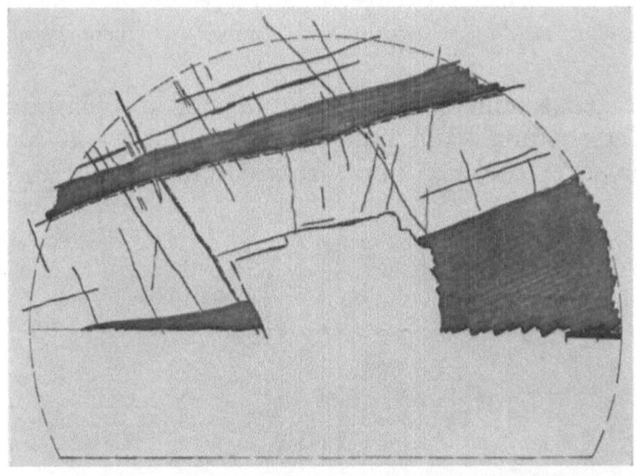

Abb. 3. Strukturtyp I im Mitterbergtunnel: bankige Ausbildung des Gebirges mit einzelnen Großklüften bzw. mylonitisch belegten Störungsflächen

Structure type I as found in the Mitterbergtunnel: Bedding schistosity cut by several major joints and fault planes, bearing slickensides and mylonite

Der zweite Typ ist gekennzeichnet durch ausgeprägte Bankung im dm-Bereich und mittelsteiles bis steiles Einfallen der *s*-Flächen. Zahlreiche Schieferungsflächen sind als Scherflächen betätigt worden, tragen mylonitisch belegte spiegelnde Harnische und wirken als bevorzugte Ablösungsflächen. Mehrere steilstehende Verwerfungen mit ausgeprägten Mylonitstreifen queren die Tunnelbrust und versetzen charakterische Bänke in der Art einer Horst-Graben-Tektonik.

Der dritte Typ ist fast ausschließlich auf die dünnschiefrigen dunklen Granatglimmerschiefer bzw. Biotitschiefer beschränkt. Blättchenförmige

Glimmermineralien prägen die Gesteinstextur. Er ist durch flache bis mäßig steile Schieferungsstellung gekennzeichnet, die durch ein ebenso flach liegendes zweischariges Netz von Störungsflächen überlagert wird. Fast alle

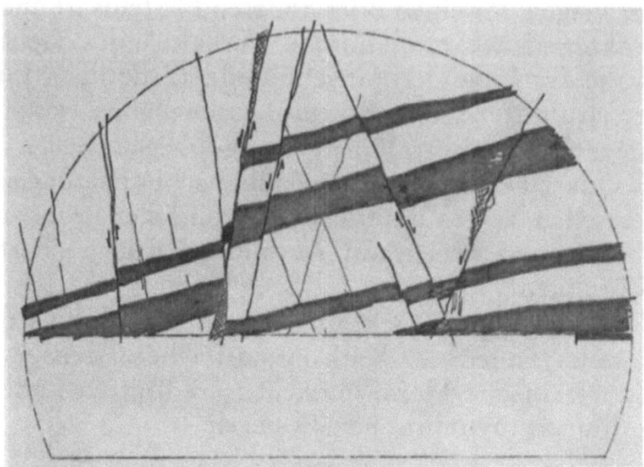

Abb. 4. Strukturtyp II: bankiges Gebirge, an Störungen deutliche Relativbewegungen
Rock mass showing bedding schistosity and notable movement along fault planes

Trennflächen tragen schmierige Bestege und sind als Harnischflächen ausgebildet. Die Zerscherung kann sich bis zur Bildung von Meter-mächtigen Bereichen tektonischer Brekzien und Grobmyloniten steigern.

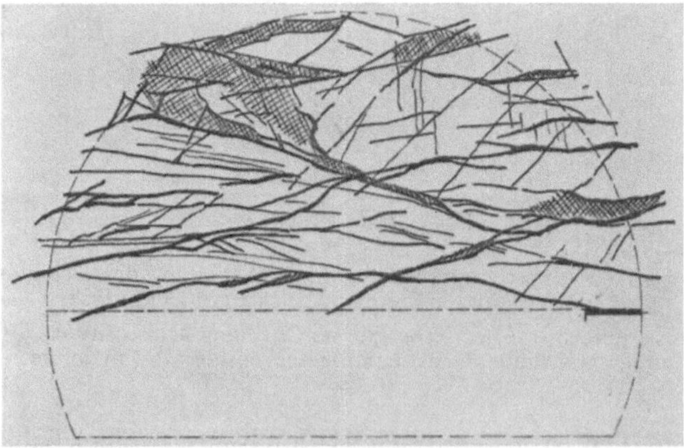

Abb. 5. Strukturtyp III: intensive Zerscherung des Gebirges, Bildung von Zerrüttungszonen und Grobmyloniten
Crushed zones and coarse grained mylonites formed by intense shearing of the rock mass

Diese drei oben charakterisierten strukturellen Grundtypen lassen sich den in der Ausschreibung bezeichneten GGKL III, IV und V zuordnen bzw. stellen eine unter Verarbeitung der im Vollausbruch gewonnenen Erfahrungen präziser gefaßte Charakterisierung dieser GGKL dar.

Die in der Ausschreibung zur Charakterisierung der GGKL herangezogenen Merkmale treffen im wesentlichen gut zu. Auch hat sich gezeigt, daß die nach den genannten Gesichtspunkten klassifizierten Strecken mit den dafür in der Ausschreibung vorgesehenen Stützmaßnahmen beherrschbar sind.

Die anfänglichen Schwierigkeiten rührten wohl davon her, daß die festgelegten geologischen Klassifizierungsrichtlinien nicht immer streng genug angewendet wurden. In ihrer Formulierung ließen sie einen gewissen Spielraum für subjektive Interpretationen zu und verleiteten so zu einer manchmal zu optimistischen Gebirgsklassifizierung.

Besonders die gebirgsentfestigenden Einflüsse einer überaus intensiven und tiefgreifenden „Verwitterung" wurden erst nach und nach in ihrer vollen Bedeutung erkannt und berücksichtigt.

Dazu kommt die Entdeckung und der Nachweis von quell- und schmierfähigen Tonmineralien wie Montmorillonit und Talk in Kluftbestegen und Myloniten des Mitterbergtunnels mithilfe von Röntgendiffraktometeraufnahmen. Die Anwesenheit solcher Mineralien, die als Umwandlungs- und Zersetzungsprodukte des mechanisch zerrütteten Ausgangsgesteins zu verstehen sind, ist von zahlreichen Beispielen aus der Literatur (z. B. Brekke u. a., 1965) als äußerst ungünstig für das Festigkeitsverhalten des Gebirges bekannt und hat sicherlich beim Verbruch in der Nordröhre eine gewisse Rolle gespielt.

Die Erkenntnis dieser zusätzlichen Faktoren zieht zwangsläufig eine schlechtere Klassifizierung nach sich und drückt sich dadurch auch im Verhältnis der prognostizierten und der tatsächlich angetroffenen GGKL aus.

Eine weitere Verschiebung dieses Verhältnisses hängt mit der Ungleichheit der Aufschlußgrößen Sondierstollen – Vollquerschnitt zusammen.

Häufig haben Grobmylonitbänke in der Kalotte die Gebirgsgüte im Vollquerschnitt bestimmt. Bei flacher Lagerung, wie sie im Mitterbergtunnel oft auftritt, konnten sie im Sondierstollen nicht erkannt werden und so auf weite Strecken ein unzutreffendes Bild vortäuschen.

Störungen, die im Sondierstollen nur als glattwandige Klüfte kenntlich waren, entwickeln sich in ihrem Fortstreichen im Vollquerschnitt zu mächtigen Zerbrechungszonen mit Myloniten.

Die Prognose vom Sondierstollen bzw. einer vorauseilend voll ausgebrochenen Röhre auf die nur 40 m entfernte benachbarte Röhre krankt am raschen Auskeilen der Gesteinskörper, an deren ungünstiger Stellung und an der Absätzigkeit selbst bestausgeprägter struktureller Elemente.

Wie die Plandarstellung der W-Hälfte des MT deutlich zeigt, sind z. B. die auffälligen steilstehenden Verwerfungen, die die Nordröhre auf mehrere Zehner-Meter begleiten, in der Südröhre nicht wiederzufinden.

So viel zum Versuch einer Erklärung der quantitativen Unterschiede in den Mengenansätzen der prognostizierten GGKL und den nun vorliegenden tatsächlichen Werten.

Außerdem ergaben sich gewisse Diskrepanzen zwischen den in der Ausschreibung für jede einzelne Klasse festgelegten Gebirgslösungsvorschriften und den tatsächlichen Möglichkeiten der Bearbeitbarkeit des Gebirges, besonders in den Klassen III, IV und V.

Bei GGKL III, für die Vollausbruch bei Abschlagtiefen von maximal 1,5 m vorgesehen ist, heißt es zum Beispiel: „Die Randschüsse müssen 0,2—0,3 m innerhalb der theoretischen Hohlraumgrenze gesetzt werden. Die Profilierung darf nur mit dem Abbauhammer erfolgen."

Ein Versuch, dieser Empfehlung entsprechend zu arbeiten, führte zu Unterprofilen, die mühsam durch umfangreiche Sprengarbeiten nachprofiliert werden mußten. Der Einsatz von Abbauhämmern wäre äußerst unzweckmäßig gewesen.

Bei GGKL IV ist das Lösen des Gesteins laut Ausschreibung „vornehmlich mittels Schrämmen zu bewerkstelligen".

Die tatsächliche Beschaffenheit des Gebirges macht eine derartige Vorgangsweise jedoch gänzlich unmöglich. Es mußte durchwegs gesprengt werden, allerdings bei entsprechend verkürzten Abschlagslängen.

In der Ausschreibung steht bei GGKL V: „Der Ausbruch erfordert Getriebezimmerung unter Verwendung von Tunnelbogen und Stahlstollendielen ..."

Wenn man einige charakteristische Aufschlußbilder der GGKL V, wie sie im Mitterbergtunnel vorkommt, betrachtet, so sieht man die Undurchführbarkeit einer solchen Abbauweise ein. Die Wechselhaftigkeit der Gebirgsfestigkeit in ein und demselben Querschnitt, wo oft m-mächtige Grobmylonithorizonte mit kompakten Bänken wechseln, erfordert Sprengvortrieb, obwohl teilweise mit dem Ladegerät abgebaut werden konnte.

In Anbetracht dessen, daß derartige Forderungen einer Kalkulation zugrundegelegt werden können, ja müssen, und daß sich daran gewisse Erwartungen bezüglich der Qualität des erreichbaren Profils knüpfen, stellt sich die Frage, ob es für den Ausschreibenden überhaupt zweckmäßig ist, derartige Vorschriften zu präzisieren. Das Beispiel MT zeigt, daß eine Verquickung von geologischen und bautechnischen Kriterien in der Gebirgsklassifizierung zu solchen Unstimmigkeiten führen kann.

Die daraus resultierenden technischen und wirtschaftlichen Probleme wiegen umso schwerer, als sie das Grundkonzept der Kalkulation betreffen.

Es sollte deshalb vielmehr versucht werden, durch Erfassen und Darstellen aller geotechnisch wesentlichen Details ein so komplettes Bild einer jeden erwarteten GGKL zu entwerfen, daß der Bieter selbst, in Zusammenarbeit mit einem baugeologischen Berater ein optimales Konzept zu deren Bewältigung in Ausbruch und Stützung ausarbeiten kann.

Eine solche Darstellung erfordert eine möglichst genaue Kenntnis des jeweils vorliegenden Gebirges und macht genaue Definitionen der einzelnen GGKL auf baugeologischer Basis notwendig — kurz, eine gebirgsgerechte Klassifizierung. Eine solche finden wir beispielhaft bei E. H. Weiss (1976).

Es mag sich in Anbetracht einer solch schlechten Bilanz von Prognose und Tatsache die Frage aufdrängen, wie sinnvoll der Bau von Sondierstollen überhaupt ist. Vom Standpunkt der Aufschließung des Gebirges ist er zu befürworten. Das Problem liegt im richtigen Umlegen der Erfahrungen im kleinen Querschnitt auf den großen. Hier könnte einiges getan werden, um die Informationslücke zwischen dem Verhalten des Gebirges im kleinen und im großen Querschnitt zu verringern.

Es wäre z. B. durchaus sinnvoll, in der Art der bei U-Bahn-Bauten üblichen Probevortriebe jede GGKL auf bestimmte Länge an einer charakteristischen Stelle im Vollausbruch zu öffnen und mit allen zu Gebote stehenden geotechnischen Messungen zu beobachten. Auch beliebig dicht anzuordnende und auf die geologischen Verhältnisse abgestimmte radiale Kernbohrungen vom Sondierstollen aus können dazu beitragen, die quantitativen Gebirgsverhältnisse im voraus zu klären und bieten die Möglichkeit für geotechnische Messungen über den Zeitraum, der zwischen Bau des Sondierstollens und der Tunnelröhre verstreicht.

Eine derartige Erweiterung der Praxis der Vorerkundung erscheint geeignet, die Kalkulation sowohl des Bauherrn als auch des Bieters auf eine bessere Basis zu stellen.

Literatur

Beck-Mannagetta, P.: Grundlagen für die wasserversorgungswirtschaftlichen Planungen in der Südweststeiermark. 2. Teil, Ber. d. Wasserwirtschaftlichen Rahmenplanung. Amt d. Steiermärkischen Landesregierung, Graz, 1975.

Brekke, T. L., Selmer-Olsen, R.: Stability Problems in Underground Constructions Caused by Montmorillonite-carrying Joints and Faults. Eng. Geol. *1* (1), 3—19 (1965).

Homann, O.: Das kristalline Gebirge im Raume Pack-Ligist. Joanneum, Mineralog. Mitteilungsblatt, H. 2, 21—62 (1962).

Weiss, E. H.: Die baugeologische Prognose für den Schnellstraßentunnel durch den Arlberg, Tirol – Vorarlberg. Rock Mechanics, Suppl. 5, 133—156. Wien – New York: Springer 1976.

Anschrift des Verfassers: Dr. Werner Fürlinger, Spitalskystraße 10, A-4400 Steyr, Österreich.

Rock Mechanics, Suppl. 7, 13—26 (1978)

Rock Mechanics
Felsmechanik
Mécanique des Roches

The Engineering-Geologic Expectation Model as a Basis for the Prognosis in Underground Construction

By

H. K. Helfrich

With 2 Figures

Summary — Zusammenfassung

The Engineering-Geologic Expectation Model as a Basis for the Prognosis in Underground Construction. An engineering-geologic expectation model develops in the following manner: The first presumption is the compilation of a specific geotechnical questionnaire for the given project (Planning model).

Thereafter, the project-adapted engineering-geologic conditions are to be worked out, where the knowledge of the regional geologic environment has to be taken into consideration, in order to be able to expect on site unexposed repeated properties and structures of geotechnical significance and to produce an initial prognosis (Geologic-tectonic expectation model).

After the conception of this model and the defining of the existing geologic (homogeneous) units, the engineering-geologic exploration is principally outlined by type and extent of the utilized methods according to the chosen exploration strategy (Exploration model). This model contains the quality of the performance, the density of the survey, the range of the results obtained, related to the direct range of validity (measuring and testing), the indirect range of validity (sphere of reference) and the broader sphere of consideration.

The exploration ought to be carried out in the following steps: Reconnaissance and pattern recognition-reconstruction, detailed and finally supplementary investigation. For each step, the correctness of the exploration model has to be examined and — if necessary — to be changed.

The content of the geotechnical expectation model (prognosis) includes the material (rock and rock mass) by quality and quantity, its classification and percental zone distribution, the construction parameters, the type and extent of required temporary and permanent support. The degree of confidence is to be designated as certain, probable, possible or uncertain.

The prognosis level of the geotechnical model depends on the correctly defined geologic units, the extent of outcropping areas, the degree of coordination, the evaluation procedures and the strength of the statement in the information content.

Finally, any engineering-geologic expectation model should be followed-up, reevaluated and verified during construction time (Geotechnical follow-up model). The engineering geologist's work ends with the analysis and interpretation of the relationship between outfall versus prognosis.

Das ingenieurgeologische Erwartungsmodell als Prognoseunterlage für den Felsbau unter Tage. Ein ingenieurgeologisches Erwartungsmodell erwächst in folgender Weise: Voraussetzung ist die Erfassung der spezifischen gebirgstechnischen Fragestellung des gegebenen Projektes (Planungsmodell).

Danach sind die projektangepaßten ingenieurgeologischen Voraussetzungen zu erarbeiten, die regionalgeologischen Erkenntnisse zu berücksichtigen, um im Planungsbereich nicht aufgeschlossene unzufällige Eigenschaften und Strukturen gebirgstechnischer Bedeutung erwarten zu können sowie eine Initialprognose zu erstellen (Geologisch-tektonisches Erwartungsmodell).

Nach Aufstellung dieses Modelles und der erfolgten Abgrenzung vorhandener Homogenitätsbereiche wird die ingenieurgeologische Prospektion gemäß der gewählten Prospektionsstrategie nach Art und Umfang der einzusetzenden Methodik prinzipiell festgelegt (Prospektionsmodell), welches die Qualität der Ausführung, die Dichte des Untersuchungsnetzes, die Reichweite der erhaltenen Resultate, bezogen auf den unmittelbaren Gültigkeitsbereich (Meß- und Testbereich), den mittelbaren Gültigkeitsbereich (Bezugsbereich) und den weiteren Betrachtungsbereich, umfaßt.

Die Prospektion sollte in den Etappen: Vorbereitende Erkundung, detaillierte und schließlich komplettierende Untersuchung durchgeführt werden. Für jede Etappe ist die Richtigkeit des Prospektionsmodelles zu prüfen und allenfalls zu ändern.

Der Inhalt des gebirgstechnischen Erwartungsmodelles (Prognose) umfaßt das Material (Gestein und Gebirge) nach Qualität und Quantität, dessen Klassifikation und prozentuale Zonenverteilung, die Konstruktionsparameter, die Art und den Umfang eventueller Sicherungs- und Ausbauarbeiten. Die Prognose ist jeweils als sicher, wahrscheinlich, möglich oder unsicher zu bewerten.

Das Prognoseniveau des gebirgstechnischen Modelles hängt von der richtigen Abgrenzung der Homogenitätsbereiche, dem Aufschlußgrad, dem Deckungsgrad, den Auswerteverfahren und der Aussagekraft des Informationsinhaltes ab.

Schließlich sollte jedes ingenieurgeologische Erwartungsmodell während der Bauausführung nach Kenntnisstand erweitert und gegebenenfalls revidiert werden (Gebirgstechnisches Analysemodell). Der Einsatz des Ingenieurgeologen endet mit dem Abschluß der Analyse und der Interpretation des Verhältnisses Wirklichkeit kontra Prognose.

1. Introduction

Practice shows that the geotechnical prognosis and the real findings often differ to a great extent.

The Swedish Rock Mechanics Research Foundation has with the Research project "Value and extent of geotechnical exploration- and documentation-methodics for underground construction" made it to its task to analyse the cause and to work out guide lines supposed to reduce the mentioned contrast.

Four essential steps should lead to success:

— A systematization of the engineering-geologic working pattern,
— An investigation quantity dictated by geotechnical homogeneity units,
— A precise statement of the prognosis level, and
— A current documentation of the geotechnical condition and a confrontation of the prognosis and outfall after completion of the construction work.

Essential causes for contradictions between prognosis and outfall can be:

- — Insufficient knowledge of the geotechnical conditions,
- — Occurrence of random geologic structures,
- — Answering of geotechnical questions asked after completion of the investigation, for which the methodics was not prepared,
- — That an investigation method is required to have a higher strength of statement than it deserves methodically,
- — Utilization of an excavation technique which is not adapted to the rock mass conditions,
- — Supplementary changes in the construction plan.

Recently, the subject of pre-investigation in underground construction has strongly moved into the foreground.

In 1974 at the Dept. of Civil Engineering at the Mass. Inst. of Technology (MIT), a tunnel cost model was developed which represents a stochastic simulation model for the underground construction, by which construction costs and time of a tunnel can be estimated (Lindner, Einstein, Vanmarke, 1975).

The International Society for Rock Mechanics (ISRM) issued in 1975 "Recommendations on site investigation techniques" in a report which presents a compilation of all methods of investigation.

Finally, in November 1976 in Johannesburg a symposium on "Exploration for Rock Engineering" was held which in numerous articles dealt with the problem of investigation procedures (Bieniawski, 1976).

It could be presumed that hereby the subject is exhausted; yet, it might be useful to look in detail at some of the causes which can influence geotechnical prognoses negatively.

The author considers a procedure on the basis of an engineering geological expectation model suitable.

The following presentation will show the development of a model which seems suitable in order to eliminate a number of the mentioned causes.

2. Geotechnical Questioning

Any rock construction has its own specific conditions which are to be stated in the planning model. Basically, a decision has to be made whether a freedom of choice in siting — as exists in the prospecting of suitable site areas for underground storage of petroleum products (Söderberg, 1977) — or local confinement for the project in question exists. At this point, the first decision has to be made whether favourable rock mass properties are to be prospected or existing ones merely to be investigated.

Necessarily, the following questions arise:

Location	Shape
Rock mass surface	Excavation properties
Depth	Excavation technique
Stress distribution	Support requirements
Orientation	Construction duration
Dimension	Construction costs

The effort of answering these questions will differ within a wide range, depending upon the kind of project. Therefore, the significance of each question and the required value of statement has to be defined as early as at this stage.

Here the first occasion is given to synchronize the geotechnical questioning with the engineering-geologic ones and to establish the requirements towards the Engineering Geology. The co-operation engineer-geologist established at this stage shall not be interrupted until the project is completed.

3. Geologic-Tectonic Expectation Model

The decisive step for the level of statement is to produce the geologic-tectonic expectation model. The motivation for this is quite clear.

In various cases, the geologic environment of construction in rock is accessible to direct observation only to a limited degree. Therefore, conclusions drawn from the regional-geologic relations have to be utilized for the material and structural elements of the planning area. Basic data for this are air photographs, topographic and geologic maps, publications, experience from other rock constructions as well as field reconnaissance.

By such data regional characteristics can be recognized, even if they are not exposed in the planning area. Therewith they can be expected and taken into consideration for the coming exploration.

However, one condition would be that the characteristics are non-accidental ones, i. e. those repeating themselves in the three dimensional geologic environment. As an ideal case, a geometric repetition (congruence) in the planning area can be mentioned.

Every unexpected characteristics, especially when interpreted incorrectly by the investigation results — after all, it is not programmed — influences the value of a prognosis.

As an example, weakness zones in the rock mass localized by refract. seismic, can be interpreted as steep standing or flat structures. It will depend on the expectation model, which category they are going to be associated with.

The space distribution of the geologic material and its structure make it possible to define the areas of homogeneity. By area of homogeneity in rock mass construction, that type of geologic unit is understood where characteristics have a similar influence on the choice of construction technique, the costs and the progress. Similarity of the rock type at a certain fabric symmetry, however, does not mean similarity of the material, when for example strength behaviour of the rock is subject to strong variations.

If such conditions exist, it is — of course — a question of different homogeneity areas. A certain homogeneity unit may contain several rock mass zone classes.

From the above definition follows the conclusion that the extension of a homogeneity unit has to be large enough in order to influence the construction activity. The excavation method cannot be changed meter by meter.

Even under consideration of all the above mentioned conditions, there will probably remain some uncertainty which, when realized, must be kept in mind.

The geologic-tectonic expectation model furnishes the base for an initial prognosis which is supposed to contain the following criteria:

Geologic parameters	*Geotechnical parameters*
Homogeneity units	Cavern orientation
Rock type distribution	Roof depth
Principle structure interpretation	Span width
Contact conditions	Wall height
Weakness zones	Rock mass classes
Water	Support roof
Gas, Temperature	Support wall
Seismicity	Surface influence

The initial prognosis contains a preliminary judgment of the mentioned characteristics, the significance of which are to be weighed against each other and evaluated according to the level of knowledge. The initial prognosis in its turn furnishes the proper basis for the following engineering-geologic exploration.

4. Exploration Strategy

Based on the geologic-tectonic expectation model, the strategy for exploration has to be established. This depends upon the existing tasks.

In order to resolve certain problems, certain methods of investigation or groups of such are utilized. It has to be decided to what extent a question can and should be answered. Does an economical relation exist between the content of information, its strength of the statement and the construction technical consequences?

A simple model can demonstrate that often the last few percents of probability are achieved with an equally high cost effort as that of the principal information (Fig. 1).

Depending upon the type of project, by area information is gathered or point by point in the field, by laboratory evaluation models, tests, finite element analyses, or mathematical simultation models (Lindner *et al.*, 1974).

It is essential to make the right choice among the rich arsenal of geologic, geophysical, drilling-technical, geomechanical and mathematical-analytical methods. The chosen methods shall, if possible, lead to quantitative results (Helfrich, 1974).

However, one should not be of the opinion that an exploration strategy — once determined — could not be changed. On the contrary, the importance of keeping the exploration flexible will be shown later on.

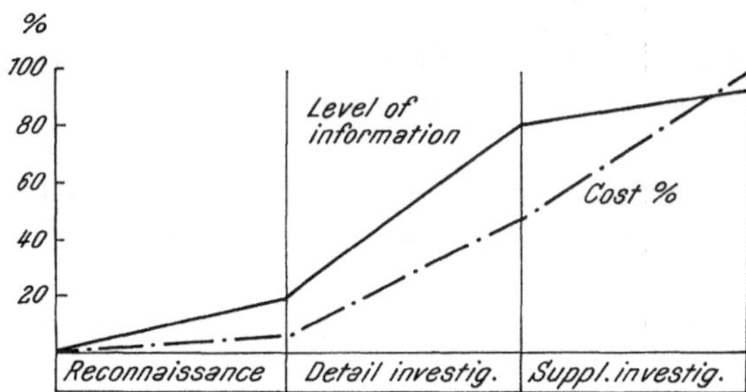

Fig. 1. Example for relation between information level and costs
Beispiel eines Verhältnisses zwischen Informationsniveau und Kosten

At this point it should be stressed once more that all decisions ought to be made in co-operation with the engineer familiar with the project, the main objective here being the joint power of decision which is characterized by the mutual belief in aiming at the project goal.

5. Exploration Model

It is the goal of each exploration to obtain data which make an estimation of cost and construction time possible. A certain exploration model has to be chosen and adapted according to those premises.

This model yet will vary depending upon the type of terrain, kind and size of the homogeneity units and the expected distribution of material properties.

After deciding upon the kind (methodics), the extent of the methods to be applied is estimated, quality of the performance is defined and the density of the investigation net is co-ordinated with the original questioning.

Of utmost importance is the definition of the validity range for each method. Principally, the range of every investigation method can be divided into three sections:

Direct validity range or measuring or test point resp. Here it is dealt with results obtained along measuring lines or from samples.

Indirect validity range or relation range into which the results can be extrapolated for geologic reasons in case of known geometric structure relations.

Wider consideration range, where the geometric structure relations only can be presumed but the results must be extrapolated in lack of sufficient investigation coverage.

These different relations of the results to the geologic space have to be stressed in particular. Namely, a different level of statement with an information content of its own and a specific strength of statement is to be co-ordinated to each range which has to be expressed distinctly in the final prognosis. A sort of scale effect is involved here, which — if interpreted incorrectly — will lead to a higher prognosis level than motivated.

If this range consideration is omitted, the quality of the prognosis is decreased, and cause for justified criticism is given.

The model contains — as indicated before — a list of quantity and the initial locations of investigation which, however, by no means are to be understood as final ones.

In our model it seems important to express the quantity in figures. When the corresponding investigation unit is defined by square dimension and volume, the investigations can be expressed in the rate of outcropping area, m^3 volume per m borehole, or m^2 square dimension per seismic profile meter, number of tests per observation unit, and others. In this way you find relation units which are comparable and make it possible to talk about the so-called coverage degree.

Full coverage is reached when the indirect validity ranges touch each other. There exists for example an 80% coverage if only 20% of the investigated unit is outside the indirect validity range of the utilized methods.

The most difficult question is that of the investigation quantity. For the geological mapping the boundries' limitations are indicated by the spatial division of the existing homogeneity units, as far as these touch the project area or are influenced by the adjoining connections. The quality of a detailed mapping will always depend upon the regional geology.

A refraction seismic investigation program starts with the display of reconnaissance profiles, which have to be placed in a way, that all the "expected" geotechnically significant structural directions can be comprehensible. Accordingly, the investigation net is to be tightened, due to the existing anisotropy in such manner that the spatial persistence can be comprehended.

To what extent this must be done is shown by the results of the partial investigation and by the requested statement.

Even there, it is essential to take into consideration the distribution of the homogeneity units and to adjust the measurements individually. If one should unnoticed exceed a homogeneity limit, this could lead to incorrect conclusions.

Similar principles apply also for drilling operations. Weaver (1976) mentions certain guidelines about the borehole spacing in relation to the depth level of a project and mentions that a minimum of 3 boreholes should be drilled. For tunnels, 5 boreholes are advised without consideration to the tunnel's length.

These guide lines should not be utilized by quantity, but in reference to the homogeneity units.

For each homogeneity unit there will be planned at least one borehole provided that the outcrop situation requires supplementary information. The more complicated such a unit is expected to be, the greater will be the amount of the boreholes within the same area. Extensive weakness zones in which a consumptive excavation or support is to be expected have to be treated individually. And they require special boreholes, if they cannot be solved in connection with other investigation objects. The above mentioned applies also to geomechanical investigations.

Frequently, there are economical limitations for exploration activities. These have to appear as a basis for the model. Otherwise, in case of poor result, the impression may arise that either too little was suggested or important factors were overlooked.

Every investigational effort serves, of course, the purpose of completing the geological background. If sufficient knowledge has been obtained by field mapping, no further efforts will be required.

6. Exploration Stages

After having set up an exploration strategy and having developed the exploration model, the course of the investigation is to be planned in different steps.

As a rule, the following stages are applied: reconnaissance, detailed investigation and finally supplementary investigation.

A step-wise procedure offers the possibility to utilize the results obtained thus far in order to examine the correctness of the model and eventually undertake modifications (Fig. 2).

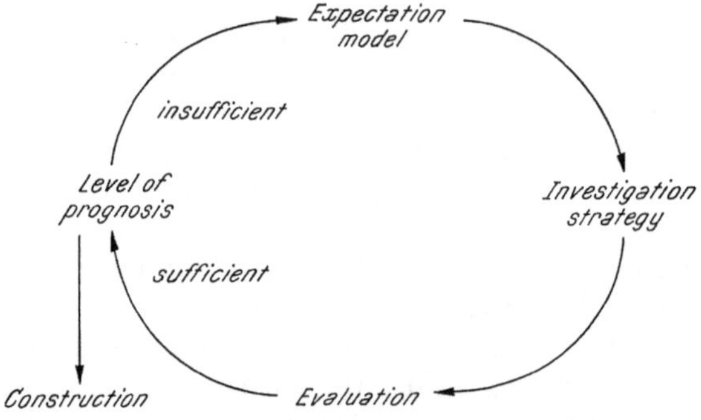

Fig. 2. Step-wise investigation procedure
Schrittweiser Untersuchungsfortschritt

This procedure will frequently have a cost-saving influence. By an earlier achievement of a sufficiently high level of statement, direct investigation costs can be saved through a reduction of the exploration. This is also true for the construction costs if by an amplified exploration unfavorable conditions can be recognized in time.

The step-wise development of an investigation program also makes a constant adjustment of operations to the state of result more possible.

L. Müller (1963: „Der Felsbau") expressed this characteristically in chapter 8. 13 „Bohrungen". This is, of course, also true for a geophysical investigation net or the mapping scale.

Utilization of adits with in situ measurements can be included in the supplementary stage, if geotechnical problems claim this for the final solution.

Even during the construction work itself, modifications can still be made. This is to say, that the investigation of a project first ends with the completion of the construction.

7. Limits of the Methods

Each method has a specific range. The geological mapping is depending on the outcrop situation. Interpolations from one outcrop to another easily lead in the inhomogeneous unit to incorrect interpretations, especially when a large scale is used. However, it is in many cases not the exact position of a feature of interest, but its principle occurrence and distribution.

Important is the question: "Is the specific feature of significance for the choice of a construction parameter or for the type of support"?

Important geological characteristics can be derived, already by using detailed geologic mapping. Of essential significance is the material character and its distribution in space. To rely upon analogical decisions may result in serious consequences. It means that analogous material has to be re-inspected for each site.

This circumstance is of particular significance if a boring machine tunnelling alternative with its particulars (Blindheim, 1976; Wanner, 1975) is considered, a fact which already has to be considered by choosing the exploration model.

As far as geophysics is concerned, the limitation is partly in the measuring range, partly in the variation of interpretation. The refraction seismic for example cannot distinguish deeply sitting flat faults. Certain seismic velocities do not necessarily mean the same rock mass or jointing frequency.

Boreholes can be without value, if their goal is the achievement of a hole. If, however, the total information is essential this will be an unquestionable help in obtaining information necessary for a requested prognosis level.

As far as geomechanic investigations are concerned, numerous measuring values are of a much greater importance than single exact ones.

The report later on should point out all deficiencies in the exploration procedure or in the data processing, in order to avoid being over-charged concerning the level of statement.

8. Geotechnical Expectation Model

The evaluation achieved and the interpretation of the investigation results obtained in the field and in the laboratory make an optimal geotechnical expectation model or prognosis possible. It contains the following parameters:

Homogeneity units
Stress distribution
Distribution of the material and its properties
Interpretation of the structures
Contact conditions
Weakness zones
Water, Gas, Temperature
Construction parameters
Support measures

Depending upon the needs for a certain project, parameters can be chosen from the above list.

For the characterization and classification of each parameter a number of systems are available (Grønhaug, 1975; Schimmer, 1975; Bieniawski, 1976) for which recently was added a general rock mass classification "Basic geotechnical classification of rock mass (BGC)", compiled by the Committee for Rock Mass Classification of the ISRM (1977).

Without going into the subject of qualification in single systems, a certain concentration on the Rock Mass Rating (RMR) — System by Bieniawski (1976) seems to exist.

Yet, regional geological aspects concerning the tectonic position of the rock mass seem not to have been considered.

Summarized, the expectation model can be built up as follows:

Rock mass classification
Zone representation by Z (RQD) in per cent (5 classes)
Construction parameter
Excavation method
Field measurements
Support requirement by type and in per cent (5 classes)

Each classification of rock mass zones (one rock mass zone is one rock mass unit which distinguishes itself through similar characteristics and behaviour; yet, the limitation of which depends upon the dimension of the respective rock construction), favourably interpreted by the aid of Z (RQD) (Helfrich, 1975, 1976), is associated with a support class, where the tunnelling method has to be considered. Applying a boring machine in the same rock mass, the support class will improve by one to two degrees possibly even more as proposed for drill and blast techniques.

9. Level of Prognosis

The most difficult point of a prognosis is the determination of the level. There are, however, certain factors, which can be pointed out for an evaluation.

The following consideration shall act as a basis (Tables 1 and 2).

Table 1

Base of evaluation	Level
Investigation	
Method	Complete
Density (Coverage)	Sufficient
Quality	Deficient
Range	Insufficient
Evaluation	
Method	In situ measurements
Density	Numerous values
Quality	Few values
Validity range	Estimation
Prognosis	
Material	
Rock mass classes	Certain
Zone distribution	Probable
Support requirement	Uncertain
Construction parameter	Very uncertain
Excavation method	

The investigation can be considered as "complete" if the degree of coverage covers the geotechnical conditions in their entirety.

Table 2

Base of Evaluation	Designation			
Degree of investigation	Complete	Sufficient	Deficient	Insufficient
Evaluation	in situ measuring	numerous measuring values	single measuring values	estimation
Level of prognosis	certain	probable	uncertain	very uncertain

A "sufficient" investigation means that the homogeneity conditions in the rock mass allow an interpolation of the results, which equals an enlargement of the indirect validity range.

An investigation is "deficient" if the degree of coverage is limited and the homogeneity or geometric conditions do not allow an extension of the indirect validity range.

A wider consideration range has to be considered, and an increase of the coverage degree would be required.

Finally, an investigation has to be considered "insufficient", if no coverage is existent but the homogeneity conditions require a coverage.

The degree of evaluation is dependent upon the existence of

— In situ measuring values,
— the amount of measuring values and calculation procedures,
— estimation.

As appears in Table 2 the prognosis level decreases with a declining number of data.

For the evaluation and presentation of the measuring values, certain aspects appear suitable. It has become customary to use the decimal system. Modern instruments give a high measuring accuracy, and calculators (even pocket size) allow an automatic decimal definition.

Yet, the geological material is not adapted to the decimal system. A non-existent accuracy would be pretended. Mostly, it should be sufficient to count with figures rounded off to five or ten. This is not only true for single measuring values, but even for statistics and percentage calculations.

Of a real significance is, however, the presentation of the limit ranges and the spreading of a parameter, respectively. Due to the inhomogeneous character and the anisotropic and heterogenic condition of the geological bodies, the features vary within a wide range.

For each evaluation, these variations and also median values are of an essential significance. At the same time it has to be said that it is more important to establish this spectrum, than to achieve absolute values.

This is especially true for the establishment of model parameters, as models should be used for the solving of the extreme cases. Even the "models" of the finite element method which by no means gives obligatory solutions should be added.

The fundamentals of the prognosis evaluation are discussed and now the values as such remain to be characterized: A certain prognosis is present, if a nearly complete conformity with outfall is found (90—120%).

If the prognosis is characterized as "probable", a 70—150% conformity can be expected.

In the case of an "uncertain" prognosis, the corresponding percentage lies between 50 and 200%.

The range below 50% and above 200% remains for a prognosis which is to be evaluated as "very uncertain".

10. Geotechnical Analysis Model — Follow up

One of the greatest problems of the engineering-geologic exploration is the frequent lack of subsequent control possibilities.

If the prognosis was correct, this is hardly mentioned. If it was wrong, however, this fact is kept alive. In both cases, a comparison of the correct result, as well as the incorrect one with the actual conditions would be more constructive. Only in this way, the number of good prognoses has a chance to grow. A documentation, giving the background for a correct or incorrect prognosis, has to be considered a necessity.

A current documentation of the geotechnical conditions, presented as a "tunnel" profile, serves as an example for this. The following data should be included:

— Space distribution of the rock,
— Geotechnically significant structures,
— Jointing (frequency, character, filling),
— Water,
— Drainage,
— Strength characteristics,
— Stand up time,
— Support measurements in m^3 and m^2 resp and intensity per running meter tunnel,
— Temporary support classes,
— Rock mass deformability,
— Permanent support classes,
— Rock mass excavation classes,
— Excavation progress,
— Disturbances (seismicity, rock burst, failure and so forth).

If the listed data are registered continuously, a priceless observation material exists. It will first serve as a basis for the temporary support work and in continuation for the permanent support work.

In many cases, the temporary support work has to be incorporated into the excavation routine. If shotcrete is used, there exists no method more effective in order to disguise the original condition of the exposed rock mass surface. Herewith the rock mass is well protected against an areal inventary later on. This can be avoided, if the shift personel could be brought to a participation in the mapping activity. The active participation in the excavation progress can be very constructive.

The documentation of the actual conditions is an absolute necessity not only for the contractor, e. g. for correct support decisions, but also for the customer. In case of eventual stability disturbances, the interacting of the disturbing factors can easily be recognized.

After the termination of the construction work, the geotechnical conditions are documented in such a manner that a synthesis is possible. The lessons to be learned from incorrect conclusions serve as the starting capital for new tasks. The payment of interest will be accordingly.

References

Bieniawski, Z. T.: Exploration for Rock Engineering. Vol. 1 and 2. Rotterdam: Balkema 1976.

Blindheim, O. T.: Borsynkindeks — DRI. Prosjektrapport 6—75. NTH Trondheim 1975.

Blindheim, O. T.: Fullprofilboring av tunneler. Prosjektrapport 1—76. NTH Trondheim 1976.

Bundesministerium der Verteidigung: Vorläufige baufachliche Grundsätze für Untertage-Bauvorhaben der Bundeswehr. Tragkonstruktionen im Fels, Standsicherheitsnachweis. März 1973.

Grønhaug, A.: Geologiske undersøkelser for tunneler. Veglaboratoriet meddelelse (Oslo) *48*, (1975).

Helfrich, H. K.: Geophysikalische und gebirgstechnische Prospektion des Gebirges als Unterlage für Ausschreibungen von Felsbauwerken. Bundesministerium für Bauten und Technik: Straßenforschung (Wien) H. 18 (1974).

Helfrich, H. K.: Gebirgstechnische Kennziffern mit petrographischem Begriffsinhalt. Rock Mechanics, Suppl. 4. Wien—New York: Springer 1975.

Helfrich, H. K.: Zon (RQD) som underlag för bergklassificering (Zonen RQD als Unterlage für die Gebirgsklassifikation). BeFo-rapport, Stockholm 1976.

ISRM: Recommendations on Site Investigation Techniques. Final Rapport, 1975.

Lindner, E., Einstein, H. H., Vanmarke, E. H.: Exploration: Its Evaluation in Hard Rock Tunneling, 16th Symposium on Rock Mechanics, Minnesota 1975.

Müller, L.: Der Felsbau. Bd. 1. Stuttgart: Ferdinand Enke Verlag 1963.

ÖNORM: Untertagebauarbeiten. ÖNORM B 2203, Vornorm 1975.

Schimmer, E. R.: Geotechnische Gebirgsklassifizierungen. Dipl.-Arbeit. Ruhr-Universität Bochum, Inst. für Geologie, 1975.

SIA: Erfassen des Gebirges im Untertagebau. Empfehlung 199. Zürich 1975.

Söderberg, L.: Practical Procedure for Preinvestigations Concerning Unlined Rock Storage Caverns for Petroleum Products. Proceedings Rockstore 77, Stockholm.

Wanner, W. J.: Einsatz von Tunnelvortriebsmaschinen im kristallinen Gebirge. Abhängigkeit des Vortriebes von geologisch-petrographischen Bedingungen. Dissertation, ETH Zürich 1975.

Weaver, J. M.: Exploratory Drilling for Rock Engineering. Proceedings of the Symposium on Exploration for Rock Engineering, Johannesburg. Rotterdam: Balkema 1976.

Address of the author: Prof. Dr. Hans K. Helfrich, Swedish Rock Mechanics Research Foundation (BeFo), Storgatan 19, S-114 85 Stockholm, Sweden.

Rock Mechanics, Suppl. 7, 27—40 (1978)

Rock Mechanics
Felsmechanik
Mécanique des Roches

Requirements of Geological Studies for Undersea Tunnels

By

A. Grønhaug

With 5 Figures

Summary — Zusammenfassung

Requirements of Geological Studies for Undersea Tunnels. The Norwegian coastline is dissected by deep fjords that create large problems for modern landbased transport communications. Although road tunneling has a tradition in Norway, and many tunnels have been driven under lakes and rivers, no road tunnel project in rock under a fjord has been studied seriously.

A project now under study may initiate a new phase in Norwegian road construction. Studies of a bridge project at Vardø on the Arctic Sea coast at the extreme North of the European continent, revealed such high costs that it was decided to look into an undersea tunnel project.

Studies of construction records from those of the Severn railway tunnel, built a hundred years ago, to the ongoing Seikan tunneling of today indicated that unforeseen difficulties tend to take place causing costs which exceed the estimates. It was realized that our present procedure for geological studies does not give sufficient guarantee against water inflow and cave-in.

A closer study of all information, possible methods and the costs of achieving them has therefore been undertaken. Firstly, mapping of the geology of the area is quite inexpensive compared to the amount of information given provided the rock is not covered by overburden. Echo soundings furnish a detailed contour map of the bottom topography, from which some geological interpretations may be drawn.

For mapping on the sea bottom frogmen geologists are employed, but the information obtained is dependent on the extent of exposed rock, and costs involved are reasonable.

Next, by means of acoustic soundings, the position of the rock surface and the thickness of the overburden may be determined with a good degree of accuracy, and at an acceptable cost.

Seismic measurements are useful to confirm the rock surface topography and to identify eventual low speed zones, but due to prevailing storms and heavy currents, such measurements are time-consuming and accordingly relatively expensive. Careful planning of the seismic profiles is therefore essential for good economy.

The only kind of investigations that provide exact and concrete information on the ground conditions are drillings. They may be started from land, bottom, platforms and barges. Drillings from barges and from land have been studied in particular. Geological mapping, acoustical and seismic soundings are just finished, and drillings will proceed in July 1977.

Anforderungen an geologische Studien für Unterseetunnel. In die norwegische Küste schneiden sich tiefe Fjorde ein, die für moderne Transportverbindungen an Land große Probleme bringen. Obwohl der Bau von Straßentunneln in Norwegen nichts neues ist und viele Tunnel auch schon unter Seen und Flüssen gebaut wurden, hat man sich bisher noch nie ernsthaft mit der Möglichkeit eines Straßentunnelprojektes in Fels unter einem Fjord befaßt.

Es ist möglich, daß ein Projekt, das zur Zeit näher untersucht wird, den Beginn einer neuen Phase im norwegischen Straßenbau darstellt. Untersuchungen haben ergeben, daß ein Brückenprojekt bei Vardø an der Küste des Nördlichen Eismeeres im extremen Norden des europäischen Kontinents mit so hohen Kosten verbunden ist, daß beschlossen wurde, die Möglichkeit eines unterseeischen Tunnelprojektes in Erwägung zu ziehen.

Untersuchungen der Bauunterlagen für den vor hundert Jahren gebauten Severn-Eisenbahntunnel bis zu den derzeitigen Bauten am Seikan-Tunnel lassen darauf schließen, daß immer mit unvorhergesehenen Schwierigkeiten zu rechnen ist, wodurch die Kostenvoranschläge überschritten werden. Es wurde erkannt, daß unser gegenwärtiges Verfahren für geologische Untersuchungen nicht genügend Garantie gegen das Einströmen von Wasser und Einbrüche bietet.

Deshalb wurde eine nähere Untersuchung aller Informationen sowie der möglichen Methoden und der damit verbundenen Kosten durchgeführt. Zunächst kann die Geologie des betreffenden Gebietes im Verhältnis zu den vermittelten Informationen auf ziemlich billige Weise kartographiert werden, sofern das Gestein nicht mit Abhub bedeckt ist. Durch die Echolotung erhält man eine genaue Höchenschichtenkarte der Bodentopographie, aus der verschiedene geologische Schlüsse gezogen werden können.

Zum Kartographieren des Meeresbodens werden Froschmann-Geologen eingesetzt aber die dadurch erhaltenen Informationen hängen davon ab, wieviel Gestein bloßliegt. Die damit verbundenen Kosten liegen in annehmbarem Rahmen.

Jetzt kann durch akustische Lotung die Position der Gesteinsoberfläche und die Dicke des Abhubs mit ziemlicher Genauigkeit und zu annehmbaren Kosten festgestellt werden.

Seismische Messungen sind zur Bestätigung der Gesteinsoberflächen-Topographie und zur Feststellung eventueller Zonen für niedrige Geschwindigkeit sehr nützlich, aber aufgrund der herrschenden Stürme und starker Strömungen nehmen solche Messungen viel Zeit in Anspruch und sind daher auch relativ teuer. Eine sorgfältige Planung der seismischen Profile ist daher für gute Wirtschaftlichkeit unerläßlich.

Die einzigen Untersuchungen, die exakte und konkrete Informationen über die Bodenart liefern, sind Bohrungen. Sie können vom Land, vom Boden, von Plattformen oder von Kähnen aus begonnen werden. Bohrungen von Kähnen und vom Land aus wurden mit besonderer Aufmerksamkeit studiert. Die geologische Kartographierung sowie die akustischen und seismischen Lotungen sind jetzt aber abgeschlossen und mit den Bohrungen wird im Juli 1977 begonnen.

Background

The population in Western and Northern Norway is concentrated along the coast. The coastline is dissected by the deep and long fjords and has high mountains rising from the shores. Traditionally communication has taken place by sea. This situation has created large problems for the modern landbased transport communication. To meet the traffic demands, about 200 ferry routes are now in service on our National Road Network (Riksveger), which comprises 25000 km of main Public Roads. The ferries serve a total distance of 2500 km of the National Roads. This service is very expensive and is subsidized by the government. Work has therefore been going on to replace the ferry routes by roades and bridges. Up to now, about 7500 bridges are in service on our National Roads, some of the newest concrete bridges have spans of up to 160 m and there are suspension bridges with main spans of 525 m. The remaining ferry routes on the National Roads require very large investments if they are to be replaced. Some of those most urgently needing replacement are at Vardø, Tysfjord, Sannessjøen, Kristiansund and Oslofjorden at Drøbak (Fig. 1).

Possibilities for Undersea Rock Tunnels

The idea of constructing undersea tunnels in rock instead of bridges is, of course not new. A milestone in the construction of such tunnels was reached with the construction of the 7 km long Severn Railway tunnel completed in 1886. At present the Japanese are constructing the 54 km long Seikantunnel under the Tsugaru Strait between Honshu and Hokkaido.

Road tunneling has a tradition in Norway, and on the National Road Network, now in service, there are about 300 tunnels with a total length of 150 km. Because the tunnels have less than 3% concrete lining, very little ventilation and good illumination, they are relatively inexpensive.

Undersea road tunnels in rock have also been considered, but never studied seriously, mainly because of uncertainty as to the consequences. Construction problems, demand for waterproof lining, and expensive installations might make this design less economically attractive. On the other hand, large investments could be saved provided such tunnels could be constructed for prices of the same order as our normal road tunnels. Many types of tunnels other than road tunnels, constructed in rock under fjords, lakes, rivers and water reservoirs indicate that this may be feasible. For example a 5 km long rock tunnel under the 3 km wide Frierfjord has recently been completed with very little rock reinforcement and water proofing other than pre-injection.

The Norwegian fjords are normally very deep. The largest fjord, Sognefjorden, is 180 km long and 1200 m at its deepest point. However the continental shelf is about 200 m deep in areas close to the coast except for a trough just outside southern Norway. For a long time it has been known that the fjords have their shallowest parts at their mouth areas (Fig. 2). Accordingly, the Sognefjord is about 200 m deep where it finishes at the coast.

The shallow mouths of the fjords have been attributed to the existence of terminal moraines. Recent studies indicate to the contrary that the soil cover is relatively thin. If this is correct, some of the fjords may be crossed in undersea rock tunnels at their outer parts. This may be the case at Tysfjord which constitutes one of the two ferry connections left on E6, the main road to Northern Norway. Plans to eliminate this ferry connection are now under way, and a normal solution calls for 80 km of new road,

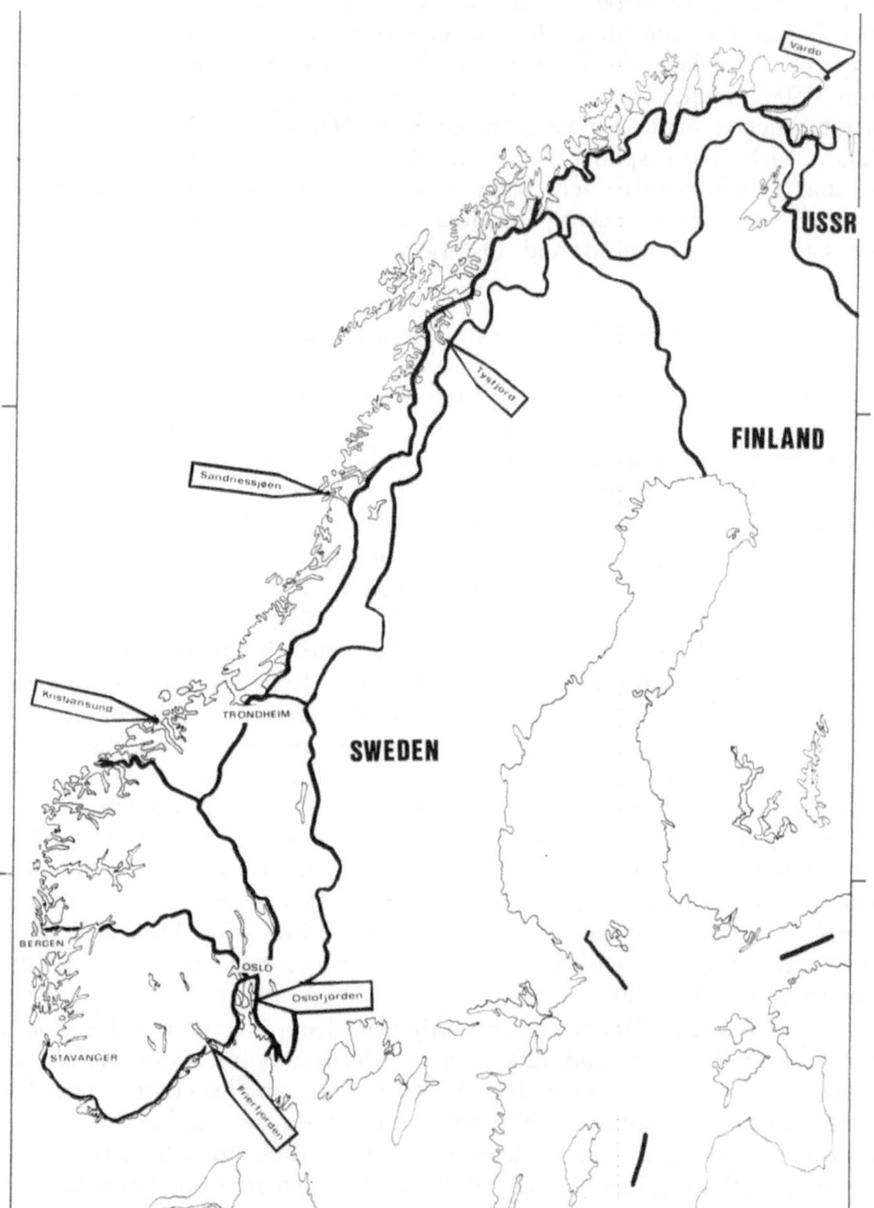

Fig. 1. The undersea road tunnel projects discussed are at Vardø, Tysfjord, Sannessjøen, Kristiansund and Oslofjorden

Die diskutierten Untersee-Straßentunnel-Projekte sind: Vardø, Tysfjord, Sannessjøen, Kristiansund und Oslofjord

including 30 km of tunnels and three very large suspension bridges. The fjord is very deep, but at its mouth rock may be found at a depth as shallow as 200 m. If this is correct, the whole normal project may be replaced by 6 km of undersea tunnel and about 20 km of new road.

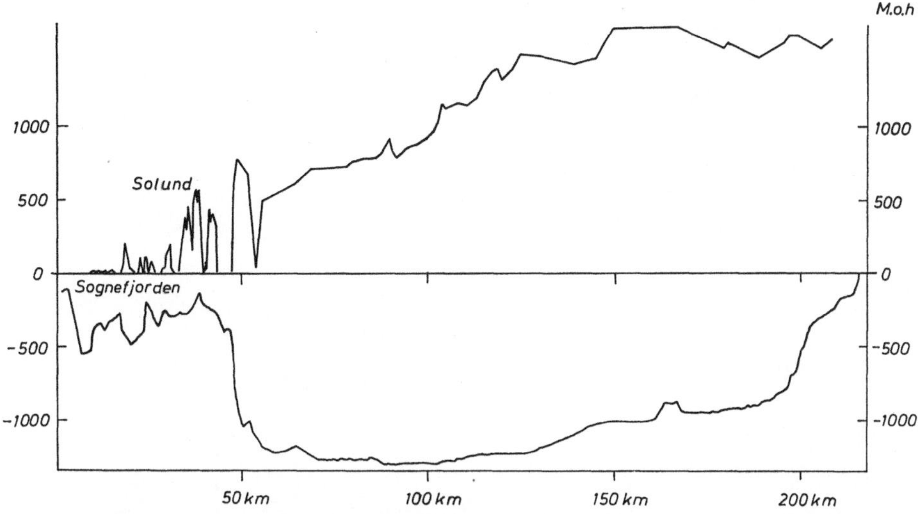

Fig. 2. Longitudinal section of the Sognefjord, showing the treshold so common in many Norwegian fjords. The level of the land area north of the fjord is also shown

Längsschnitt des Sognefjordes mit der Schwelle, welche in vielen norwegischen Fjorden angetroffen wird. Man erkennt auch das Niveau des Landstriches nördlich des Fjordes

The Vardø Project

The city and port of Vardø is situated on the island of Vardø lying in the Arctic Ocean just outside the extreme north of the European Continent. The city has 3000 inhabitants, but is an important sea port, as well as a centre for fisheries and sea communication. Because of prevailing storms the road on the mainland is closed on an average of 12 days a year, and the ferry connection to Vardø even more often. A new breakwater has been constructed on the mainland, and the future expansion of the city is planned here.

A road project for replacement of the ferry has been studied for 20 years. From the beginning, cost estimates for a bridge were so low that the prospect of a bridge connection was accepted as a matter of course. Cost estimates made recently have, however, revealed extremely high prices for a bridge, partly because of poorer ground conditions than anticipated, and partly because of the soaring cost situation. Consequently the Authorities began to look for a tunnel project in 1977. The Authorities have decided to make a choice between tunnel and bridge in the beginning of 1978, and construction will start to take place in 1979 (Fig. 3).

If an eventual tunneling at Vardø is successful, this experience may initiate a new phase in Norwegian road construction. The possibility of

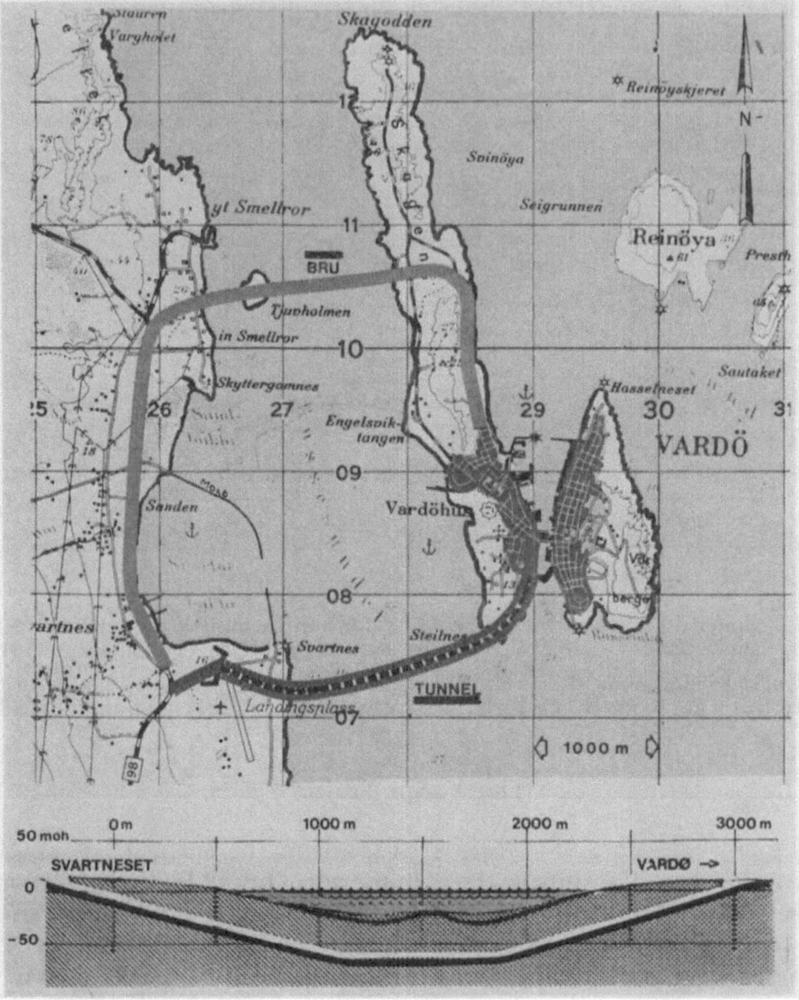

Fig. 3. The Vardø tunnel project
Das Vardø-Tunnelprojekt

constructing undersea road connections as reasonably as road tunnels on land will be received with great interest by the many small communities in areas waiting for a road connection.

Required Studies

Studies of undersea rock tunnels should of course include all aspects of the project such as installation, maintenance and service. In the following outline, only the scheme of the geological studies, consequences of tunneling and plans for investigations during driving are dealt with. In the Vardø case, a tunnel project should have a cost estimate that does not exceed that of the bridge project. In addition it should offer some advantages because it is a new design considered by many to be somewhat risky. Consequently, at the planning stage, all risks involved in the project should be forseen and measures to meet them be worked out. This situation enforces strict requirements on the studies, and extra care must be taken to make an account of the project as complete and carefully worked out as possible.

Firstly, the studies must give a reliable outline of the position of the bedrock, because it is the deepest lying surface of the bedrock that determines the depth, and consequently the length, of the tunnel.

Secondly, a great deal of attention must be paid to detect the presence of any gaps in the rock surface. Even though geological studies indicate even and homogeneous topography, experience has shown that extrapolation of geological information has had some shortcomings. In the case of undersea tunneling in rock, driving unexpectedly into overburden which is filling a gap in the rock surface would be catastrophic. Thirdly, stability of the rock is essential for the tunneling costs, and in undersea tunnels special care must be taken to avoid cave-in and to prevent a situation that cannot be controlled.

Fourthly, the permeability of the rock should be mapped to decide the extent and method of water inflow control during tunneling and a method of permanent leakage prevention and waterproofing. Normally the upper crust of the bedrock is weathered and permeable, and it gradually changes downwards to more sound and less permeable rock. Consequently a tunnel depth must be found at which the tunnel length is weighed against costs of water inflow prevention measures.

Location of the Bedrock

The rugged topography of Norway make it unreliable to assess bedrock depth from soundings or echo soundings for a judgement of an undersea tunnel project. The physical contrasts between bedrock and overburden is normally so sharp that the accoustical soundings have supplied quite reliable data on the thickness of the overburden and the depth of the bedrock.

Deviations tend to appear in areas where the bedrock surface slopes steeply, or where overburden is thick. Consequently, accoustical soundings should supply reliable information on average bedrock depth. The accoustical soundings are therefore useful at the early planning stages when only rough estimates of trace and costs are needed. When the amount of information given by the accoustical soundings is considered, the expenses are low.

More reliable information has to be supplied by other, more exact, methods. Seismic refractive measurements is a step further in this direction, firstly because the equipment is stationary and a more exact methods of location have to be applied. Experience from measurements taken on sea bed in the Vardø area reveal that in areas with sloping rock surface, seismics have resulted in larger depths to bedrock and thicker overburden than the accoustics. It is therefore unjustified to rely only on these methods. Confirmation of the results are required at least in selected areas. However, the method give a good basis for working out a drilling programme.

Vertical drillings are normally used for a reliable verification of bedrock depth. Such drillings may be accomplished from barges, platforms or from special drilling rigs that are made for underwater operation from the bottom. At great depths, drilling from the bottom is the only alternative. Such equipment exists but operational records are scarce. It is believed that this procedure is expensive and dependability still uncertain.

Drilling from platform is favourable in shallow seas and the drilling may proceed with little influence form currents and storms. Removal of the platform from one point to the next requires better weather. Drilling from platforms has an important advantage in that inclined drillings are possible, and that closely spaced holes are drilled quickly without moving the platform.

Drilling from barges is normally the least expensive in shallow and sheltered waters. Drilling from barges has been accomplished even at Vardø this year at sea depths of 25 m and where the current goes up to 3 knots, and where storms may be strong. When the waves became higher than 1.5 m, the drilling had to be stopped.

Fig. 4. The Odex drilling method. Drilling position at left, retraction at right
1. Pilot bit, *2.* reaming bit, *3.* steering, and *4.* casing (2)

Das Odex-Bohrverfahren. Links: Bohrposition, rechts: Einziehen
1 Stufenmeißel; *2* Nachnehm-Bohrmeißel; *3* Lenkung; *4* Gehäuse (2)

Experience has demonstrated that the drilling has to be accomplished with casings of 10″, 5″ and 3″ to avoid breaking the drill rods. The 3″ casing is provided with the Odex bit, an eccentric rotating bit that may penetrate rock so that the casing is tied 2—3 m in the rock (Fig. 4). Through this casing normal percussion or core drilling was undertaken. The rock surface was determined by observing the drilling and by automatic recording of drilling speed, rotation speed and feeding pressure. The rock was drilled to a depth of 10 m and a core of 1 m length was taken at the bottom of the holes in good rock. The rate of drilling gives some indication of rock strength when feed pressure is taken into consideration. Variation in the rate of drilling held together with the variation in rotational speed indicates joints in the rock. For every 3 m, some of the drill mud was sampled.

Mapping Rock Stability

Progressive rockfall and cave-in is not common in Norwegian tunneling, because the bedrock is formed by sound crystalline rocks or old, competent sedimentary rocks. Some faulting occurs, and faulted or strained zones with heavy jointing may cause instabilities.

Regional geologic maps give indications on the main structures in the bedrock. More detailed studies are required to locate rock quality and zones that may be unstable. The stratigraphy and rock types classified in units pertinent to the project should be carefully drawn on detail maps. Areas classified in terms of frequency of fracturing should also be indicated, and diagrams showing fracture directional statistics be worked out.

In the Vardø case, frog-men geologists detected very few outcrops of bedrock. Consequently information on geology could mainly be extrapolated from outcrops on land and from regional studies.

From the geological mapping, prospects of rock quality may be extrapolated. More detailed studies are necessary to obtain the information needed. Held together with the seismic information, the outlined geology gives a good basis for working out a drilling programme that will supply a maximum of information from a minimum of effort.

Mapping Rock Permeability

Water leakages constitute the most expensive feature in Norwegian road tunneling: The rock permeability, or by a more precise term, rock conductivity, have to be paid special attention dealing with an undersea tunnel project. In most road tunnels, leaks originate from limited sources. Water pressure decreases after tunneling and water inflow is stabilized or become periodical when the ground water reservoir is partly drained.

In undersea tunnels, water pressure must be expected to be the same as in underland tunnels, and dependent on the distance to the water table. However, water inflow must be assumed to be constant or even increase with time. Tunneling costs are therefore very much dependent on the type and extent of the water-proofing. A prospect of the leakage conditions is essential for a well prepared driving and water-proofing method.

Improved driving efficiency may be obtained in areas where no water proofing measures are needed, and therefore information on sections with

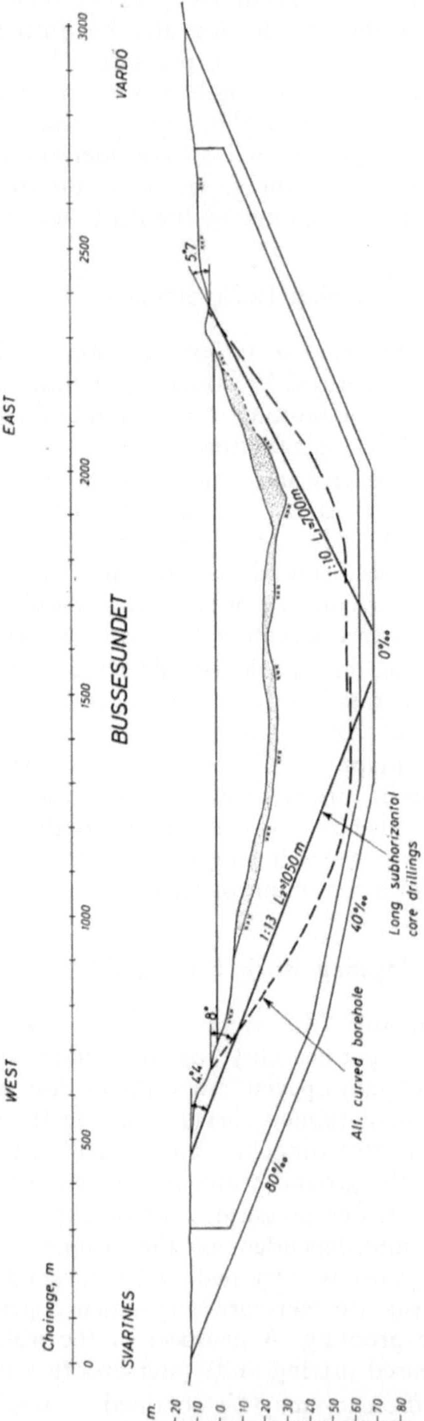

Fig. 5. Possible courses for long exploratory drillings
Möglicher Verlauf für lange Erkundungsbohrungen

large inflow hazards is needed at the earliest possible stage during planning. Water pumping or pressure tests undertaken in drillholes in the rock provide a good indication of permeability and conductivity. An average estimation of the permeability may be obtained by performing such tests during the vertical drilling programme. By studying all the information from the investigations, and extending the drilling programme a good average idea of the leakage conditions may be reached.

Detection of Gaps in the Rock Surface

Normally, seismic measurements should give some indication of marked topographical features buried below overburden. However, the method has its limitations, for instance in the detection of narrow clefts or clefts directed at small angles to the seismic cable.

By performing vertical drillings, chances of striking such features are very small, even though drilling on seismic indications. The only way of obtaining complete and dependable detection of such clefts is by inclined or horizontal drillings. Inclined drillings may be executed from the shore, islets, platforms or by underwater drillings rigs. If possible, the best answer is to drill long holes in the rock above the projected tunnel. A study on the possibility of this has recently been undertaken. There seem to be few documentations on such drillings. Core drillings with length of 1000 m have been recorded, but the critical factor is the drilling deflection.

In Vardø case, two drillings are required, each of 1000 m from the opposite shores (Fig. 5). To be of use, the drillings should have a total deflection of less than 20 m. Indications have been received from drilling companies that this might be possible.

To be successful, such drillings have to be controlled by deflection measurements and course corrections. Many factors tend to deflect the drilling from keeping a straight course. These factors originate from gravitation, rock structure, equipment and drilling procedure. Normally, gravitation tend to deflect a 1000 m long borehole 10—20%. Possibilities are good for controlling deflections caused by equipment and operation, but controlling the influence from gravitation and rock structure may prove to be difficult. Several methods have been applied or suggested to direct the drilling in the wanted position.

The simplest method is to map the deflective tendency by drilling, and start new drillings in the most favourable direction. At the same time the deflective capability of equipment and operation should be adjusted to counteract the deflection from gravitation and rock structure. If deflection still becomes to large, the course may be corrected by installation of wedges that force the bit in a more favourable direction.

The most sophisticated method is to apply guided mud-turbine drilling machines.

Long, horizontal drillings with directional control are expensive. However, the quality of the rock may change over very small distances, so that the closer the drilling is performed to the projected tunnel, the more reliable the results. Shorter and less expensive drillings may of course be undertaken at intervals from the tunnel face. At that time little can be done to avoid bad ground conditions.

Driving Method

The purpose of the exhaustive studies of the project is to arrive at a cost estimate and a scheme of the driving conditions that is reliable. At Seikan and Severn the situation was different in that the projects had to be completed whatever the costs, and that driving conditions had to be found in front of the working face because a thorough study done beforehand was insuperable. In such cases a driving method must be undertaken that includes drillings and special operations to prevent disasters.

In the Seikan project the hazard measures are threefold, firstly by driving two pilot tunnels, secondly by continuous exploration of the rock ahead of the face by drillings always undertaken 350 m ahead of the working faces, and thirdly by continuous pre-injection of 70 m long sections. In addition, water gates are constructed at certain intervals. Even with this procedure severe water inbursts have happened four times.

A procedure like this is timeconsuming and expensive. For this reason, the Norwegian undersea tunnel projects have to include a minimum of operations other than driving. On the other hand, a complete and perfect knowledge of the ground conditions is not obtained until the project has been completed. The studies of the project supply assurance of the ground conditions that is increasing with the efforts, but the last pieces of information are going to be extremely expensive. It is consequently reasonable that the driving method should include some sort of exploration and/or security operations.

The risks involved in the project are those of water inbursts and striking into the overburden. It is important to find a scheme of drilling in front of the face that will in the least possible degree interfere with the driving; for instance, drilling every week-end. It may be difficult to detect all zones of water inburst by drilling only a couple of holes in front of the face. The drilling of about 100 holes in the round would therefore serve as a last check on water leakages. When such water conductive zones are found, preinjection has to be accomplished immediately. In order to drive through overburden, freezing machinery should be on hand.

There may be a more simple driving method, at least of sections in the best quality rock. In case of cave-in or inflow, and when the inflow exceeds the pumping capacity, it will probably be very difficult to have work done by divers at a water depth of 60 m at the face of a 1 km long tunnel. If however, a transportable freezing machine could be placed at the face, it could freeze a clog of ice and permit water to be pumped out and injection work started. Such a procedure would allow for considerable savings if water leakage problems appear to be small.

Discussion on Study Schemes

There should be no doubt that drillings are necessary to establish reliable information in order to assess the ground conditions for undersea tunnel projects. Some information is needed in order to work out a rational drilling programme.

The mapping of geology may give indication of disturbance zones made up of poor rock. Soundings and accoustical soundings are relatively inexpensive and supply important information on average bedrock depth and thickness of overburden. Seismic measurements are by far more expensive, but supply better locational control, better detailed results and classification of the bedrock.

If vertical drillings are planned, seismic measurements are considered desirable for the working out of a reasonable drilling programme. There may still be undetected occurrences of poor rock, at least in narrow zones.

To obtain the highest degree of ground condition control, there are no alternatives other than inclined or horizontal drillings. Such drillings may be accomplished in selected areas indicated by previous studies. For instance, short range drillings through fault zones may supply important information about rock quality and thickness of the disturbed zone. The conditions in rock, and especially in zones of disturbance, may vary over very short distances. It is essential therefore, that drillings for ground condition control should penetrate the poor rock areas as close to the tunnel as possible.

This may not be feasible by means other than long subhorizontal drillings started from the shores. If the drillings are directed along or above the crown of the projected tunnel, they may also give continuous recording of the rock quality, as a reliable detection of harmful buried subductions or gaps in the bedrock surface, and lastly map the permeability and water conduction in the rock. Such drillings will penetrate the predominant rock in the project and eventually leave undetected small, local problem areas. Parallel structures may also be undetected for a shorter distance but this possibility should be taken into consideration when planning the drillings. The drillings are expensive, but they supply information of a more exact nature than seismic measurements. If long subhorizontal drillings are considered, seismic measurements should be required only to plan the location of the drilling. If the drillings may be planned on the basis of the accoustic soundings, expenses for seismic measurements may be saved.

Provided long horizontal drillings are possible it seems that, for feasibility and cost estimation, such drillings should be the most reasonable way of obtaining the required information.

Long horizontal core drillings are at present limited to about 1000 m in length, deviational control may not be satisfactory, and the information given may be limited if the tunnel project is curved. If the drilling strikes into overburden filling gaps and subductions in the bedrock surface, other drillings may be required to re-locate the tunnel. This procedure may therefore be expensive and parallel disturbance zones may not be detected.

If such a situation may be foreseen, a more reasonable approach would be to undertake a thorough mapping of the ground conditions by means of seismic measurements, and subsequently to apply a reduced drilling programme. The risk involved for tunneling into local water inbursts and cave-ins is considered higher by this procedure than by drilling long subhorizontal holes. For the Vardø project, discussions so far have resulted in a step by step progress in the studies, to have the opportunity to stop them immediately if and when high cost ground conditions are found. This means that the last procedure is pursued, and that eventual long horizontal drillings are postponed.

Where the sea is wide and deep, neither of the discussed procedures may be feasible for a complete exploration of the project. Large risks have to be accepted when dealing with tunnel projects in such areas. The only way of dealing with the risks of accidents is by the undertaking of continuous drillings ahead of the face. However, with this procedure, large costs may arise if the tunnel has to pass areas with very poor rock or overburden.

References

[1] Andersen, B. G.: Randmorener i Sørvest-Norge. Norsk Geografisk Tidsskrift. *XVI*/5-6, (1954).

[2] Holtedahl, H.: Notes on the Formation of Fjords and Fjord-valleys. Geografiska Annaler *49 A*/2-4, 188—203 (1967).

[3] Holtedahl, H.: The Geology of the Hardangerfjord, West Norway. Norges Geologiske Undersøkelse No. 323, Bull. 36, 1975.

[4] Atlas Copco: Odex-metoden. Jordborrning enligt excentermetoden AHB 33-16, No. 15490, 1975.

Address of the author: Arne Grønhaug, Veglaboratoriet, Gaustadalleen 25, Blindern, Oslo 3, Norway.

Rock Mechanics, Suppl. 7, 41—51 (1978)

Rock Mechanics
Felsmechanik
Mécanique des Roches

Geomechanische Erfahrungen beim Bau und Betrieb eines Pumpspeicherwerkes mit gedichtetem Unterbecken in Buntsandstein

Von

A. Roloff

Mit 7 Abbildungen

Zusammenfassung — Summary

Geomechanische Erfahrungen beim Bau und Betrieb eines Pumpspeicherwerkes mit gedichtetem Unterbecken in Buntsandstein. Das Pumpspeicherwerk Langenprozelten liegt im Spessart westlich Gemünden/Main. Es versorgt die Deutsche Bundesbahn mit Spitzenstrom. Bauherr ist die Donau-Wasserkraft Aktiengesellschaft. Die Planung wurde von der Rhein-Main-Donau A. G. in Zusammenarbeit mit Siemens A. G., Erlangen, durchgeführt.

An den Flanken des bis zu 300 m tiefen Sindersbachtales sind hangparallele Klüfte bis 0,50 m geöffnet. In der Talsohle enden diese Aufweitungserscheinungen an einer Schichtfläche und es kam durch den auf das Unterlager wirksamen Schub der gleitenden Hangkluftkörper zu Aufwölbung und Faltung der Wechsellagerungsschichten in der Talmitte. Dieser mehrere Meter mächtige Felshorizont besitzt Lockermaterialeigenschaften, er wurde im Unterbecken und Hauptdamm belassen.

Während der Bauzeit des Kraftwerkes war die geringe Reibung auf den Schichtfugen registrierbar an langen Vorspannwegen bei Ankern, Versatz von Perforationsspuren und an der schichtigen Umläufigkeit bei Felsvergütungsarbeiten.

Der Einbruch der Beckendichtung nach Inbetriebnahme erforderte den Bau einer begehbaren Wasserfassung unter der Dichtung. Die intensive Beobachtung der Bergwasserverhältnisse wird durch ihre Verknüpfung mit elastischen unf plastischen Verformungen im Untergrund notwendig. Synchron mit der Zusammendrückung bzw. der Rückfederung des Felsuntergrundes läuft ein ruckartiger Anstieg bzw. Abfall in Wasserbeobachtungsrohren in diesem Schichtpaket. Die Empfindlichkeit (0,50 m Wasserspiegelschwankungen sind in 11 m Tiefe registrierbar) und Gleichzeitigkeit der Reaktion des Felsverbandes auf Betriebswasserspiegel-Änderungen belegt Porenwasserdruckwirkung in einzelnen Schichten und Klüften.

Die elastische Verformungs-Komponente in vertikaler Richtung, gemessen in einem Schichtpaket von 11 m Dicke, erreicht bei Belastungsänderung von 10 m Wassersäule einige mm. Der Absolutbetrag spielt eine geringere Rolle als die Beobachtung der spontanen Wirksamkeit des Druckausgleichs, wodurch in Zwischenstadien größere Endausschläge als bei Stau- und Absenkziel erreicht werden.

Die dabei auftretenden horizontalen Mobilisationen, auch auf alten Bewegungsflächen, wurden nur geodätisch erfaßt. Sie sind für das Verhalten der Kluftfüllungen an den Hängen zusammen mit den beobachteten starken Bergwasserschwankungen von großer Bedeutung.

Geomechanical Experience During the Construction and Operation of a Pumped Storage Power Station With a Lined Lower Basin in "Buntsandstein". The pumped storage scheme Langenprozelten is situated in the Spessart west of Gemünden/Main. It supplies the German Railway with peak energy. Client is the Donau-Wasserkraft A. G. Consulting engineers were the Rhein-Main-Donau A. G. in cooperation with the Siemens A. G.

On the slopes of the up to 300 m deep Sindersbach valley are joints running parallel to the slope of as much as 0.5 m in width. The phenomenon ends at the valley bottom as the thrust of the sliding slope material on the underground causes compression and folding of the clay and sandstone layers in the middle of the valley. This rock formation, which is several meters thick, possesses the characteristics of loose material and was retained in the lower basin and main dam.

During the construction time of the power station the low shearing strength in the strata joints was registered in the length of the elongation of the prestressed anchors, the displacement in the vertical drilling traces and the percolation during rock-grouting.

The collapse of the basin lining after the commencement of operations necessitated the construction of an inspectable gallery under the lining. The intensive observation of the ground water conditions was necessary because of its close association with the elastic and plastic deformation of the underlaying material. Synchronous with the compression of the rock underground was a sudden rise or fall in the water observation holes in this formation. The sensitivity (0.5 m water level differences are measurable at a depth of 11 m) and at the same time the reaction of the rock joints to the change in the operational water load documents the pore-water pressure in the individual layers and cracks.

The elastic deformation component in the vertical direction, measured in an 11 m thick strata, can reach several mm with a change in water load of 10 m. The absolute value is not so important as the observation of the spontaneous effectiveness of the pressure balance, whereby in an intermediate stage the movement reached is greater as by the full and empty situation.

The horizontal movements which occur also in the fossil movement joints, were surveyed. They are for the behaviour of the joint filling in the slope together with the observed large ground water variation of great importance.

1. Einleitung

Durch den Bau des Pumpspeicherwerkes Langenprozelten wurde der Blick in den tieferen Felsuntergrund eines Buntsandsteintales freigegeben. Schon die Voruntersuchung vom Geotechnischen Büro Heitfeld hatte auf die Notwendigkeit der Wannendichtungen im Unter- und Oberbecken hingewiesen. Die Zone der Hangzerreißung reicht von der Talsohle bis 70 m unter die Hangoberfläche, ihre Untergrenze verläuft zunächst nahezu horizontal. Erst allmählich, auf Schichtflächen treppenartig aushebend, steigt der durch die Talbildung unbeeinflußte Felsuntergrund an. Sperrenbauwerke müßten sicher in diesen Horizont einbinden und bis weit in die Hangflanken ausgeführt werden, umgekehrt würden Wasserversorgungsanlagen auf diesem Niveau das Wasser eines gesamten Bergmassivs über dem Talniveau erschließen.

2. Beobachtungen beim Bau

Bei der Voruntersuchung wurden im Nahbereich des späteren Krafthauses zwei Schrägbrunnen von 40 m Tiefe abgeteuft. Die zu erwartende Pumpwassermenge wurde mit dem Absenkversuch erfaßt, nicht der Durch-

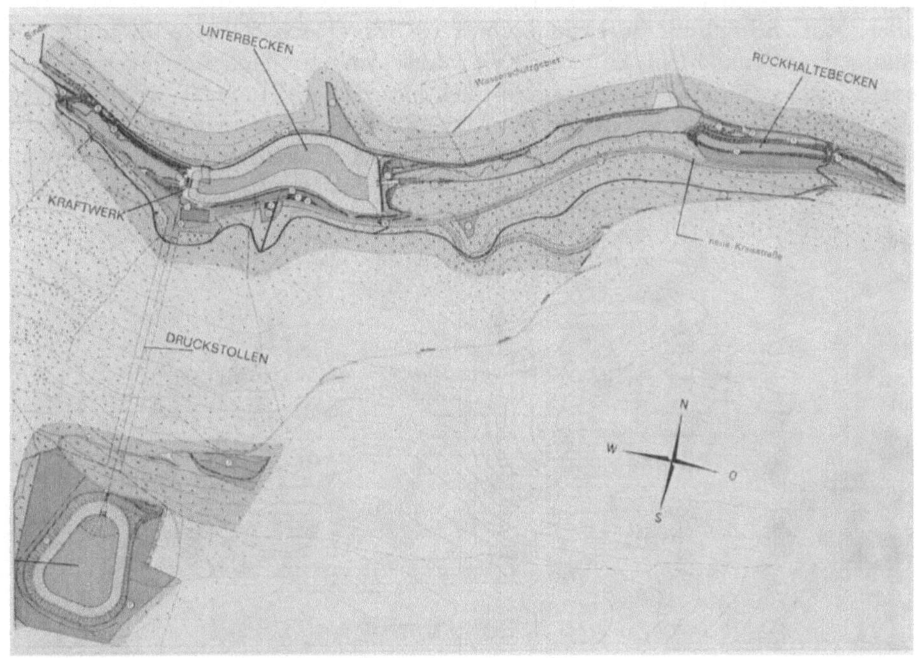

Abb. 1. Landschaftsplan
Topography

strömungsmechanismus. Aus den Pegelständen im Nahbereich, die tatsächlich trichterförmige Absenkung zeigten, wurde auf ein durch Klüftung und Schichtung homogen durchströmtes Gebirge (Modell: Kies) geschlossen. Der Ausbruch zeigte aber, daß in der tieferen, eingespannten Sandstein-Tonstein-Wechsellagerung, die hier der sogenannten Miltenbergfolge angehört, unter der Talsohle keine vertikale Wasserbewegung stattfindet. Die einige Meter weitständige Klüftung durchtrennt meist nur einzelne Bänke. Tonsteinlagen wirken als Stauhorizonte, in der Regel ist das Grundwasser gespannt.

Durch leichtes Nordeinfallen (3—5^0) ist die gesamte Südseite des Sindersbachtales von weit geöffneten Klüften (bis ca. 0,40 m) zertrennt. Das Abgleiten der Hangkluftkörper übt auf das Unterlager einen Schub aus, der schließlich in der Talmitte von der oberen Felszone aufgefangen wird. Die Aufbeulung und Verfaltung der Wechsellagerungsschichten wurde in einem bis zu 4,00 m mächtigen Horizont beim Bau des Krafthauses und des Tosbeckens beobachtet. Die Einmessung der beobachteten Verformungselemente, ebenflächige Abscherungen und faltenartige Aufbeulungen, zeichnen eindeutig die Talkrümmung nach. Der Betrag der Einengung erreicht einige Meter. Erstmals wurden derartige Schichtdeformationen in Sandsteinen von Eissele und Link (1968) beschrieben.

Die hier beobachtete schichtparallele Schwäche, die geringe Reibung des Buntsandsteins auf der Schichtfläche, ist das „Unglück“ für den Felsbauer. Auf die Schwierigkeiten beim Bau des Druckstollens soll nicht eingegangen werden, beim Bau des Krafthauses machte sich diese Eigenschaft bemerkbar bei nach oben geneigten Zugankern mit bis zu 40 cm Spannweg. Schichtparallele Verschiebungen wurden auch sichtbar an vertikalen Perforationsbohrungen. Die Umläufigkeit auf Schichtflächen verlangte weiterhin

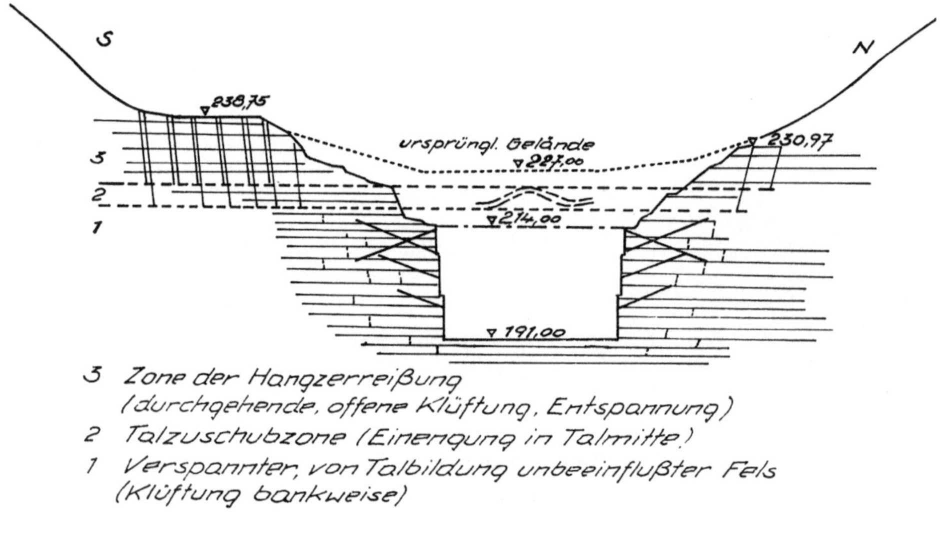

Abb. 2. Felsausbruch Krafthausgrube
Rock excavation for the power station

das gleichzeitige Packersetzen für die Felsinjektionen von bis zu 6 Ankern in einer Lage. Eine generelle Verlängerung der Haftstrecken beim Ankern brachte ebenso Verbesserungen wie die Auswahl der Haftstrecken in Sandstein-Dickbänken, die durch einfache geologische Lagerung und vorauseilenden Ausbruch bei nach oben geneigter Lage der Anker möglich war.

In der bis ca. 35 m tiefen Krafthausbaugrube, ca. 40 × 40 m, traten durch umfangreiche Sicherungen kaum Bewegungen an Extensometern auf. Oberhalb der senkrechten Wände wurde eine abgeräumte Felstreppe beobachtet. Ihre Gesamtneigung war gegenüber der alten Hangneigung kaum verändert. Zwei Systeme von offenen Klüften (hang- bzw. baugrubenparallel und senkrecht dazu) wurden mit paarigen Meßpunkten und einem Setzungsdehnungsmeßgerät überwacht. Im Juli 1973 ergab eine Messung eine gleichzeitige

Aufweitung aller Spalten bis zu 2,5 mm bei 14° C und Regenwetter gegenüber einer Messung bei 28° C und trockenem Wetter. Die Bewegungen eines solchen Baukasten-Verbandes über dem Bergwasserspiegel können verglichen werden mit denen eines Schuppenpanzers beim Atmungsvorgang, wobei der Zusammenhang des Verbandes durch unterlagernde Tonsteinlagen hergestellt wird.

Ein solcher Felsverband dient unter geringfügiger Hangschuttüberdeckung (0,40—ca. 2,00 m) als Unterlager für die Unterbeckendichtung. Die Hänge wurden vor Aufbringen der Dichtung nur vom Oberboden befreit, geringfügig eingeebnet und mit einer Drainschotterschicht aus Mineralbeton (Muschelkalk) abgedeckt. Die ursprüngliche Neigung blieb weitgehend erhalten. Angetroffene offene Klüfte wurden beim Böschungsplanieren mit Beton verfüllt.

3. Dichtungsschaden, Ursachen, Sanierung

Nach mehrmonatiger Probeinbetriebnahme traten bei hohem Grundwasserstand im Nahbereich einer Quellfassung in der Beckendichtung Einbrüche auf. Die Schadstellen liegen im unteren Böschungsbereich, nahe dem Übergang zur Sohlendichtung und vor dem Hauptdamm. Als Auslösemechanismus für den Einbruch kommt in irgendeiner Form der Übertritt von Wasser aus dem Lockermaterial in die offene Felskluft oder in die Spalte Fels/Hohlraumfüllung in Frage. Eine umfassende Darstellung der Schadensursache kann hier nicht gegeben werden.

Das Aufgraben der Schadstellen zeigte folgendes Bild: Rückschreitend baut sich der freigespülte Spalt im Fels in den Hangschutt vor. Dies führt zu Nachsackungen des Drainschotters, zur Bildung von „Fuchsgängen" und schließlich zum Durchbrechen der Dichtung. Veränderungen der Grundwasserführung legten nahe, die Fassung der Hauptquelle vor dem Damm begehbar auszuführen. Die Wirkung der Galerie kann bei niedrigen Bergwasserständen an Grundwasserisohypsenkarten gut erkannt werden. Vor Errichtung der Galerie waren Grundwassergefälle im Nahbereich der Galerie um 20—40 ‰ verteilt vorhanden. Nach Fertigstellung stellt sich oberstromig eine Verebnung auf ca. 10 ‰ ein, unterstromig stieg das Gefälle an einer scharfen Linie um bis zu 80 ‰ an. Bei ruckartigem Anstieg in der Hangzerreißungszone verschwindet die Verebnung. Größere Abgrabungen im Nahbereich der Wasserfassung zeigten eine flexurartige Aufbiegung der Schichten, so daß wirksame flach geneigte Drainagebohrungen vom Eckbauwerk der Galerie bis auf die Länge des Hauptdammes ausgeführt werden konnten und den Zufluß, vor allem bei hohen Bergwasserständen, erheblich verstärken und ein Überstauen verhindern.

An einzelnen Quellen wurden Abflußmengenschreiber mit Grenzsignalgeber in der Schaltwarte installiert. Die Wasserstände werden zum großen Teil mit Schreibpegeln aufgezeichnet. Vollautomatische Temperaturschreiber sind in den Drainagen des Oberbeckens installiert, am Unterbecken werden die Temperaturen an Meßwehren festgestellt.

4. Grundwasserbeobachtungen, Verformungsverhalten

Zwischenzeitlich läuft nun das Pumpspeicherwerk seit 1 1/2 Jahren ohne weitere Unterbrechung. Es soll hier über die Meßeinrichtungen und über die Ergebnisse berichtet werden. Im Nahbereich des Unterbeckens und des

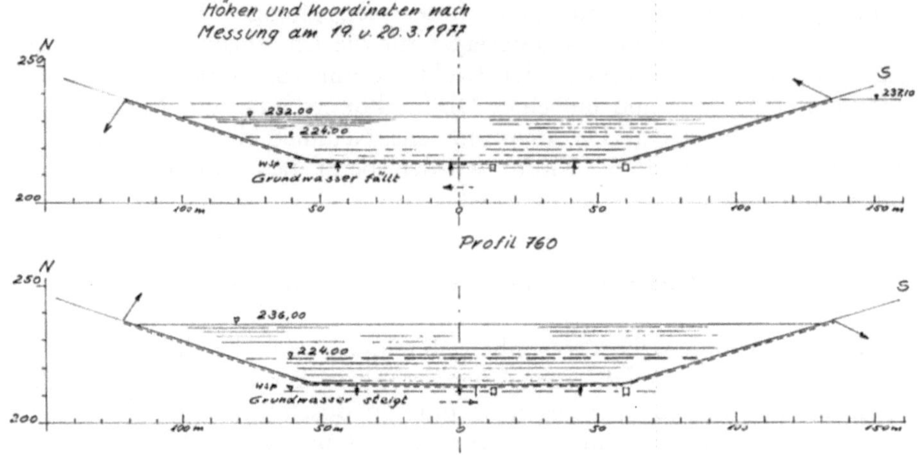

Abb. 3. Schematischer Bewegungsablauf im Unterbecken
Schematic movement in lower basin

Hauptdammes wurden zahlreiche Grundwassermeßrohre eingebaut, die einige überraschende Gegebenheiten erbrachten.

1. Die Grundwasserstände am unteren Hangfuß und im Tal schwanken sehr stark, in einzelnen Pegelrohren um bis zu 9,00 m. Die schnellen Anstiege erinnern an Karstverhältnisse, aus den Hohlräumen wird bei Anstiegen Pilzgewebe abgespült, wie es aus Karstkalken bekannt ist. Das starke Auftreten von schaumigen Pilzstoffen spricht für turbulente Durchströmung in Zeiten hoher Bergwasserstände.
2. Die Fließrichtungen ändern sich um bis zu 90°.
3. Grundwasserströmungen sind über die Verformung von Belastungen abhängig.

Die Verformung des Untergrundes durch die Betriebswasserbelastung führt zu Änderungen im Grundwasser unter dem Becken, im Damm und abgeschwächt an den Hängen. Besonders beachtenswert erscheint, daß an den Hängen eine ähnliche Reaktion entsteht, wie sie bei ungedichteten Talsperren bekannt ist. Diese Änderungen sind gleichartig über den gesamten Nordhang vorhanden und reichen sogar nach Westen über das Krafthaus hinaus. Der Grundwasserspiegel sinkt dabei auf 800 m in Talrichtung um 17 m ab.

Die generellen Grundwasserströmungen und einige Anomalien werden in einer noch andauernden Studie des Bayerischen Geologischen Landesamtes untersucht. Suffosionsvorgänge sind in die Untersuchung einbezogen. In diesem Zusammenhang ist zu erwähnen, daß innerhalb von einigen

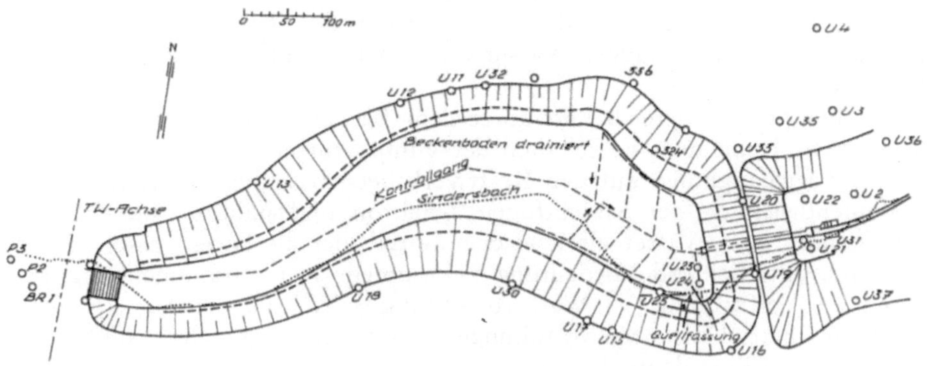

Abb. 4. Lageplan des Unterbeckens
Layout of lower basin

Pegelstelle: U. 19 1.6.77 bis 1.7.77

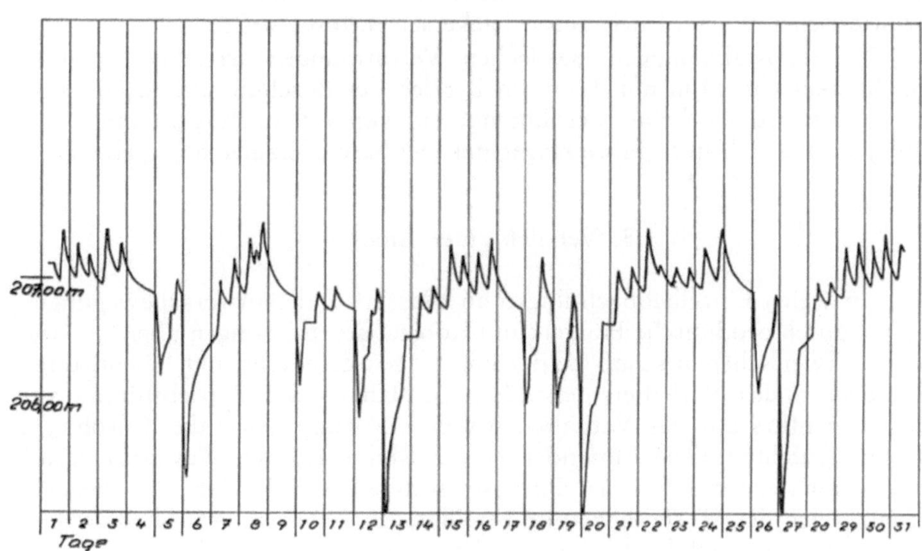

Abb. 5. Pegelganglinie U 19
Water level in piezometer U 19

Wochen sich um Absenkbrunnen offene Wasserwege bilden, in denen das Wasser nach Markierungsversuchen im Fels so schnell fließt wie in einem Rohrgerinne.

Der Zusammenhang Belastung — Verformung — Wasserbewegung soll näher betrachtet werden. Da die Beobachtung der gegenseitigen Beeinflussung im Nahbereich des Hauptdammes besonders stark ist und an ihm nach unterstrom endet, wird der Dammauflast eine nach der Tiefe wirkende Abminderung der schichtparallelen Durchströmung im Buntsandstein zugeordnet. Diese Durchflußabminderung erreicht aber keine Sperrenfunktion, da zwischen Dammschüttung und Felsuntergrund der Talschotter eine Entwässerung zuläßt und andere Wasserwege im Fels auflastunabhängig einen Abfluß bewerkstelligen.

Im Hauptdammbereich und unter dem Becken steigen bzw. fallen die Pegel praktisch mit dem Einschalten der Pumpenturbinen, so daß man vom Bild der Pegelganglinien auf den Kraftwerksbetrieb schließen kann. Da die Zusammendrückung der Schichtfugen synchron und nahezu voll elastisch verläuft, muß dem Wasser als Zwischenmittel oder als Porenwasser in den Füllstoffen entscheidende Bedeutung zukommen. Der Zusammenhang Belastung — Verformung — Wasserbewegung wird aus der gemeinsamen Auftragung Betriebswasserschwankungen, Extensometerganglinie und Wasserbeobachtungslinie deutlich.

Neben diesen ruckhaften Anstiegen bzw. Abfällen in den Pegeln und dem dann nach abgeschlossener Belastungsänderung sich asymptotisch einstellenden Ruhewasserspiegel sind auflastabhängige Schüttungen aus Vertikalbohrungen im Beckenuntergrund, Quellen und horizontalen Drainagen zu erwähnen. Beim Abteufen der Vertikalbohrung wurde nach unten zunehmender artesischer Wasserdruck durch Packersetzen festgestellt.

Die auflastabhängigen elastischen Verformungen erreichen in einem Schichtpaket von 11 m nur 3 mm im Bereich des Böschungsfußes. Die Hüllkurve zeigt jedoch keine Beruhigung, sie verläuft flach geradlinig nach unten, so daß sich über Jahre betrachtet meßbare „Setzungen" ergeben.

5. Versuch einer Analyse

Mit vielen Einzelbeobachtungen an Wasser- und Bauwerksbewegungen, ergänzt durch geodätische Höhen- und Koordinatenmessungen über 1,5 Jahre wird in zwei schematischen Darstellungen bei gefülltem und teilentleertem Unterbecken das Verhalten der Böschungsflanken auf Betriebsniveau verglichen. Einfluß auf das Verhalten beider Böschungen hat die Talsohle, in der eine schichtparallele Abminderung der Reibung durch Wasserdruck auftritt. Mindestens teilweise ist Porenwasserüberdruck in den zwischengelagerten Tonen der Wechsellagerung beteiligt. An Bohrkernen in diesem Horizont konnte mehrfach beobachtet werden, daß intern feingeschichtete Schluffstein-Tonsteinlagen stark zerbrochen und das noch erkennbare Interngefüge um bis zu 45° rotiert ist. Obwohl die über- und unterlagernde Sandsteinbank keinerlei Anzeichen einer Durchbewegung trägt, erhalten die mechanisch zerstörten Tonsteine Lockermaterialeigenschaften. Durch diesen Sachverhalt tritt bei jedem Lastwechsel eine Mobilisierung der seitlichen Einspannkräfte des Felses in der Talsohle auf. Nach den geschilderten Untergrundgegebenheiten kann diese Funktion nur der tiefere, eingespannte, durch Talbildung unbeeinflußte Felsuntergrund übernehmen.

An der Südseite begünstigen entlastete Schichtfugen in der Talmitte horizontale Verschiebungen, entsprechend dem Schichtfallen sind sie nördlich (talwärts) gerichtet und erreichen bis zu 40 mm, dabei tritt eine Hebung am Betriebsweg mit bis zu 20 mm auf. Die Hebung kann möglicherweise durch das elastische Verhalten der schluffig-tonigen Füllstoffe im völlig durchtrennten Kluftkörperverband der Südböschung erklärt werden.

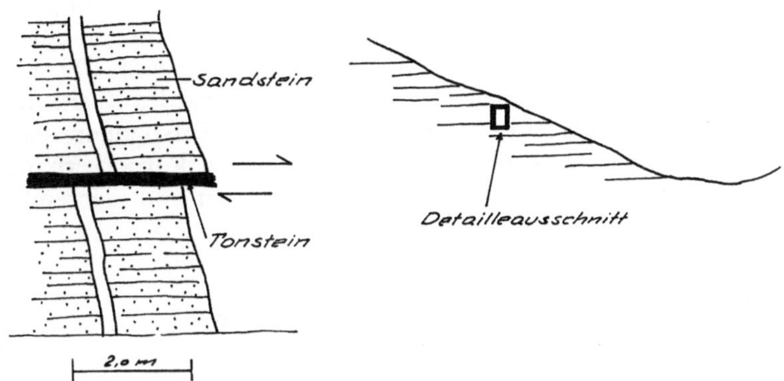

Abb. 6. Kluftkörperverschiebung am Hang
Fissures and sliding in the slope

Die nördliche Talseite zeigt gegensätzliches Verhalten, bei Entlastung tritt überwiegend eine Verschiebung der Böschung bergeinwärts um bis zu 10 mm ein. Gleichzeitig ergibt sich eine Senkung auf dem Betriebsweg um

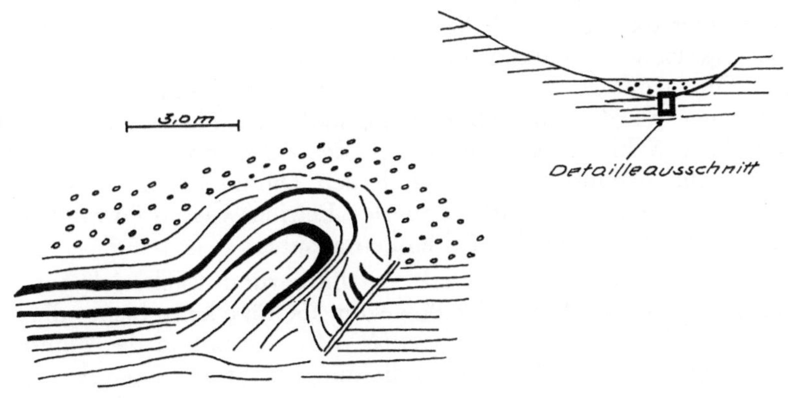

Abb. 7. Aufbruch in Talsohle
Compression and folding in the valley bottom

bis 15 mm. Die Entlastung an der Südböschung verursacht die Übernahme des nach Norden gerichteten Schubes durch das Felsgefüge. Insgesamt fällt das entsprechend der Schichtlagerung „felsartige Verhalten" der Nordböschung auf. Sie bildet das Felswiderlager für die aktiven Bewegungen der Südböschung. Sinngemäß gleich ist an beiden Böschungen die Verknüpfung von bergwärtiger Verschiebung mit Senkung und talwärtiger Bewegung mit Hebung.

Die auslösende Größe der Belastungsänderung, die das Umsetzen des Abstützungsmechanismus hervorruft, ist sehr gering. Der betriebsbedingte Durchsatz von 60 m^3/sec auf eine gemittelte Fläche bringt eine Auflaständerung von 0,02 kp/cm^2 in 6 min, oder auf labormäßig geläufige Dimensionen übertragen, bedeuten die Betriebswasserspiegeländerungen eine jeweils viertelstündige Laststeigerung um 5 kp auf eine Fläche von 100 cm^2. Dies führt bereits zu einer momentanen Druckbeaufschlagung des Kluftfüllsystems an der Südseite. Erst nach Abklingen der Änderung (Entspannen des Porenwasserüberdruckes) werden die Kräfte wieder vom Festkörperverband übertragen, wobei sich dann die Endzustände bei gefülltem und „leerem" Unterbecken nicht wesentlich unterscheiden.

6. Auswirkung auf das Bauwerk

Zusammenfassend wirkt einer natürlichen Belastung des „Talsohlgewölbes" durch Schub eine betriebsbedingte Entlastung durch Wasserdruck im völlig durchtrennten Kluftkörperverband der Südseite entgegen. Das Wechselspiel verursacht eine horizontale Verschiebung mit Kippbewegung und letztlich ein Einrütteln der Schlüssel. Im Verzahnungsbereich der sich mechanisch unterschiedlich verhaltenden Talböschungen liegen bis zu 5,00 m mächtige Talschotter. Sie bilden eine Ausgleichsschicht über der „Scharnierzone". Die Dicke der Talschotter und ein bis zu 4,00 m mächtiger Aufwölbungshorizont verlegen die Übertragung der Einspannkräfte in etwa 10 m Tiefe unter das ursprüngliche Talniveau. Über diese horizontalen Verformungen erklärt sich zwanglos die Tiefenwirkung der geringen Belastungsänderung. Die größten, geodätisch gemessenen horizontalen Nord- bzw. Südbewegungen wurden mit 60 mm im südlichen Hauptdamm ermittelt. Eine zusätzliche Sohlenaussteifung durch den 30 m hohen Damm ist also weder durch die Erdauflast noch durch den Dammkörper selbst gegeben. Daher würde auch die Erdauffüllung des Betriebswasser-Totraumes keine stabilisierende Wirkung auf die Talflanken ausüben.

Das Zusammenwirken Bauwerk — Untergrund spielt sich auf großer Fläche ab. Die Größenordnung der Bewegungsbeträge kann bisher die Sicherheit des Bauwerks nicht in Frage stellen. Mit Ausnahme des Nahbereichs der Quellfassungen steht die Asphaltdichtung nicht im Kontakt mit dem durchströmten Felsverband. Er bildet durch seinen Gebirgsbau und seine Talgeschichte die Ursache für die Bewegungen. Eine technische Lösung für die Verhinderung der Durchströmung zeichnet sich nicht ab. Daher soll weitere Beobachtung die Veränderungen und Kriecherscheinungen in einem äußerst sensiblen Felsverband aufzeigen, um Gefahrenquellen zu erkennen. Hierzu mögen die Schlußfolgerungen beitragen, ohne Anspruch auf endgültige Klärung zu stellen.

Literatur

Bayerisches Geologisches Landesamt: Gutachten über die hydrogeologischen Verhältnisse im Bereich des Pumpspeicherwerkes Langenprozelten — Unterbecken, 1976.

Eissele, K., Link, G. B.: Schichtdeformationen im Buntsandstein des nördlichen Schwarzwaldes. Jh. geol. Landesamt Baden-Württemberg, *10,* S. 157—173, Freiburg i. Br. 1968.

Fuchs, H.: Das Pumpspeicherwerk Langenprozelten — Entwurf und bauliche Gestaltung. Elektrizitätswirtschaft *74,* H. 6, 139—147 (1975).

Geotechnisches Büro Dr. Heitfeld: Endbericht über die ingenieur-hydrogeologischen Untersuchungsarbeiten zum Hauptprojekt Pumpspeicherwerk Langenprozelten, vom 15. 1. 1972.

Grüner, J.: Das Pumpspeicherwerk Langenprozelten. — Maschinen, elektrische Einrichtungen und Stahlwasserbau. Elektrizitätswirtschaft *24,* (1975).

Hollingworth, S. E., Taylor, J. H., Kellaway, G. A.: Largescale Superficial Structures in the Northampton Ironstone Field. Quart. Journ. of the Geol. Soc. of London, S. 1—34, London 1944.

Rümelin, B.: Schäden am Unterbecken des Pumpspeicherwerks Langenprozelten. Vortrag am 9. 9. 1976 in Klagenfurt.

Anschrift des Verfassers: Dr. Achim Roloff, Materialprüfstelle der Rhein-Main-Donau A. G., Schallershofer Straße 86, D-8520 Erlangen, Bundesrepublik Deutschland.

Rock Mechanics, Suppl. 7, 53—65 (1978)

Rock Mechanics
Felsmechanik
Mécanique des Roches

The Reliability of Estimating Rock Excavation Cost in Tunneling Specifications

By

E. Kurzmann

With 5 Figures

Summary — Zusammenfassung — Résumé

The Reliability of Estimating Rock Excavation Cost in Tunnelling Specifications. The statistical evaluation of the distribution function of quantities and qualities of excavated rocks in 14 (10) tunnels of different lengths and profiles, crossing the Central Alps of Austria leads to the conclusion, that their distribution-function has the form of a Gauss-bellgraph, which shows a remarkably small scattering-range, provided, that the data of the final invoices are taken into account.

The total cost of the excavation works, which were finally charged in these projects, exceeded heavily the sums of the bid, when the standard deviation of the rock class quantities-qualities-distribution function, related to the weighted average unit price, or to the most frequent unit price, of the bid surpassed 35%. Thus the standard deviation of the distribution function of rock classes of the bid is a measurement for the reliability of the offered price.

Improvements in the interpretation of measured deformation velocities and the applied support means, the standardization of the lists of quantities, and the further publications of offered and charged prices in connection with the quantities and prices of support means will contribute to the most economical development of the New Austrian Tunnelling Method (NATM). A method to ease this task without revealing company secrets is shown and applied in the paper.

Die Zuverlässigkeit der Kostenschätzung für Ausbrucharbeiten. Die statistische Auswertung der Kollektive von Einheitspreisen und Mengen von Felsausbruchsklassen in 14 (10) Tunneln von verschiedenen Längen und Querschnittsformen durch die österreichischen Zentralalpen berechtigt zur Schlußfolgerung, daß ihre Häufigkeitsfunktion die Form einer Gauss-Glockenkurve hat, die ein bemerkenswert geringes Streumaß aufweist, soferne man von den Mengen der Schlußabrechnungen ausgeht.

Die Abrechnungssumme für den Felsausbruch überschritt ganz wesentlich die Anbotsumme, wenn die Standardabweichung, d. i. das Streumaß der Häufigkeitsverteilung über 35% betrug, sowohl bei einer Beziehung auf den gewogenen mittleren Einheitspreis der Felsausbruchsklassen als auch auf den Einheitspreis der im jeweiligen Hohlraumbau am häufigsten vorkommenden Felsausbruchsklasse. Das Streumaß der Häufigkeitsfunktion ist somit ein Maß der Zuverlässigkeit der Kostenschätzung für Ausbruchsarbeiten.

Die Wirtschaftlichkeit der Neuen Österreichischen Tunnelbaumethode NATM ließe sich noch steigern durch eine zutreffendere Inbeziehungsetzung der gemessenen Konvergenzmessungen zu den jeweils angewandten Stützmitteln, ferner durch eine Vereinheitlichung der Leistungsverzeichnisse für Untertagearbeiten und schließlich durch die Veröffentlichung der Anbotspreise und der zugehörigen Abrechnungspreise, worin die Ausbruchsklassen und zugehörigen Stützmitteln samt Einheitspreisen und Mengen bekanntgegeben werden. Eine Methode zum Vergleich, bei deren Anwendung keine Firmengeheimnisse verraten werden, wird im Aufsatz gezeigt.

La sûreté de l'évaluation des prix du creusement inhérante aux cahiers de charges. Une évaluation statistique de la fonction de distribution des quantités et des qualités de rochers en excavations souterraines de 14 (10) tunnels, de longueurs et de gabarits différents, transversant les Alpes Centrales Autrichiennes, mène à la conclusion que cette fonction a l'apparence de cloche-Gauss, distinguée d'une dispersion remarquablement serrée, pourvu que les valeurs du calcul soient prises du règlement de compte définitif.

Le coût total des travaux de creusement qui fut facturé à la fin de l'éxecution du projet excédait sensiblement le prix total offert dans les exemples, dont la dispersion standard excédait 35%, si bien que le calcul de distribution des qualités et des quantités de rochers fut basé sur le prix moyen pesé d'un certain offre de prix, de même s'il était basé sur le prix de la classe de rocher la plus fréquente. Ainsi la déviation standard de la fonction de distribution des prix unitaires et des quantités de classes de rochers d'un certain offre de prix pour un creusement souterrain s'avère d'être un étalon pour la sûreté d'un prix offert.

L'amélioration de rentabilité de la Nouvelle Méthode Autrichienne de Creusement des Tunnels NATM serait garantie par une meilleure interprétation des vitesses de déformations arpentées en relation aux moyens de support appliqués, l'utilisation des cahiers de charges standardisés, la publication des prix offerts en comparaison avec les prix mis en charge définitifs incluant les quantités et les prix des moyens de soutien relatives aux classes diverses de rocher. Une méthode qui allégérait cette tâche sans trahir des secrets de gérance est demonstrée et employée dans la thèse.

1. Scope of the Problem

Tunnels are very expensive investments. All decisions influencing the cost of their construction and maintenance should be based upon reliable estimates. As 30% to 50% of the total tunnel cost are evoked by the cost of rock excavation, the reliability of estimating the qualities and quantities of the different rock classes within one certain project is an important problem [1, 2].

The overwhelmig influence of rock excavation cost upon the total investment cost may be shown in the following examples.

The impact of the cost of rock excavation plus support means on the total investment cost of this tunnel is:

$$83.42\% \times 1.1 \times (0{,}2326 + 0.1969) = 39.41\%$$

which is well situated between the average of 30% to 50% reported in the technical papers. The reliability of estimating the total cost of a tunnel will be influenced by an eventual error of $\pm 10\%$ concerning the cost of

Table 1. Composition of Total Investment Cost of a Tunnel With Extremly Low General Cost (Austrian Schilling 1962—1967) [4]

Structural investments		
Buildings for ventilating	29 026 588	
Boxes for toll receivers	389 815	
Control posts	144 348	29 560 751
Tunnel North	171 121 380	
Tunnel South	163 851 055	
		334 972 435
Structural investments		364 533 186
Mechanical and electrical Equipment		
Lighting	5 813 950	
Installations for water supply	1 197 968	
Installations for ventilating, traffic control, noxious outfall	8 036 637	
Power supply and distribution	22 231 634	
Mechanical and electrical equipment		38 280 189

General cost

Investment	Interests	Overhead	Engineering	
Buildings	1 291 180	3 295 096	2 096 089	
Tunnel	22 267 904	—	—	
M & E-Equipment	1 427 377	—	160 000	
General cost	24 986 461	3 295 096	2 256 089	
				30 537 646

Total Investment Cost

Investments	Net cost	General cost	Total	%	
Buildings	29 560 751	2 382 402	31 943 153	7.37	
Tunnel	334 972 435	26 567 867	361 540 302	83.42	
M & E-Equipment	38 280 189	1 587 377	39 867 566	9.21	
Total (besides access roads)				100.00%	
					433 351 021

Table 2. Composition of Different Cost Elements of Tunnelling Construction (Austrian Schilling 1962—1967) Tunnel North. (Final invoice) [4]

Cost element	AS	%
Site plant	13 101 205.60	10.00
Rock excavation	30 459 005.09	23.26
Support with welded meshes, shotcrete, steel beams, arches, steel bars, woven wires, rock bolts	25 791 529.53	19.69
Drainage	3 787 976.53	2.89
Bore holes for and injections, dewatering, sealing, insulation, concrete of walls and vault, forms, joints, joint sealing strips	28 511 924.60	21.77
Structural investment for ventilation	18 116 075.49	13.83
Pavement and footways	6 646 864.97	5.07
Geologically provoked supplementary works	1 186 557.96	0.90
Supplementary construction, swinging cavern	3 342 271.44	2.59
Construction cost of Tunnel North, contract prices without inflation compensation	130 933 401.21	100.00 %

rock excavation and support with $\pm 10\% \times 0.4 = \pm 4\%$ of the total investment cost.

As the influence of the support means cost is of the same amount that is shown of excavations cost, i. e. 19.69% to 23.26%, and as a pessimistic evaluation of the rock qualities provoques the same uncertainty in the cost of support means, it is justified, to restrain this research to the specific cost of rock excavation. This limitation is also imposed by the diversity of specifications philosophies, where the imputation of a certain amount of support cost to a specific cost of excavation is rarely made.

It has to be kept in mind, that the average deviation of the definite values from the estimated cost of rock excavation will have nearly the twofold impact upon investment cost, as consequence of the error in the cost of support means, which will be of a similar scale.

The composition of the components of tunnel construction cost of a bid for another tunnel is shown in Table 3.

Table 3. Components of the Tunnel Construction Cost [5]

Construction works	% of construction cost	
Rock excavation	42.31	
Shotcrete	9.85	Support means
Welded meshes, support with steel beams and arches	5.30	Support means
Steel bars	2.14	Support means
Woven wires and rock bolts	5.25	Support means
Bore holes and injections	1.34	
Concrete lining, forms, joint sealing strips	15.17	
Dewatering, supplementary provisions against water influx	2.98	
Sealing, drainages	6.34	
Walls and floors of air channels wall plates, placing, ducts	9.32	
Total cost of construction works	100.00%	} 22.54%

The purpose of this paper is to provide the answer to the following problems:

1) Reliability of geological prognostic in tunnelling;

2) Frequency and distribution of rock classes in underground excavations;

3) Geotechnical rock classification;

4) Reasonable investment cost portion for the geological investigations and the technical survey of the support means;

5) Statistically based prognostics of rock classes.

2. Solution of the Problems Mentioned Above

The following computations are established upon the hypothesis that so many causes of geological nature are influencing the qualities and quantities of rocks along every tunnel axis that their frequency and distribution are subjected to the law of hazard. The second assumption of the author, which justifies this statistical approach to this problems, is, that the cost of rock excavation integrate the influence of rock quality parameters much more realistically than the mechanical properties of rocks utilized for the rock classification. The strength of rocks depends upon specimen preparation, technology of laboratory tests, rate of loading, environment, moisture, liquids etc. and especially on the confining pressures, that are changing on every spot and at every time in situ. These parameters are dynamic, time dependent, influenced by the tunnel profile and the procedure of tunnelling and therefore not comparable constants. Every rock may be elastic, plastic, brittle, its failure behaviour is not caused by the mineralogical and structural nature of the rock but by the kind of stressing.

These ideas of the author could be easily proven within a few tunnels. The quantification of the data derived from different sites, situated in diverse rocks, and originated under various costlevels and performance conditions, and procedures, had to be chosen under the handicap that uniform specifications are nonexistent. It must be kept in mind that a statistical approach to these problems will never be able to predict the objective rock class on a certain spot of a tunnel axis but may only estimate the relative frequencies of a collectif of rock classes. The geological expertise is the starting point of the statistical prognosis. Therefore a thorough investigation of the geological conditions and the geotechnical behaviour of the rocks that are perforated by the tunnel are never to be replaced by a statistical calculation.

The verbal classification of rocks which is used in this paper is also applied to bids and final invoices for performances of the New Austrian Tunnelling Method (NATM). Within the collective mass of rock quantities and their specific unit prices for rock excavation it was possible to eliminate the differences in site conditions, basic price levels, inflation, by utilizing relative prices, in reference to the weighted average unit price and to the most frequent unit price. The Histographs were drawn with the specific unit Price S_i upon the abscissa and frequency F_i upon the ordinates [6].

Rock class	Quantity Q_i (m, m^3)	Unit price P_i (S/m, S/m^3)
I Sturdy	Q_{I}	P_{I}
II Friable	Q_{II}	P_{II}
III Very friable	Q_{III}	P_{III}
IV Squeezing	Q_{IV}	P_{IV}
V Very squeezing	Q_{V}	P_{V}
VI Extremly squeezing ...	Q_{VI}	P_{VI}

Average unit price

$$P_A = [P_i \cdot Q_i]_{\text{I}}^{\text{N}} / [Q_i]_{\text{I}}^{\text{N}} \quad (1)$$

Total Quantity

$$Q_T = [Q_i]_{\text{I}}^{\text{N}} \quad (2)$$

Specific unit price (percent)

$$S_i = 100\, P_i / P_A \quad (3)$$

Relative frequency (percent)

$$F_i = 100\, Q_i / Q_T \tag{4}$$

Standard deviation

$$s_A{}^2 = [\,(P_A - P_i)^2\, Q_i]/(100-1) \text{ resp.}$$
$$s_F{}^2 = [\,(P_F - P_i)^2\, Q_i]/(100-1) \tag{5}$$

Most frequent unit price P_F.

The statistical computation was made for every tunnel with the unit-prices of the bid which got the award. The quantities were taken from the bid, and than from the final invoice; the qualities (unit prices) were taken from the bid, so that the results of final price corrections are not taken into account, but only the differences of quantities of rock classes as they

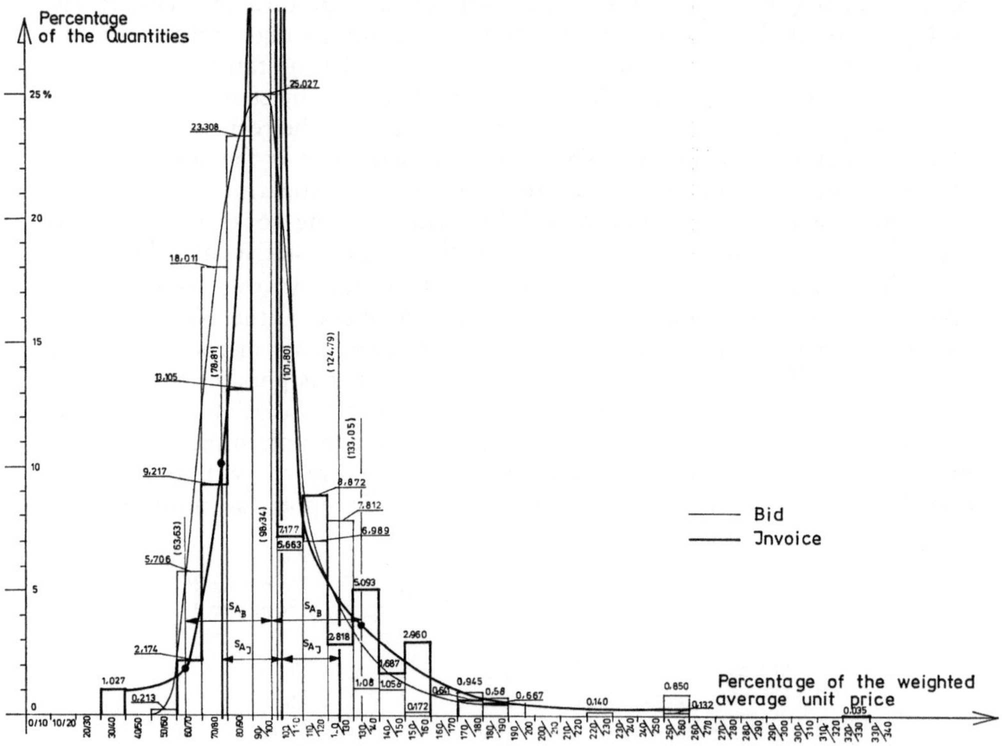

Fig. 1. Distribution of rock classes. Comparison of Bids and Invoices. Abscissa: Percentages of the weighted average unit price. Ordinates: Percentage of Quantities
——— Bid ——— Invoice

Häufigkeitsverteilung der Felsausbruchklassen. Vergleich der Anbote und Schlußabrechnungen. Abszisse: Prozente des gewogenen mittleren Einheitspreises. Ordinaten: Prozente der Mengen
——— Anbot ——— Schlußabrechnung

Distribution des classes de rocher. Comparaison des offres de prix et des comptes rendus. Abscisse: Pourcentage de la valeur moyenne pesée des prix unitaires. Ordonnée: Pourcentage des quantités
——— Offre de prix ——— Compte rendu

were foreseen and as they were really stated in situ. The standard deviation in the bid is S_{A_B} (bid), S_{A_I} (final invoice) is calculated for the average unit price S_A and for the most frequent unit price S_F. A comparison was also made

Table 4. Standard Deviation of the Distribution Function of Rock Classes in Bids and Invoices. Deviation of the Final Price From the Offered Price in Relation to the Standard Deviation of Rock Classes Predicted and Realized

Tunnel	Length m	Profile m²	s_A % B	I	s_F % B	I	%
Malta*	8998/9001	6.65—23.0	8.43	5.62	8.74	5.73	−1.64
Göss*	5910/5901	6.65—23.0	5.90	28.87	5.98	30.01	+2.55
Klamm	1584	66.5	28.22	25.43	32.55	29.65	−4.74
Mauth	207.3	66.5	76.36	27.41	105.03	32.58	+106.66
Tauern, r, N, S	5482	88.7	57.30	38.60	73.41	25.44	+101.17
Katschberg, r, N	3590	88.7	43.98	18.91	62.86	20.81	−18.93
Katschberg r, S	1784	88.7	23.84	29.12	25.26	39.78	+0.04
Katschberg v, N	1600	29.12	3.62	0.74	50.39	0.75	+19.48
Katschberg v, S	3813	29.12	20.35	28.79	24.57	—	—
Wolfsberg**	292	62.0	28.35	28.83	19.05	39.77	−26.28

* Enlargement to a pilote tunnel.

** Enlargement to 2 pilote tunnels, not representatif, delay 1942—1970 between first and final construction works.

between the preliminary cost and the final cost, I/B, calculated with the unit prices of the bid, but with the quantities of the offer and of the final invoice

$$d\ \% = 100 \cdot \frac{[Q_i P_i]_I - [Q_i P_i]_B}{[Q_i P_i]_B} \tag{6}$$

The collectifs of quantities/weighted average unit prices and resp. most frequent unit prices were established as follows:

$$\frac{\text{Unit prices}}{\text{Histograph of weighted unit price / rock quantities}}$$

Bid: $98{,}34 \pm 34{,}71\%$

Invoice: $101.08 \pm 22{,}99\%$

$$\frac{\text{Unit prices}}{\text{Histograph of most frequent unit price / rock quantities}}$$

Bid: – –

Invoice: $106.64 \pm 21.97\%$

In order to establish the histrograph of relative unit prices/weighted average unit price and quantities the different price groups were subdivided in prices of 10% to 10%; the histrograph of relative unitprices/most frequent price the subdivision was made in 20% to 20%.

3. Conclusions

3.1. Reliability of the Rock Classification in Bids

As it has been shown in Point 2 of this paper, the collectif of unit-prices/weighted unit price and relative quantities of rock excavation within every underground project has a surprisingly small standard deviation, when the histograph is established from the quantities of the final invoice. The

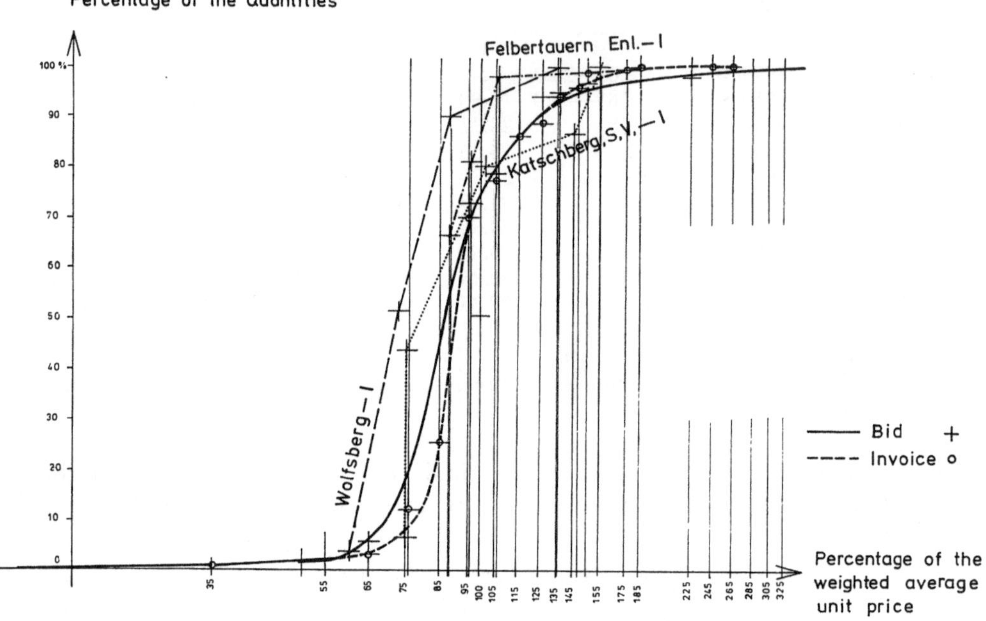

Fig. 2. Cumulative distribution of rock classes. Bids and Invoices. Abscissa: Percentages of the weighted average unit price. Ordinate: Percentage of the quantities

Summenhäufigkeitsfunktion der Felsausbruchsklassen. Anbote und Schlußabrechnungen. Abszisse: Prozente des gewogenen mittleren Einheitspreises. Ordinate: Prozente der Mengen

Courbe cumulative de la fréquence des classes de rocher. Offre des prix et comptes rendus. Abscisse: Pourcentage de la valeur moyenne pesée des prix unitaires. Ordonnée: Pourcentage des quantités

standard deviation of the collectif of unitprices/most frequent unit price and relative quantities is also surprisingly small, when it is calculated from the quantities of rock excavation, realized during the performance of the works and taken from the final invoice. The same statement may be drawn from the totality of 10 tunnels that were perforated through the Central Alps of Austria.

A comparison of the deviations of the final prices from the offered prices with the standard deviation of the collectif of relative unit prices/relative quantities of rock excavation shows that a standard deviation calculated from the data of the bid greater than ±35% will lead to a considerable augmentation of the total price for rock excavation in the final invoice. This statement is useful, if the weighted average unit price, or the most frequent unit price, is taken to calculate the relative unit prices.

In order to prove this hypothesis there will be made a comparison of the geological forecast, the price calculation and the rock classes that were ascertained in situ in one of the projects, taken by chance, where the author prepared the tender papers, that were based upon a very thoroughly founded geological examination:

Rock classes	Prognosis (%)	Tender (%)	Invoice (%)	Unit price (S/m)	Bid length (m)
Sturdy	75	71.42	77.8	7649.70	2600 m
Yielding	15	17.85	8.0	7474.90	650
Friable	7	7.41	9.0	9330.00	270
Very friable	2	2.19	4.0	11046.00	80
Squeezing	1	1.09	1.0	18825.00	40
Very squeezing	0	0.04		42374.00	(10)

Standard deviation of the Bid: 16.66%
Total price of excavation of the invoice: 29 579 587.59 S
tender: 28 903 685.00 (+2.33%)
Calculated from geological forecast: 28 910 847.00 (+2.31%)

In establishing the tender papers the geological forecast was slightly altered in order to get also an unit price for "very squeezing rock" and to round up the quantities. The geological prognosis was established by an expert who has a broad experience in NATM [7].

3.2. Frequency and Distribution Function of Rock Classes in Tunnel

The relatively small collectif of rock classes and unit prices which was available for an investigation, unveils that an almost normal distribution of quantities/qualities seems to be existent. It is deplorable that publications of costs of underground excavation, peculiarly of the comparison of the prices of bids with the prices of invoices are nearly nonexistant. Given the fact, that there is no uniformity in the geometry, the methods and equipments for excavation and support in tunnelling, the costly gained experiences are dissipated without a reporting based upon quantified cost parameters. A neutral kind of comparison, which will not betray business secrets, has here been shown and its usefulness has been proven.

3.3. Geotechnical Rock Classification

The verbal description of rock classes utilized by geologists is mainly derived from their previous experience in tunnelling. The prices for advancing depend form the speed in excavating (boring, tamping, shooting, ventilating, application of security means, loading, hauling). The supplement time necessary for the support works (shotcreting, bolting, meshing, arching etc.)

are heavily influencing the speed of advancing, and this is true not only for the usual excavation works with explosifs but also for mechanical continuous mining. Therefore the list of quantities should specify the different rock classes with the allocation of the necessary support means to every class of rock excavation. There are two methods in forecasting the amount of support means: a statistical evaluation, drawn from former experiences in similar conditions, or an analytical approach. The definitif means of support are calculated in accordance with the results of deformation measurements, e. g. from model tests in the scale 1 : 1. The author is convinced, that the qualification of the support means-classes through their specific costs and

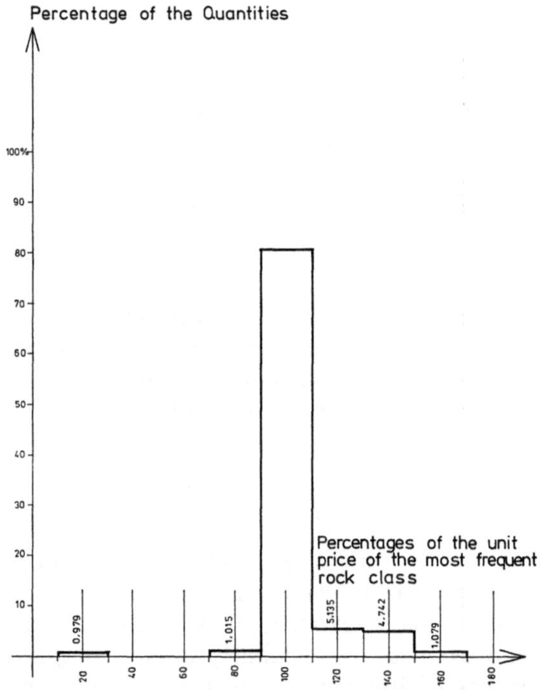

Fig. 3. Distribution of rock classes. Abscissa: Percentages of the unit price of the most frequent rock class. Ordinate: Percentage of the quantities

Häufigkeitsverteilung der Felsausbruchsklassen. Abszisse: Prozente vom Einheitspreis der häufigsten Felsausbruchsklasse. Ordinate: Prozentsätze der Mengen

Distribution des classes de rocher. Abscisse: Pourcentage du prix unitaire de la classe de rocher la plus fréquente. Ordonnée: Pourcentage des quantités

a statistical approach to this problem will be at least as valuable as any analytical calcul, which leads to a simplification, whereas every price is a synthesis of all rock data. The statistical evaluation would be facilitated by the standardization of the specifications and by the publication of rock parameters, deformation velocities and support means with a comparison of the bid and invoice total price.

3.4. Investment Cost Portion for the Geological Investigations and for the Survey of the Support Means

The choice of the optimum route within the variants of a project depends heavily on the total cost of the different tunnels, and their cost are mostly influenced by the cost of underground excavation and support works. As a geologist will be able to elaborate a prognosis of the underground, based upon the data that are available on the surface, he may

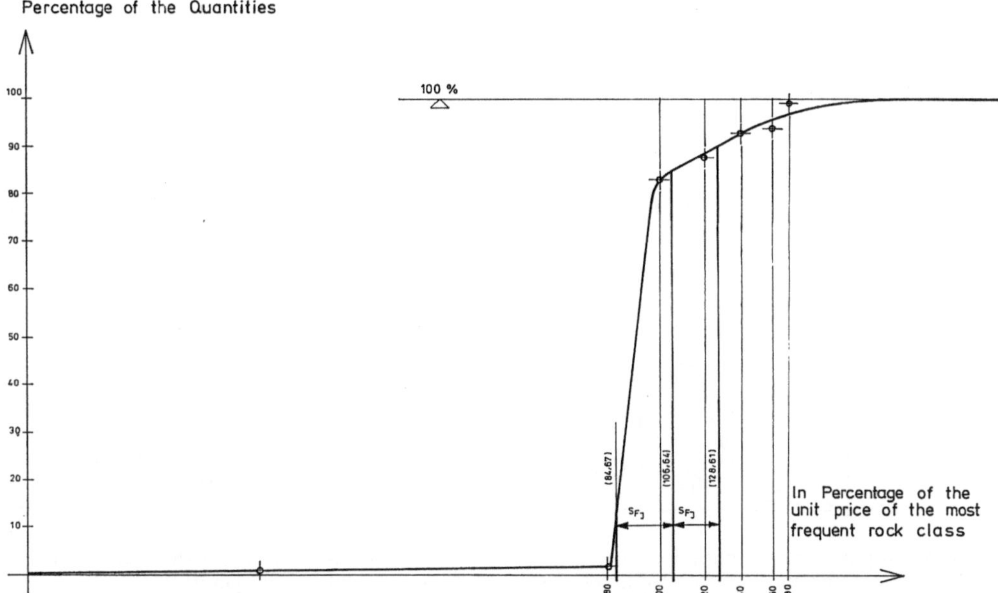

Fig. 4. Cumulative distribution of rock classes. Abscissa: Percentages of the unit price of the most frequent rock class. Ordinate: Percentage of the quantities. Invoices.

(Scale of the Abscisse: log nat of the percentage of the unit price of the most frequent rock class)

Summenhäufigkeitsfunktion der Felsausbruchsklassen. Abszisse: log nat des Prozentsatzes vom Einheitspreis der häufigsten Felsausbruchsklasse. Ordinate: Prozentsatz der Mengen. Schlußabrechnungen

Courbe cumulative de la fréquence des classes de rocher. Abscisse: log nat du pourcentage du prix unitaire de la classe de rocher la plus fréquente. Ordonnée: Pourcentage des quantités. Compte rendu

sketch a trustful picture of the rock classes and their relative distribution along the tunnel axis. Thus the one leading coordinate of the distribution function, the quantities are given. The other coordinates, the qualities, are the unit prices for excavation and support, they are the result of the submission of bids. One offer may be proven as to be reliable, if the standard deviation does not exceed $\pm 35\%$. A supplementary criterium is shown by [8].

The cost of survey-measurements of deformations during the optimization of support means are 2% to 3% of the construction works of tunnelling.

As every tunnel is a prototype, with more or less available geological maps, data from previously performed underground projects in the vicinity etc., an average percentage of the cost of geological investigation may not be

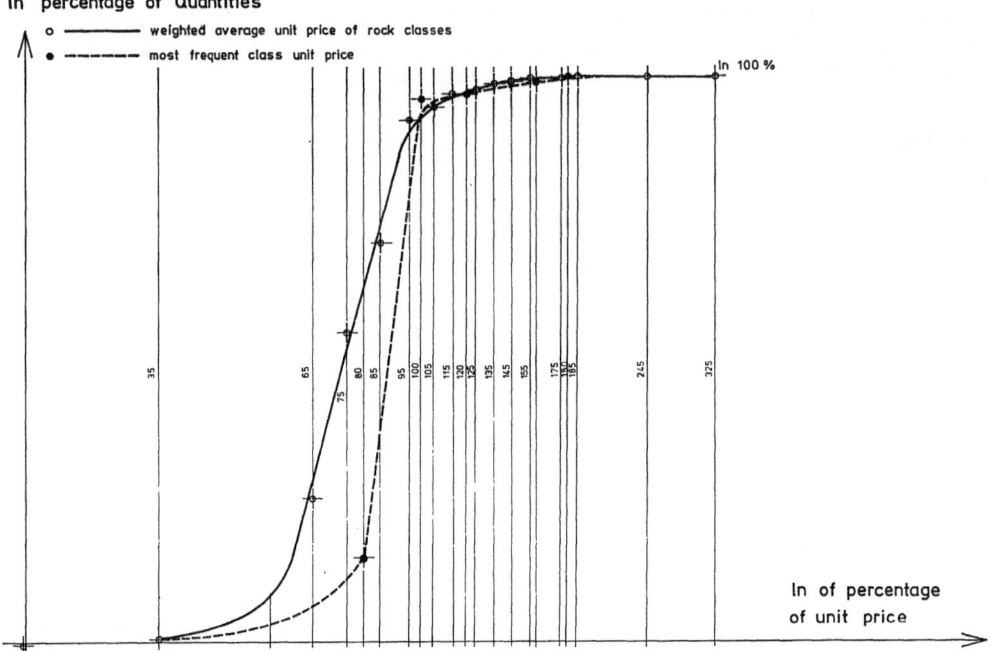

Fig. 5. Cumulative distribution of rock classes (Invoices). Abscissa: log nat of the percentage of the unit price. Ordinates: log nat of the percentage of quantities

——— weighted average unit price of rcok classes; – – – most frequent class unit price

Summenhäufigkeitsfunktion der Felsausbruchsklassen (Schlußabrechnungen). Abszisse: log nat des Prozentsatzes vom Einheitspreis der Felsausbruchsklasse. Ordinate: log nat des Prozentsatzes der Mengen

——— log nat des Prozentsatzes vom gewogenen, mittleren Einheitspreis für den Felsausbruch; – – – log nat des Prozentsatzes vom Einheitspreis der häufigsten Felsausbruchsklasse

Courbe cumulative de la fréquence des classes de rocher (Comptes rendus). Abscisse: log nat du pourcentage du prix unitaire. Ordonnée: log nat des quantités

——— log nat du pourcentage du prix unitaire moyen pesé des classes de rocher; – – – log nat du pourcentage du prix unitaire de la classe de rocher la plus fréquente

excepted rationally. A geological prognosis is undispensable. Table 4 of this paper may show the amount of reliability of the rock class prognosis. Unforseen complications are here excluded, because they cannot be eliminated by the most elaborate expertise.

3.5. Statistically Based Prognostics of Rock Classes

The existence of an almost normal distribution of excavation-rock-classes leads to the expectation, that there will also be a normal distribution of support-classes. In order to get a high reliability of cost estimates, the publication of the frequency of support classes and their relation to the total sum of cost in the bid and invoices is suggested, being much more useful than any refined statical calcul, which will ever be an oversimplification of the dynamical behavior of rocks.

References

[1] Kurzmann, E.: Die Zuverlässigkeit von Kostenschätzungen für Großinvestitionen. Forschungsförderungsfonds der gewerblichen Wirtschaft, Wien, 1974.

[2] Ministère de l'Equipement et du Logement, Service des Tunnels: Dossier pilote des Tunnels. Organe technique régional de Lyon/France, 1970.

[3] Dokumentation österreichischer Straßentunnels. Forschungsförderungsfonds des Österreichischen Bundesministeriums für Bauten und Technik, Wien, Österreich (to be published 1979).

[4] lo [3].

[5] lo [3].

[6] Kreyszig, E.: Statistische Methoden und ihre Anwendungen. 4. Aufl. Göttingen: Vandenhoeck & Ruprecht 1973.

[7] Demmer, W.: Geologische Prognose und Tatsache. Unpublished Note from 75-11-24 to Österr. Draukraftwerke, Austria.

[8] Kurzmann, E.: Strukturanalytische Feststellung des Billigstbieters auf Bauleistungen. ÖIZ 11. Jg., H. 10 (1968).

Address of the author: Dipl.-Ing. Dr. Ernst Kurzmann, Zivilingenieur für Bauwesen, A-9535 Schiefling 135, Österreich.

Rock Mechanics, Suppl. 7, 67—85 (1978)

Rock Mechanics
Felsmechanik
Mécanique des Roches

Dreidimensionale Spannungsumlagerungsprozesse im Bereich der Ortsbrust

Von

L. Müller-Salzburg, G. Sauer und M. Vardar

Mit 11 Abbildungen

Zusammenfassung — Summary

Dreidimensionale Spannungsumlagerungsprozesse im Bereich der Ortsbrust. Die Probleme des Tunnelbaues in nicht standfestem Gebirge werden maßgeblich durch das dreidimensionale Geschehen in der Umgebung der Ortsbrust — der Vortriebs- und Sicherungsart sowie der Vortriebsgeschwindigkeit in bezug zur Standzeit — bestimmt. Diese Erkenntnisse sind an sich sehr alt, doch hat wohl die komplexe Natur des Verhaltens des Gebirges in der Tunnelumgebung die Forschung von globalen Untersuchungen in dieser Richtung weitgehend abgehalten. Erst in jüngster Zeit wurden einige Schritte unternommen, die räumlichen Probleme und den „Zeiteinfluß" analytisch und numerisch mit Hilfe der Kontinuumsmechanik zu untersuchen.

Der vorliegende Artikel stellt eine Zusammenfassung zweier umfangreicher, experimenteller Arbeiten dar, die auf teilweise unkonventionellen Wegen zu einigen bemerkenswerten Ergebnissen gelangt sind. Es wurden sowohl dreidimensionale Spannungsumlagerungen als auch der Zeiteinfluß auf das Tragverhalten untertägiger Hohlräume meßtechnisch an Tunnelbauwerken und an vergleichbaren äquivalenten Modellen untersucht.

Eine spannungsentlastete Zone im Vortriebsbereich, gefolgt von Gebirgsdruckkonzentrationen oder Konvergenzsprüngen im Abstand von 1 bis 2 Durchmessern hinter der Ortsbrust, eine wellenförmige Spannungsverteilung im Gebirge sowie eine sprunghafte zeitabhängige Umlagerung von Spannungsspitzen im Zusammenhang mit der Stehzeit und der Vortriebsgeschwindigkeit stellen einige Ergebnisse dieser Untersuchungen dar.

Die Grenzen unseres hauptsächlich verwendeten Gedankenmodells — der Kontinuumsmechanik — werden sichtbar und es war mit ein Ziel dieser Arbeit, zu neuen Denkmodellen anzuregen.

Three-dimensional Stress Rearrangement Processes in the Area of a Working Face. The problems of tunnelling in unstable rock masses are essentially determined by the three-dimensional behaviour in the vicinity of the working face — the mode of excavation and of supporting as well as the speed of excavation with regard to the standup time. These findings in themselves are very old, but the complex behaviour of the rock mass around the tunnel has up to now considerably prevented the researchers from global investigations in this direction.

The present paper is to summarize two extensive experimental works, which have led to some remarkable results on partly unconventional ways. As well three-dimensional stress rearrangements as also the influence of time on the bearing behaviour of underground excavations had been measured at tunnel sites and at comparable equivalent models.

Some results of these investigations are: A stressrelieved zone in the heading area, succeeded by concentrations of ground pressure of leaps of convergency in a distance of 1 to 2 diameters behind the working face, an undulatory stress distribution in the rock mass and a jerky rearrangement of stress peaks in connexion with standup time and driving speed.

The limits of our mainly applied abstraction — the theory of the continuum — become obvious, and it was an aim of this work to suggest new models.

1. Problemstellung

Der weitaus größte Teil aller wissenschaftlichen Untersuchungen im Tunnelbau beschäftigt sich mit dem Problem der gelochten Scheibe in allen Variationen. Nichtlineare Spannungs-Dehnungsbeziehungen, Post-Failure-Verhalten, Anisotropie und rheologische Modellvorstellungen bilden die Marksteine dieser Entwicklung.

Die Grundlagenforschung bezüglich des Materialverhaltens von Boden und Fels überrascht die angewandte Forschung mit immer neuen Zusatzgliedern des konstitutiven Tensors. Nachdem die Analytiker zum großen Teil das Feld geräumt haben, sind gegenwärtig eine Vielzahl von Numerikern mit dem Einbau dieser Erkenntnisse in ihre Stoffmatrix beschäftigt.

Man sieht sich einer schillernden Vielfalt von dermaßen komplizierten Ansätzen für das zweidimensionale Problem gegenüber, daß echte dreidimensionale Untersuchungen im Felshohlraumbau sowohl an mathematischen wie auch an physikalischen Modellen kaum gewagt oder doch nur sehr zögernd in Angriff genommen werden. Nun zeigen aber gewisse Ergebnisse der neueren Grundlagenforschung, vor allem aber systematisch ausgewertete Erfahrungen vor Ort, daß eine Reihe vor allem geometrischer Faktoren, wie Teilausbruchsquerschnitt, Abschlagtiefe, Vortriebsgeschwindigkeit und wirksame Ringschlußzeit, mit Ausnahme des ersten alles Dinge, die nur in der dritten Dimension, in der Tunnelvortriebsrichtung, zur Geltung kommen, doch auch das Spannungs- und Deformationsfeld der zweidimensionalen „gelochten Scheibe“ ganz entscheidend bestimmen! Da die meisten Spannungsumlagerungen im Boden und Fels mit irreversiblen Prozessen verbunden sind, bei denen das physikalische Verhalten des Materials bleibend verändert wird, ist bei dreidimensionalen Studien von vereinfachten, etwa elastizitätstheoretischen Ansätzen nicht viel zu erwarten. Es liegt aber nahe, die leistungsfähigeren Modellversuche heranzuziehen (Abb. 1). Von solchen Studien soll hier berichtet werden.

Die dreidimensionalen Umlagerungsprozesse in der Umgebung einer Ortsbrust zeigen je nach Vortriebsweise hinsichtlich Spannungs- und Deformationsverteilung teilweise geradezu das Gegenteil von gerechneten Ergebnissen. Theoretisch ungeklärt war bisher das in gewissen Zeitabständen auftretende, bruchartige Einstellen neuer Gleichgewichtszustände, die bei entsprechenden Voraussetzungen zu Gebirgsschlägen führen können. Dreidimensionale Betrachtungen auf analytischem sowie auf numerischem Gebiet sind in den letzten Jahren mehr und mehr versucht worden, können aber (aus Kapazitätsgründen) nur relativ einfache Stoffgesetze berücksichtigen.

So versucht Smoltcyk (1969) den achsialen Erddruck auf einen Tunnelschild über die Lösung einer elastischen Differentialgleichung zu ermitteln.

Egger (1973) gibt aufgrund der „Pacher-Kennlinie" (1964) die iterativ zu lösende Differentialgleichung für das Post-failure-Verhalten an, die von Panet (1976) nachempfunden wurde. Lombardi (1974) zeigte anhand von

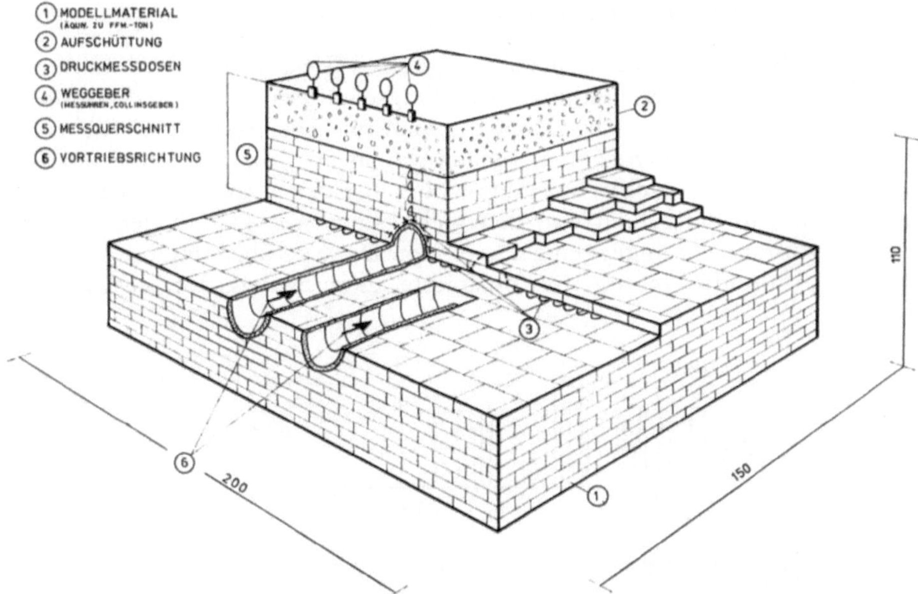

Abb. 1. Modellkasten
Equipment for model tests

Überlegungen über den Verlauf der Hauptspannungstrajektorien Einfluß und Stützwirkung der tunnelparallelen Spannungen und ging auf den Zeiteinfluß ein, während Egger (1974) ein analytischer Ansatz unter Annahme einer halbkugelig geformten Tunnelbrust gelang.

Rziha schrieb bereits 1874: „Es ist für die Sicherheit eines Tunnelbaues nicht gleichgiltig, in welcher Art und Weise man das Gebirge aushöhlt, dieses ‚abbaut' oder ‚aushaut'"; und Heim bereits 1904: „Eine sehr wichtige Rolle spielt der Faktor ‚Zeit' in allem ..."

Anwendungsbezogen zeigten Feder und Arwanitakis (1976) nach dem Prinzip des „schwimmenden Ringes" (Kompatilitätsbedingung nur für Radialspannungen) von Egger eine einfache dreidimensionale Betrachtungsweise unter Zugrundelegung der Mohrschen Bruchhypothese.

Die numerischen (fast ausschließlich FEM) Untersuchungen, wie sie von J. Aalto und E. M. Salonen (1975) sowie von Rodatz, Wallner

(1974) u. a. vorgestellt werden, zeichnen sich noch durch relativ lange Rechenzeiten bzw. sehr stark vereinfachte Ansätze aus und unterliegen wie die meisten FE-Programme einer algorithmusabhängigen Fehleranfälligkeit (Gudehus, Goldscheider, Winter, 1975), siehe auch Maidl und Geisler (1976).

Physikalische Modelle mit im Vergleich zur Natur äquivalenten Eigenschaften gibt es mit einer unserem Thema entsprechenden Zielsetzung kaum.

2. Ein Traggewölbe als Summe mehrerer Tragschalen: Zwiebelschalenartige Tragzonen nach Sauer

Das Gebirge reagiert auf eine Störung seines Gleichgewichtes durch einen Tunnelvortrieb mit der Bildung von Tragstrukturen (auch Schutzzonen, Traggewölbe, Wissmannsche Zonen genannt). Die bisherigen Lehrmeinungen (Fenner, Kastner u. a.) gehen davon aus, daß es nur einen Gebirgstragring um den Hohlraum gibt, wobei Ort und Größe der Spannungskonzentration in demselben von den Gebirgskennwerten und dem Primärspannungszustand abhängig gemacht werden. Diese gefühls- und überlegungsmäßig (Heim und Wissmann) konzipierte, dann analytisch nachgewiesene

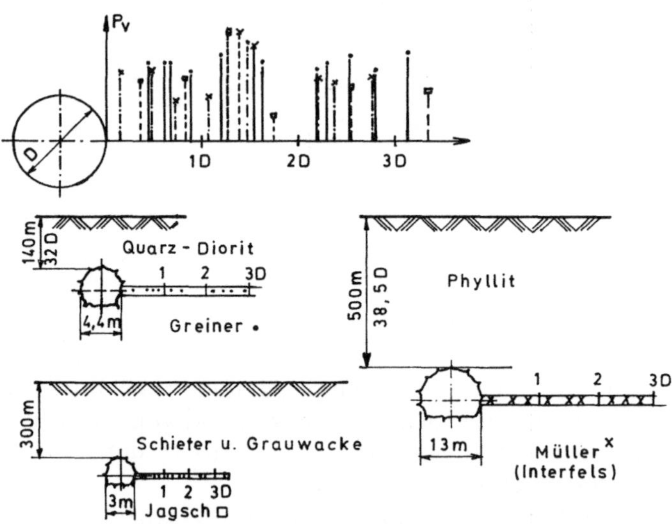

Abb. 2
Zusammenstellung verschiedener Vertikaldruckmessungen in unterschiedlichen Tunneln
Examples of vertical stress measurements in different tunnels

(Schmid, Fenner, Kastner) Schutzzone wird auch durch inzwischen zahlreich vorliegende numerische Berechnungen bestätigt — nur gemessen wurde in dieser Richtung wenig und noch weniger veröffentlicht.

Die In-situ-Messungen von Müller, G., 1976; Greiner, G., 1976, und Jagsch, D., 1974, in einem Tunnel und zwei Stollen (Abb. 2) zeigen mehrere hintereinander angeordnete Druckmaxima mit dazwischenliegenden Minima. Interessant ist dabei, daß sich trotz der sehr unterschiedlichen felsmechanischen und geometrischen Gegebenheiten die Maxima bzw. die Minima, auf den jeweiligen Tunneldurchmesser bezogen, bei den verschiedenen Tunneln in etwa decken. Die Messungen im Quarzdiorit (Greiner) und im Phyllit

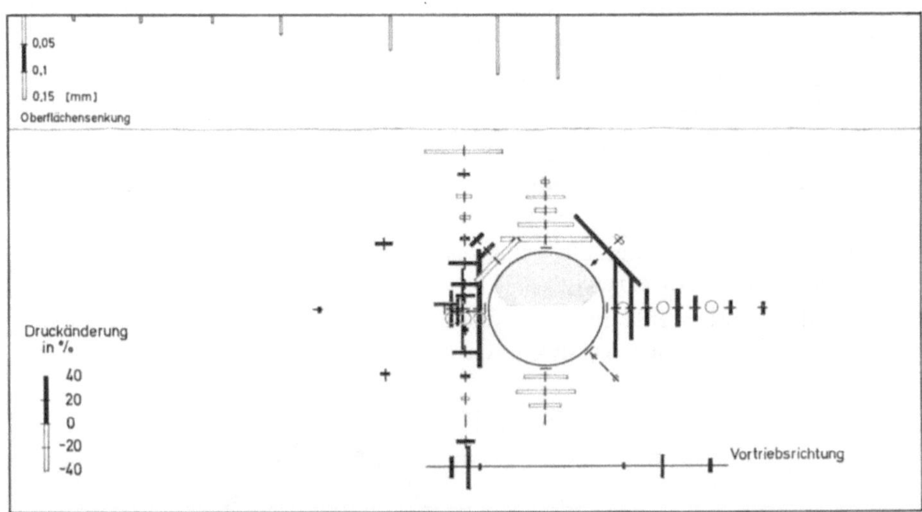

Abb. 3. Oberflächensenkung und Spannungsänderungen in einem Querschnitt nach dessen Durchörterung mit einer Tunnelröhre in einer Entfernung von 3D, bezogen auf die Werte, welche im Abstand von 3D vor Erreichen dieses Querschnittes gemessen wurden

Settlements of surface and changement of stresses in a section after its excavation. The stress values are related to the measurements before starting the excavation

(Müller) wurden mittels „Dorstopper-Methode" durchgeführt, letztere zusätzlich durch Ultraschallmessungen nachgeprüft und bestätigt. Die Messungen im Schiefer-Grauwacken-Gebirge erfolgten mit dem „Interfels-Meßstern". Auch bei weiteren Messungen (Obert, 1967; Kovari et al., 1972) sowie bei den Modellversuchen zeigte sich ein ähnliches Bild, bei letzteren jedoch ist das erste Maximum verschiedentlich noch unterteilt in einzelne Spannungsspitzen. Diese für die Vertikalspannungsverteilung in einem horizontalen Schnitt durch die Röhrenachsen geltenden Aussagen, welche an sich schon überraschend sind, werden noch verwirrender, betrachtet man die Ringspannungsverteilung in einem Bereich von der Breite 1 D im Gebirge um die Röhre herum (vgl. Abb. 3, Sauer 1976). Unter der Annahme von Gebirgstragringen von der Form stehender Ellipsen lassen sich in der Umgebung der Röhre im ersten Maximum noch große Spannungsdifferenzen feststellen.

Ganz eindeutig lassen sich jedoch aus den In-situ-Messungen und den Modellversuchen (Sauer, 1975) mindestens zwei hintereinanderfolgende Abstufungen (die erste zwischen 1,5 und 2, die zweite etwa 1 D vor der Ortsbrust) erkennen (Abb. 4). Das Spannungsumlagerungsgeschehen im unmittelbaren Bereich des Ausbruchsquerschnittes während eines Vortriebes läßt sich folgendermaßen charakterisieren:

— Etwa drei Ausbruchsdurchmesser (D) vor der Ortsbrust tritt im Gebirge in Tunnelachsenhöhe eine leichte Entlastung ein;

— 2 D vor der Ortsbrust ein merklicher Druckanstieg;

— 1 D vor der Brust ein nochmaliger starker Druckanstieg mit dem Maximum etwa 2/3 D vor der Ortsbrust;

— danach ein starker Tangentialspannungsabfall im unmittelbaren Bereich des vorübergehend unausgekleideten Hohlraumes vor Ort;

— Etwa 1 D nach Vorbeifahren der Ortsbrust (nach dem Ringschluß also) steigt die Tangentialspannung im hohlraumnahen Bereich wieder an und erreicht ein erstes Maximum etwa 1,5 D hinter der Ortsbrust:
  · In der Firste ist dies ein relatives Maximum, welches den Eingangswert nicht mehr erreicht.
  · In der Sohle fehlt in den Modellversuchen (nicht bei den In-situ-Messungen) dieser Anstieg nach dem Ringschluß.

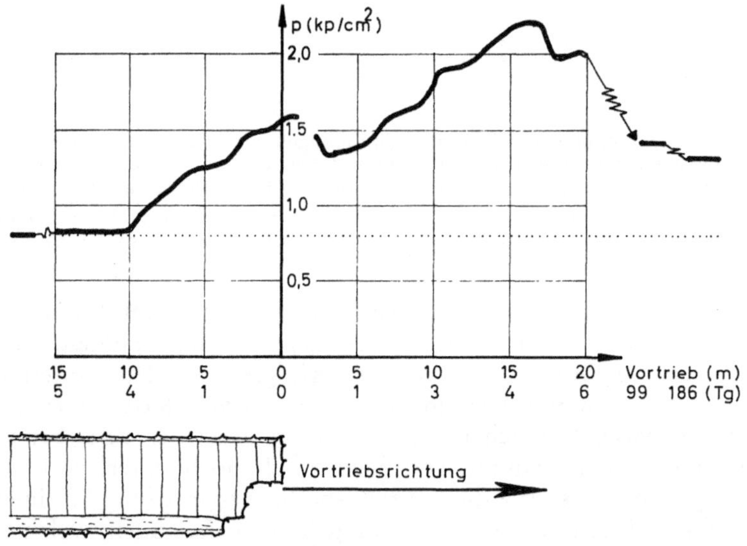

Abb. 4. Vertikalspannungsverlauf im Gebirge in Höhe der Tunnelachse während des Vortriebes

Course of vertical stresses in the rock mass at the level of the tunnel axis during excavation

Zusammenfassend scheint sich folgendes herauszukristallisieren:

— Bei der Herstellung eines untertägigen Hohlraumes reagiert das umgebende Gebirge mit der Bildung dreidimensionaler (räumlich ineinander übergreifender) Tragschalengebilde bzw. schalenförmigen Tragzonen (Zwiebelschalen). Für einen Tunnelvortrieb bedeutet dies ein Mitlaufen bzw. Vorsichherschieben der dreidimensionalen Tragschalen mit fortschreitendem Ausbruch (Abb. 5). Unabhängig vom Einbau einer Auskleidung versuchen sich die hohlraumnahen Schalen ab etwa 1 bis 2 D

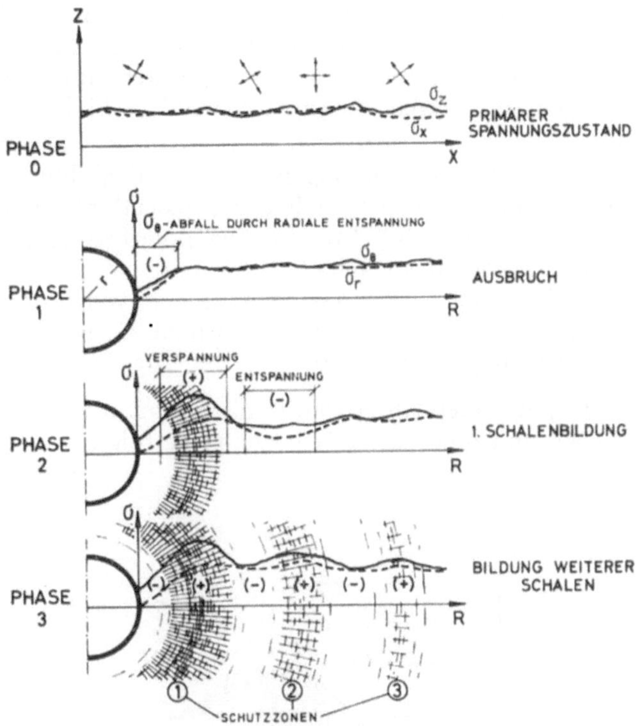

Abb. 5. Modellvorstellung von der Entstehung räumlich einander übergreifender Tragschalen bei der Herstellung eines untertägigen Hohlraumes

Model conception of the constitution of three-dimensional stress shells during construction of an underground excavation

hinter der Ortsbrust um die Röhre zusammenzuschließen und verursachen dort einen erhöhten „Gebirgsdruck", der zum Verbruch führen kann, wenn die Gebirgsfestigkeit für die Bildung der zweidimensionalen Tragringe [Wissmannsche Zone(n)] nicht ausreicht bzw. noch keine satt anliegende ringförmig geschlossene tragfähige Abstützung vorhanden ist.

— Der Abstand dieser tragenden Zonen oder Tragschalen voneinander, welche geometrische Orte erhöhter Belastungsaufnahme sind, scheint den bisherigen Messungen zufolge im Mittel in der Größenordnung des Ausbruchsdurchmessers zu liegen.

· Die Entstehung der Tragschalen ist ein zeitabhängiger dynamischer Vorgang; sie hängt mit Materialverfestigung bei Volumenkontraktion und mit Entfestigung bei Dilatation zusammen und nutzt den Unterschied zwischen Haft- und Gleitreibung.

— Die Spannungsumlagerungen sowie die Deformationsprozesse gehen meist ruckartig vor sich, wie aus fast allen kontinuierlich durchgeführten In-situ-Messungen zu ersehen ist. Diese sprunghaften Änderungen resultieren aus örtlich begrenzten, umlagerungsbedingten Bruchvorgängen bzw. aus einem Zusammenbruch der Haftreibung (Föppl, 1955), welche großräumige elastische Gleichgewichtsreaktionen zur Folge haben. Die Spannungsumlagerungen, hervorgerufen durch die einzelnen Ausbruchphasen und durch die anschließenden Einbauten, bedeuten zweierlei:
  - sofortige elastische Umlagerungen (Tragschalenbildungen, begleitet von materialbedingten örtlichen Brüchen [pseudoplastische (L. Müller, 1963) bzw. plastische (Jacobi, 1957) Verformung] durch Überschreiten der Materialfestigkeit;
  - Viskose Umlagerungen, die das anfangs entstandene Spannungsfeld laufend ändern und dadurch ausbruchsunabhängig weitere örtliche Überbeanspruchung hervorrufen.

Vardar (1977) hat diese ihrer Natur nach dynamischen Vorgänge im Gebirge rheologisch und modelltechnisch weiter verfolgt.

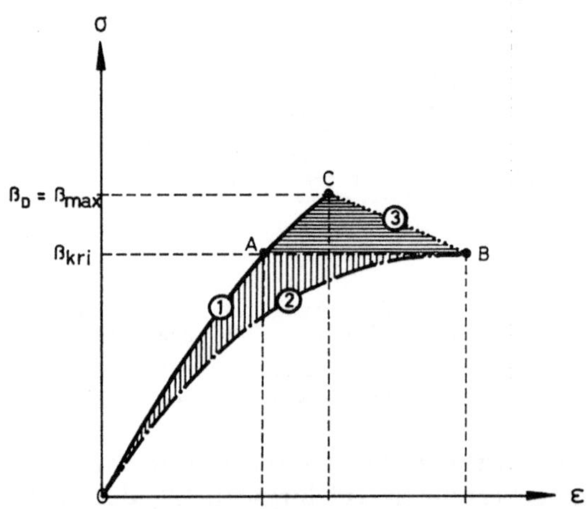

Abb. 6. Kennlinien eines elasto-plasto-viskosen Körpers

1 = σ-ε-Kurve der schnellsten Belastung (Charakteristik 1)
2 = Stabilisationskurve (Charakteristik 2)
3 = Bruchkennlinie (Charakteristik 3)

Characteristic curve of an elasto-plastic viscose body

1 = σ-ε-curve of the fastest loading
2 = curve of stabilization
3 = curve of failure

3. Gesteins- und Gebirgseigenschaften

Das zeitabhängige Verformungsverhalten des Gebirges ist auf seine elasto-plasto-viskosen Eigenschaften zurückzuführen.

Je nach der Belastungsgeschwindigkeit erhält man für ein und dasselbe Material einen anderen Verlauf der σ-ε-Kennkurve und voneinander abweichende Bruchwerte. Eine konstant gehaltene Spannung oberhalb einer für

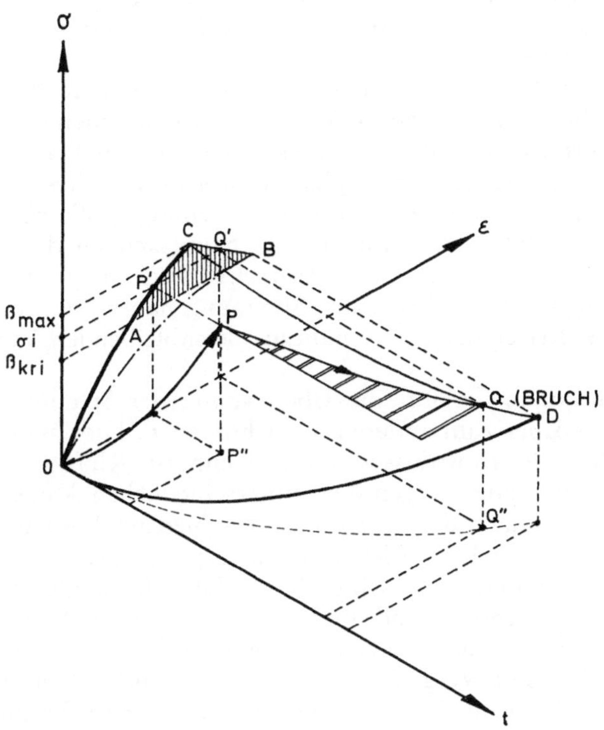

Abb. 7. Räumlicher Verlauf der Zeit-Verformungs-Kurve unter konstanter Belastung $\beta_D > \sigma_i > \beta_{kri}$

Three-dimensional characteristic of time-extension behaviour during constant loading $\beta_D > \sigma_i > \beta_{kri}$

dieses Material charakteristischen kritischen Größe führt nach einer bestimmten Belastungsdauer zu Bruch. Kleinere Spannungsgrößen dagegen lassen das Material ohne Bruchverhalten bis zur Stabilisation kriechen. (König, 1964; Dreyer, 1967; Szechy, 1971; Wittmann, 1971). Dieses Verhalten läßt sich in einem Diagramm (Abb. 6) durch drei Kennlinien charakterisieren:

- Charakteristik 1: Die σ-ε-Kurve der schnellsten Belastung
- Charakteristik 2: Die Stabilisationskurve
- Charakteristik 3: Die Bruchkennlinie

Wird die Zeit (als dritte Komponente) mit dargestellt, so lassen sich die Zusammenhänge gemäß Abb. 7 aufzeigen. Alle $\sigma = \phi$ (ε, t)-Kurven, für die $\sigma_i < \beta_{kri}$ ist, liegen im ungefährdeten Bereich, da mit der Zeit nur eine Stabilisation auftreten kann.

Das mit $\Delta\sigma/\Delta t$ beanspruchte Material folgt erst dem Kurvenast *OP*. Unter der Belastung $\sigma_i = \text{const.}$ kommt es dann bis zum absoluten Stabilisationspunkt *Q*. Die Projektionen dieser beiden Kurvenäste in der σ-ε-Ebene gehen durch die Punkte O, *P* und *Q*, das Material kommt zur Ruhe. Das Stabilisationskriechen des Materials unterhalb der kritischen Spannung besteht nur aus der ersten Kriechphase, wo die Deformationsgeschwindigkeit mit der Zeit permanent abnimmt.

Alle $\sigma = \phi\ (\varepsilon, t)$-Kurven, für die $\beta_D > \sigma_i > \beta_{kri}$ ist, beschreiben Kurvenäste, die nicht die Stabilisationsgrenze, sondern die für den Bruch charakteristische Grenzlinie CB erreichen (Abb. 7). Dabei enthält der zeitliche Verformungsverlauf alle drei Höferschen Kriechphasen. Welche von diesen Phasen überwiegt, hängt in erster Linie von der Beanspruchungsgröße ab. In der Nähe der maximalen Bruchfestigkeit findet man hauptsächlich die dritte Kriechphase mit beschleunigter Deformationsgeschwindigkeit. Je mehr die wirksamen Spannungen sich der kritischen Spannung nähern, umso ausgeprägter wird die zweite Kriechphase mit nahezu gleichbleibender Verformungsgeschwindigkeit.

Im Diskontinuum verläuft die Übertragung der Spannungstrejektorien von einem Kluftkörper zum anderen nicht homogen. Hier ändern sich Richtung und Dichte des Kraftflusses, da benachbarte Kluftkörper nie genau zusammenpassen und auch wegen ihrer unterschiedlichen Abmessungen unter gleicher Beanspruchung oder Belastung verschiedene Formänderungswerte erfahren. Dadurch entstehen in jedem Kluftkörper je nach seiner Anpassung an seine Umgebung und nach seiner Form Bereiche unterschiedlich hoher Spannungen, d. h. Richtung und Betrag des Tensors ändern sich von Ort zu Ort. Die Modellversuche zeigen deutlich, daß die Spannungsverteilung in geklüftetem Fels im Vergleich zu einem monolithischen (homogen-isotropen), durch viele kleine Spannungskonzentrationsbereiche innerhalb jedes Kluftkörpers charakterisiert ist; dennoch weist das gesamte Spannungsbild eine näherungsweise regelmäßige Verteilung der Kraftflußlinien auf.

Die Modellversuche an unterschiedlich hohen Proben mit gleichen Längen und Breiten, aber mit verschiedener Oberflächenbeschaffenheit zeigen, daß die Bruchlast unter anderem auch von der Oberflächenform und von dem Verhältnis Amplitudenhöhe der Rauhigkeiten zur Probenhöhe abhängt.

An den erhöhten Stellen der Oberfläche erreichen die Spannungen eine überdurchschnittliche Größe und die Vertiefungen wirken wie Kerben, wodurch der Gesamtkörper zum Bruch gezwungen wird. Diese Ergebnisse entsprechen der Annahme von Griffith (1920), daß für den Bruchprozeß Elastizitätsmoduli, Oberflächenenergie des betreffenden Stoffes und die Rißlänge maßgebend sein müsse. Bezogen auf das Gebirgsverhalten bedeutet dies, daß das Ausmaß der Zerklüftung die Bruchfestigkeit des Gebirges herabsetzt. Durch die Spannungsspitzen in den Kontaktpunkten erhöht sich die durchschnittliche Abweichung der lokalen Spannungen mit kleiner werdender Kluftkörpergröße. Das Verhältnis der inhomogenen Spannungsbereiche zum gesamten Kluftkörper wird mit zunehmender Anzahl der Kluftkörper (innerhalb eines betrachteten Bereiches) immer größer.

Ein derartiges Versagen des Gebirges wird mit dem Bruch eines Kluftkörpers, der unter den konzentrierten Spannungen entweder sofort oder mit der Zeit zusammengebrochen ist, ausgelöst. Nach jedem Ausfall eines solchen Kluftkörpers brechen weitere, im kritischen Zustand befindliche Kluftkörper. Neue Unstetigkeiten bringen neue Spannungskonzentrationsbereiche

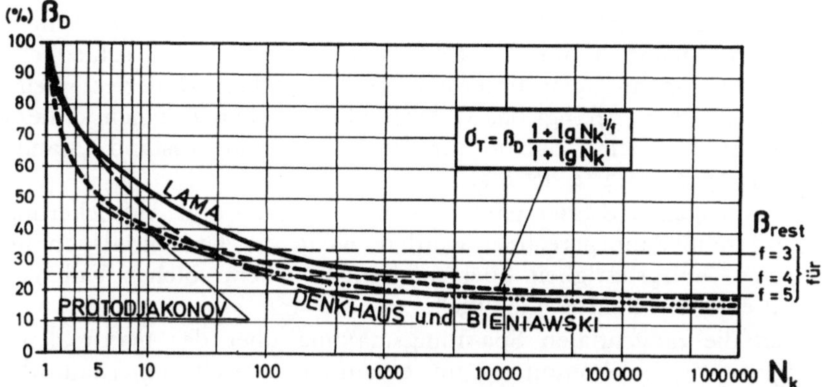

Abb. 8. Beziehungen zwischen der Gesteins- und Gebirgsfestigkeit in Abhängigkeit der Klüftigkeit und empirisch abgeleitete Festigkeitsgröße σ_T nach Vardar (1977)

σ_T = die einachsige Druckfestigkeit des Kluftverbandes; β_D = die einachsige Druckfestigkeit jedes einzelnen Kluftelementes; N_k = Anzahl der Kluftkörper im Verband; i = Inhomogenitätsfaktor; f = Festigkeitsfaktor $= \frac{\beta_D}{\beta_{rest}}$

Correlation between rock and rock mass strength, depending on jointing, empirically developed of the magnitude of strength σ_T according to Vardar (1977)

σ_T = uniaxial strength of rock mass; β_D = uniaxial strength of single joint blocks; N_k = number of joint blocks in the observed region; i = factor of inhomogeneity; f = factor of strength $\frac{\beta_D}{\beta_{rest}}$

mit sich und die erreichte Festigkeit dieses zerklüfteten Felses nähert sich somit asymptotisch der Restfestigkeit des Gebirges.

Diese Phänomene sind in der Literatur bekannt (Denkhaus und Bieniawski, 1968; Protodjakonov, Lama, 1973) und ermöglichen durch Maßstabüberlegungen die Umrechnung der Gesteinsfestigkeit auf Gebirgsfestigkeit, wobei die Kluftdichte, der Zerlegungsgrad und die Verbandanisotropie (L. Müller, 1968) mitberücksichtigt werden können (Abb. 8).

4. Rückschreitende Auflockerungszonen nach Vardar

Vor dem Auffahren eines Hohlraumes befindet sich das Gebirge in einem stabilen Gleichgewichtszustand; längere Zeit nach Beendigung des Tunnelbaues gleichfalls. Dazwischen aber durchläuft das System Gebirge plus Verbau oder Ausbau eine Reihe von Zuständen labilen Gleichgewichts. Der

Übergang vom primären Spannungszustand in den sekundären nach dem Ausbruch erfolgt im Gebirge nicht plötzlich. Zwischen den beiden im wesentlichen statischen Spannungszuständen befinden sich je nach der Größe der primären Spannungen und dem Gebirgsverhalten eine Serie von wechselnden Spannungen, die sowohl ihre Lage als auch ihre Größe um den Hohlraum ändern. Diese „lebendige Phase" ist jedem Tunnelbauer mehr oder weniger als „Spannungsumlagerungsphase" bekannt.

Das Gebirge verhält sich je nach der Höhe der Tangentialspannungen, die nach dem Ausbruch um den Hohlraum entstehen, unterschiedlich. Im Falle, daß alle Spannungen unterhalb der kritischen Dauerfestigkeit des Gebirges (σ_{kri}) liegen, gelangt das Gebirge ohne Ausbau zu einem dauernden Gleichgewicht; das heißt, daß es für diesen Spannungszustand „standfest" ist (Müller, 1963; Egger, 1973).

Mit sofortigen und/oder nachträglichen Brüchen und Auflockerungserscheinungen muß dann gerechnet werden, wenn die Tangentialspannungen die kritische Gebirgsfestigkeit überschreiten. In einem solchen Fall handelt es sich um ein „druckhaftes Gebirge".

Liegen die tangentialen Spannungsmaxima über der maximalen Gebirgsfestigkeit σ_T, so kommt es im gesamten Bereich, innerhalb dessen $\sigma_\theta > \sigma_T$ ist, ohne zeitliche Verzögerung zum Bruch und das Gebirge lockert sich dadurch auf. Außerhalb dieses Bereiches (d. h. für $\sigma_\theta < \sigma_T$) verformt sich das Gebirge zeitabhängig. Das selbe gilt auch für die Fälle, in denen die Spannungsmaxima nach dem Auffahren des Hohlraumes innerhalb des kritischen Bereiches ($\sigma_T > \sigma_t > \sigma_{kri}$) liegen.

Diese zeitabhängigen Spannungsumlagerungen lassen sich in drei Phasen einteilen.

1. Phase (Entstehung des virtuellen Traggewölbes)

Nach dem Ausbruch verspannt sich unter tangentialer Belastung eine Zone um den Hohlraum, bzw. um den Bereich sofortiger, zeitunabhängiger Auflockerung. Wenn in diesem Bereich die Spannungen größer sind als σ_{kri}, so muß hier das Material unter gegebener Bruchinformation bis an die Festigkeitsgrenze σ_{kri} fließen. Dabei entstehen im Mikrogefüge Trenn- und Scherbrüche. Im übrigen Gebirgsbereich, in dem die Spannungen unter σ_{kri} liegen, erfolgt eine Stabilisierung. Dieses Bruchfließen und Stabilisationskriechen des Gebirges sind auf Grund der viskosen Eigenschaften des Verbandes zeitabhängige Vorgänge.

2. Phase (Danymische Phase mit rückschreitenden Bruchzonen)

Wird im verspannten Bereich irgendwo die Bruchgrenze erreicht, so versagt plötzlich das gesamte Traggewölbe in seiner vollen Stärke s, so daß hier die Gebirgspartien sich auflockern und entfestigen. Dabei springen die Spannungsmaxima von der Tunnelwand zum Gebirgsinneren, wo früher die Grenze zwischen Bruch- und Verfestigungsvorgängen lag. Nach demselben Prinzip wiederholen sich diese Fortpflanzungsvorgänge so lange, bis der Maximalwert der Spannungen unter σ_{kri} liegt (Abb. 9).

3. Phase (Bildung der endgültigen Schutzhülle)

Wenn die Spannungen nur Werte unter der Dauerfestigkeit des Gebirges (σ_{kri}) erreichen, tritt infolge der Viskosität des Gebirges einerseits, eine Stabilisierung des Gebirges andererseits, ein Ausgleichsvorgang ein, der

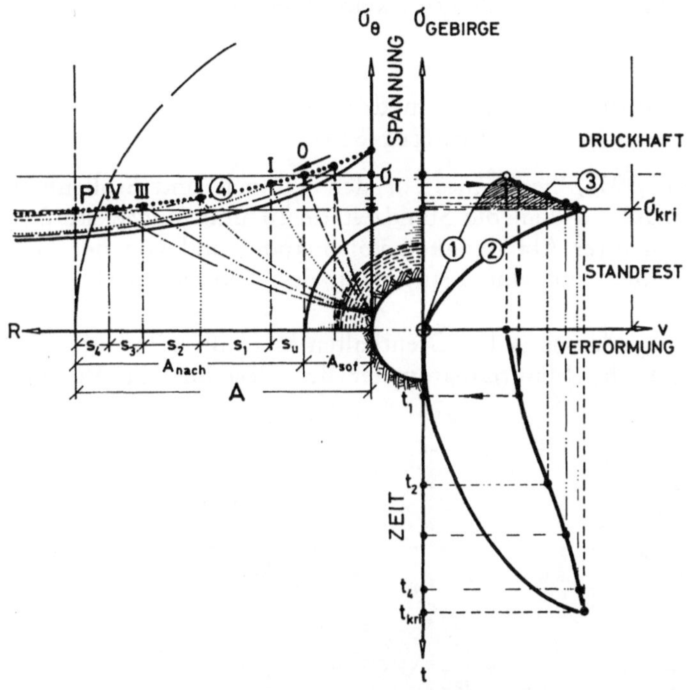

Abb. 9. Sprunghafte Entstehung von Bruchzonen und Bildung der Auflockerungszone bis zum Erreichen eines stabilen Zustandes

Stepwise development of failure zones and formation of loosening zones up to reaching the state of consolidation

die Spannungsspitzen im Gebirge abbaut. Das Gebirge bleibt in diesem Falle eine Zeit lang noch in einer allmählich abklingenden Bewegung, was jedoch nicht mehr zu weiteren Spannungssprüngen führt.

Daß das Gebirge irgendwann zum Stehen kommt, liegt daran, daß die plastifizierten und aufgelockerten Bereiche noch immer einen Teil der Kräfte aufnehmen, obwohl sie mit ihrem Gewicht gleichzeitig die Auskleidung belasten. Die Intensität des Kraftflusses im unaufgelockerten Bereich nimmt mit der Ausbreitung der Entfestigungs- und Auflockerungszone ab. Deswegen schreitet die Bruchentwicklung nicht unbegrenzt ins Gebirgsinnere fort. Die Auflockerungszone ist deshalb definiert als der Bereich, innerhalb dessen die tangentialen Spannungen während der Umlagerungsprozesse die Dauerfestigkeit (σ_{kri}) überschritten haben.

In der Abb. 9 ist dieses Phänomen dargestellt, wobei der Bereich der Auflockerung und der Gebirgsentfestigung mit A bezeichnet ist. Die nach jedem neuen Bruchvorgang erhaltenen maximalen Spannungen im Gebirgsinneren bezeichnen eine Spannungsgrenzkurve, welche die Einhüllende dieser Spannungsmaxima ist (Kurve (4).

Der Schnittpunkt *P* dieser Einhüllenden mit der kritischen Spannungslinie ergibt deshalb die Grenze dieser Auflockerungs- und Entfestigungszone, da außerhalb dieses Punktes keine Gebirgspartien mit Bruchinformation liegen.

4.1. Zusammenhang zwischen der ermittelten Standzeit und der Lauffer-Standzeit

Lauffer (1958) bestimmte für den Tunnel- und Stollenbau ein Diagramm (Abb. 10), in dem die Standzeit (in h) und Stützweite *l* (in m) sehr übersichtlich in ihrer Abhängigkeit voneinander und von der Gebirgsklasse dargestellt werden und eine Beurteilung des Gebirges nach Gebirgsklassen möglich ist.

Die Gebirgsklassen A bis G enthalten hier, allerdings ohne quantitative Aussagen, u. a. die Materialparameter des Gesteins, die Verbandfestigkeit

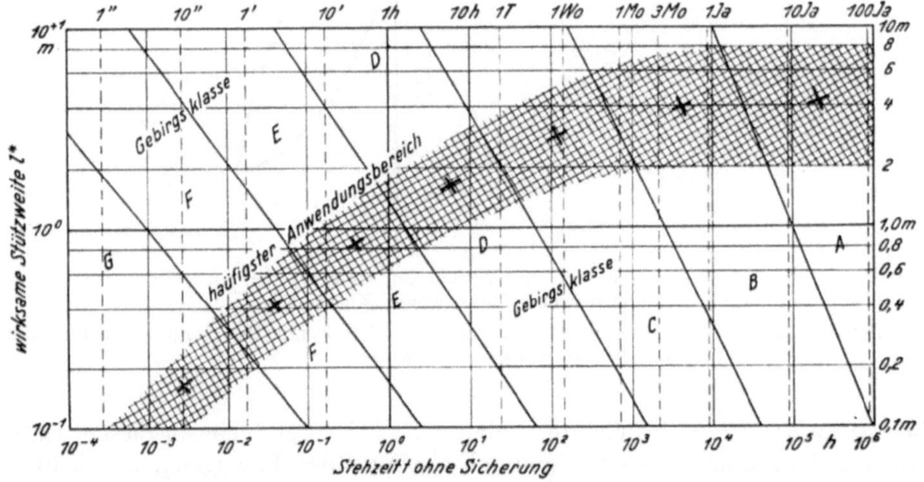

Abb. 10. Gebirgsfestigkeit beim Stollenvortrieb in Abhängigkeit von der Standzeit *t* (h) eines ungesicherten Stollenabschnittes mit der freien Spannweite 1 (m)

Rock mass strength in correlation to the standing time in an unsupported gallery with 1 m diameter

und den Spannungszustand des Gebirges. Sie sind Erfahrungswerte, die z. T. aus den erforderlichen Sicherungsmaßnahmen gewonnen worden sind.

Das Lauffer-Diagramm hat den Nachteil, daß es nur dann angewendet werden kann, wenn im voraus eine Einschätzung der Standzeiten für die einzelnen erwarteten Gebirgsarten gegeben werden kann oder wenn der Tunnel (Stollen bzw. Versuchsstollen) im Gebirge bereits vorgetrieben wird. Es ist nicht ohne weiteres möglich, es in der Entwurfsphase einzusetzen.

Die vorliegende Arbeit ermöglicht quantitative Aussagen über die Festlegung und Begrenzung der Laufferschen Gebirgsklassen. Die Abb. 11 zeigt schematisch diesen Zusammenhang zwischen der hier ermittelten Standzeit und der Lauffer-Standzeit. Hier ist für einen 3 m großen Tunneldurchmesser in einem mit seinen Charakteristiken bekannten Gebirge und unter

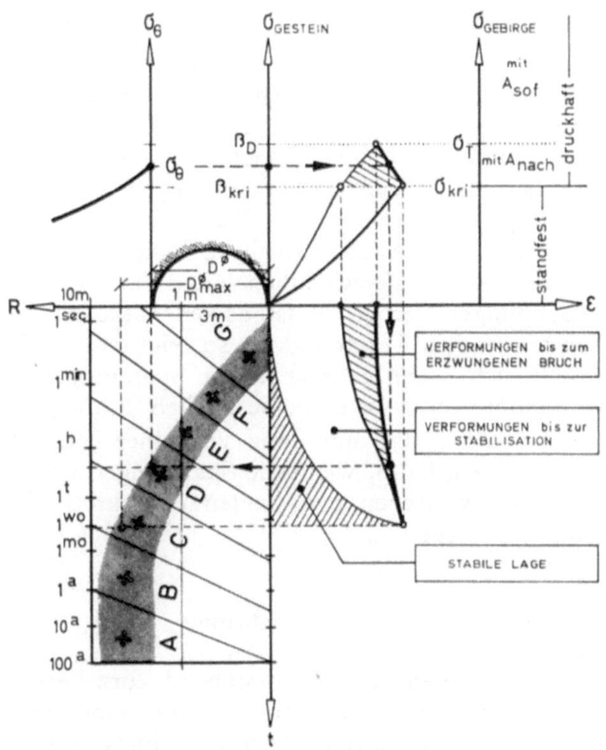

Abb. 11. Korrelation des ermittelten Standzeitwertes mit dem Lauffer-Diagramm
Correlation between stand-up time and the Lauffer diagram

einem bekannten Spannungszustand die Standzeit ermittelt. Die rechte Hälfte des Diagrammes stellt wie bisher die zeitlichen Spannungs-Verformungsbeziehungen des Gebirges dar. Im oberen linken Viertel ist der Verlauf der Tangentialspannungen eingetragen. Im unteren linken Viertel erfolgt der Einbau des Lauffer-Diagrammes in das σ-ε-t-R-Achsensystem. Bei einer zweiachsigen Betrachtung gilt für die freie Stützweite der Tunneldurchmesser. Die Pfeile zeigen den Bestimmungsweg der Standzeit und damit den der Gebirgsklasse. In diesem Fall beträgt die Standzeit ca. 6 Stunden; das Gebirge gehört also zur Klasse D.

Es ist aus dem Diagramm ferner noch zu entnehmen, daß dieses Gebirge mit gleichen Eigenschaften je nach seiner Beanspruchung zu verschiedenen Gebirgsklassen von C bis G gehören kann.

Dieses Diagramm widerlegt somit auch die weitverbreitete Meinung, daß die Gesteins- und Verbandsparameter genügen würden, die Laufferschen Gebirgsklassen festzulegen. Es zeigt vielmehr den entscheidenden Einfluß des Spannungszustandes auf die Zuordnung des Gebirges in die jeweilige Klasse. Erst mit der Berücksichtigung des Faktors „Spannungszustand" in Form von σ_θ-und σ_r-Verteilung kann das Verhalten des Gebirges richtig bestimmt werden. Die Bestimmung dieses Spannungsverlaufes sollte durch wirklichkeitsnahe Berechnungen oder/und äquivalente Modellversuche erfolgen, welche durch In-situ-Messungen zu überprüfen sind.

Ein und derselbe Gebirgsaufbau zeigt unter verschiedenen Beanspruchungen unterschiedliche Verformungs- und Bruchverhalten. Die höheren wirksamen Spannungen des kritischen Bereiches lassen das Gebirge schneller verformen und brechen als die niedrigen Spannungen, unter denen das Gebirge längere Zeit bis zum Bruch braucht. Das heißt, daß dasselbe Gebirge je nach Größe der gleich nach dem Ausbruch entstandenen Spannungen dieser neuen Beanspruchung unterschiedlich lange ohne Einbau standhalten kann. Sind die Spannungen gering, so ist dessen Standzeit lang, liegen sie in der Nähe seiner maximalen Tragfestigkeit, so zeigt das Gebirge eine kurze Standzeit; es verhält sich „sehr gebräch". Für Spannungen unterhalb der kritischen Beanspruchung (σ_{kri}) ist dasselbe zeitlich „unbegrenzt standfest".

Es ist also festzuhalten, daß nur dann von einer endlichen „Standzeit des Gebirges" die Rede sein kann, wenn die Spannungen am Hohlraumrand oder in der Schutzzone zwischen den beiden charakteristischen Gebirgsfestigkeiten (σ_T und σ_{kri}) liegen.

5. Schlußbetrachtung

Die Untersuchungen, über welche vorstehend kurz berichtet wurde — ausführlichere Darstellungen mit zahlreichen Literaturangaben enthalten die im Schrifttumsverzeichnis angeführten Veröffentlichungen —, scheinen uns mit großer Deutlichkeit zu zeigen, daß rheologische Betrachtungen, auch wenn sie zunächst noch nicht als geschlossene oder numerische Lösungen in mathematischer Form ausgewertet werden können, die tatsächlichen Geschehnisse vor Ort besser beschreiben und mit Beobachtungen und Messungen im Tunnel besser in Einklang gebracht werden können als noch so verfeinerte statische Berechnungen, welche den so wichtigen Faktor „Zeit" nicht berücksichtigen.

Dies zeigt sich schon bei zweidimensionaler, aber die rheologischen Eigenschaften der geklüfteten Felsmassen berücksichtigenden Betrachtungen; noch viel augenscheinlicher werden die Entsprechungen zwischen den erhaltenen Versuchsergebnissen sowie den daraus zu ziehenden Schlüssen und den Beobachtungen vor Ort bei dreidimensionaler Behandlung, auch wenn diese zur Zeit nur in der Form von Modellversuchen möglich ist.

Das Geschehen in der Umgebung der Tunnelbrust ist eben de facto ein dreidimensionales und zeitabhängiges. Die Spannungs- und Materialumlagerungen, welche sich in der dritten Dimension, nämlich parallel zur

Tunnelachse, abspielen, sind nicht minder bedeutungsvoll als das, was in der Querschnittsebene vor sich geht; sie sind außerdem von entscheidender Bedeutung für den endgültigen Gleichgewichtszustand der fertigen Tunnelröhre. Umlagerungsvorgänge in der dritten Dimension sind es nicht zuletzt, welche es dem Mineur erst ermöglichen, den während des Vortriebes empfindlich gestörten Gleichgewichtszustand des Gebirges wieder in einen stabilen Zustand überzuführen, also den Raum und die Zeit zwischen dem ersten Aufmachen der Ortsbrust und der endgültigen Auskleidung zu überbrücken.

Nur in Stichworten seien abschließend einige Erkenntnisse aufgezeigt, welche sich aus diesen neuen Untersuchungen ergeben haben und dem Praktiker schon in der heutigen Form bei der Wahl seiner tunnelbautechnischen Maßnahmen als Richtlinie dienen können.

Die Existenz dreidimensionaler, einander räumlich übergreifender Gewölbe hat Einfluß auf

- Formgebung der Ortsbrust
- Vortriebsmethoden
  - Vollausbruch
  - Teilausbruch

Der erhöhte Gebirgsdruck zwischen 1 und 3 D hinter der Ortsbrust beeinflußt

- Sicherungsmaßnahmen
- Ringschlußzeit
- Verspannung von Vollschnittfräsen

Die Tatsache einer spannungsverminderten Zone im Vortriebsbereich ist maßgebend für

- Stehzeitüberlegungen
- Sicherungszeit
- Abschlagtiefen
- Synchron-Asynchron-Vortrieb bei Doppelröhrentunneln
- Wahl der Mittelwanddicke (Achsabstand)

Die Dicke der Gebirgstragringe gibt Entscheidungshilfen für

- Festlegung von Ankerlängen
- Synchron-Asynchron-Vortrieb
- Lage späterer Vortriebe zu vorhandenen Tunneln
- Mittelwanddicke

Die Erkenntnisse hinsichtlich der Mittelwandbelastung zeigen

- Auswirkungen des Verhältnisses Tunneldurchmesser zu Mittelwanddicke
- Einfluß von Synchron-Asynchron-Vortrieb

Hinsichtlich der Oberflächensenkungen sind (besonders im urbanen Tunnelbau) zu beachten

— ruckartig sich ausbreitende Bug- und Heckwellen
— Aufgliederung in einzelne Komponenten bezüglich
  · Vortriebsbeurteilung
  · Vortriebsoptimierung

Literatur

Aalto, J., Salonen, E. M.: An Iterative Solution Scheme For Three-Dim. FEM Problems in Soil Mechanics. NMSR Symposium, Karlsruhe 1975, S. 325—337.

Denkhaus, H. G., Bieniawski, Z. T.: Festigkeit von Gestein und Gebirge. Bericht über das 10. Ländertreffen Leipzig, 1968.

Dreyer, W.: Neuere Forschungsergebnisse auf dem Gebiet der Gesteinsphysik. Bergb. Wiss. *14*, Nr. 8, 1967.

Egger, P.: Einfluß des Post-Failure-Verhaltens von Fels auf den Tunnelbau. Veröffentl. d. Inst. f. Boden- und Felsmechanik, TU Karlsruhe, Heft 57, 1973.

Egger, P.: Zur Abschätzung des Gebirgsdruckes aufgrund des Post-Failure-Verhaltens, 1. Nat. Tagung über Felshohlraumbau, Essen 1974.

Feder, G., Arwanitakis, M.: Zur Gebirgsmechanik ausbruchsnaher Bereiche tiefliegender Hohlraumbauten. BHM Jg. 121, H. 4, 103—117 (1976).

Föppl, L.: Der Übergang von der Haftreibung zur Gleitreibung, Geologie und Bauwesen Jg. 21, H. 4, 145—148 (1955).

Greiner, G.: Messungen von Primärspannungen untertage. Vortrag beim SFB 77 — Kolloquium, UNI Karlsruhe, April 1974.

Gudehus, G., Goldscheider, M., Winter, H.: Mechanische Eigenschaften von Sand und Ton und numerische Integrationsverfahren; einige Fehlerquellen und Genauigkeitsgrenzen. NMSR-Symposium, Karlsruhe 1975, S. 289—304.

Jacobi, D.: Die Bewegung zerbrochener Gesteinsschichten um bergmännische Hohlräume. Glückauf Jg. 93, H. 45/46 (1957).

Jagsch, D., Müller, L., Hereth, A.: Bericht über Messungen und Meßergebnisse beim Bau der Stadtbahn Bochum BL A2. Interfels Meßtechnik, Information, S. 11—13, 1974.

König, H.: Dauerstandversuche im Labor zur Bestimmung rheologischer Eigenschaften von Gesteinen. Bergakademie Jg. 16, H. 1 (1964).

Kovari, K., Amstad, Ch., Grob, H.: Ein Beitrag zum Problem der Spannungsmessung im Fels. Int. Symp. für Untertagebau, Luzern, Sept. 1972, S. 501—512.

Lama, R. D.: The Uniaxial Compressive Strength of a Jointed Rock Mass. Festschrift Leopold Müller-Salzburg, pp. 67—78, 1974.

Lombardi, G.: Berücksichtigung der räumlichen Einflüsse im Bereich der Ortsbrust. 1. Nat. Tagung über Felshohlraumbau, Essen 1974, S. 75—92.

Maidl, B., Geißler, E.: Methoden zur Berechnung von Bauzuständen im Tunnelbau. 2. Nat. Tagung über Felshohlraumbau, Aachen 1976, S. 211—230.

Müller, G.: Felstests zur Ermittlung von Spannungen in-situ. Vortrag am Institut für Boden- und Felsmechanik, UNI Karlsruhe, Januar 1976.

Müller, L.: Der Felsbau. Stuttgart: Enke Verlag 1963.

Müller, L.: Der Einfluß von Klüftung und Schichtung auf die Trompeter-Wiesmannsche Zone. Bericht über das 10. Ländertreffen des Int. Büros für Gebirgsmechanik, Leipzig 1970. Berlin: Akademie-Verlag 1970.

Obert, L.: Deformation of Stress in Rock, A State-of-art Report ASTM STP 429, Am. Soc. Testing Mats., pp. 46—53, 1967.

Pacher, F.: Deformationsmessungen im Versuchsstollen als Mittel zur Erforschung des Gebirgsverhaltens und zur Bestimmung des Ausbaues. Felsmechanik und Ingenieurgeologie, Suppl. 1. Wien – New York: Springer 1964.

Panet, M.: Analyse de la stabilité d'un tunnel creusè dans un massif rocheux en tenant compte du comportement aprês la rupture. Rock Mech. *8*, 209—224 (1976).

Rodatz, W., Wallner, M.: Untersuchungen des räumlichen Spannungs- und Verformungszustandes an der Ortsbrust eines Tunnels nach der FEM. 1. Nat. Tagung über Felshohlraumbau, Essen 1974, S. 107—118.

Rziha, F.: Lehrbuch der ges. Tunnelbaukunst. Berlin: Verlag Ernst & Korn 1874.

Sauer, G., Jonuscheit, P.: Kräfteumlagerungen in der Zwischenwand eines Doppelröhrentunnels im Zuge eines Synchronvortriebes. Rock Mechanics *8*, 1—22 (1976).

Sauer, G.: Spannungsumlagerung und Oberflächensenkung beim Vortrieb von Tunneln mit geringer Überdeckung. Veröff. des Inst. für Bodenmech. u. Felsmechanik der Universität Karlsruhe, Heft 67, 1976.

Smoltzyk, U.: Earth Pressure Reduction in Ground of a Tunnel Shield. Int. Symp. Mexico 1969, S. 473—481.

Szechy, K.: Tunnelbau. Wien – New York: Springer 1969.

Vardar, M.: Zeiteinfluß auf das Bruchverhalten des Gebirges in der Umgebung von Tunneln. Veröff. des Inst. für Bodenmech. und Felsmechanik, Univ. Karlsruhe, Heft 72, 1977.

Wittmann, F., Zaitsev, J.: Versuche zur Bestimmung der Dauerfestigkeit des Zementsteines. Zement – Kalk – Gips *84*, 119 (1971).

Anschrift der Verfasser: Prof. Dr. Leopold Müller-Salzburg, Paracelsusstr. 2, A-5020 Salzburg; Dipl.-Ing. Dr. Gerhard Sauer, Ingenieurbüro Dr. Pacher, Franz-Josef-Straße 3, A-5020 Salzburg, Österreich; Dr.-Ing. Mahir Vardar, I. T. Ü. Maden Fakültesi, Tatbiki Jeoloji Kürsüsü, Tesvijiye-Istanbul, Türkei.

Rock Mechanics, Suppl. 7, 87—102 (1978)

Rock Mechanics
Felsmechanik
Mécanique des Roches

Gebirgsbeherrschung durch Ausbauwiderstand und Nachgiebigkeit des Ausbaus

Von

H. O. Lütgendorf

Mit 14 Abbildungen

Zusammenfassung — Summary

Gebirgsbeherrschung durch Ausbauwiderstand und Nachgiebigkeit des Ausbaus. Unterirdische Hohlräume im Abbaubereich des Bergbaus sind über viele Jahre mehrmals wechselnden Gebirgsdruckänderungen unterworfen. Der Ausbau muß den immer wieder auflebenden Querschnittsverlusten des Hohlraumes entsprechend verformbar sein.

Der Ausbau macht den zerbrochenen Gebirgsmantel um ein Mehrfaches des radialen Ausbauwiderstandes höher belastbar. Der radiale Ausbauwiderstand preßt die Gesteinsteile zusammen und weckt Reibungswiderstände auf den Scherbewegungsflächen.

Ein wirtschaftlicher Ausbau hat ausreichend hohen Ausbauwiderstand und ausreichende Nachgiebigkeit.

Tragfähigkeit und Wirtschaftlichkeit des Ankerringausbaus und des geschlossenen Ausbauringes werden auf physikalischer Grundlage erläutert.

In Bergbaustrecken sind drei Ausbauklassen zu unterscheiden. Die Ausbauklassen unterscheiden sich durch den Ausbauwiderstand. Mit großem Ausbauwiderstand kann ausreichendes Hohlraumvolumen gehalten werden, wenn hoher Gebirgsdruck eintritt.

Formeln und Diagramme über Tragfähigkeit, Nachgiebigkeit und Wirtschaftlichkeit der Ausbausysteme begleiten die Ausführungen.

Strata Control by Support, Chiefly by Resistance and Capacity to Yield. Excavations in the zone of strate movements around a moving face of mining are subject to alternating rock pressure during many years. Support must be yielding in conformity with the again and again reviving deformation of excavation.

The support makes crushed surrounding rock able higher to load a multiplicity of the radial resistance of support. The radial resistance of support compresses the rock pieces and awakes frictional resistance on the planes of shear moving.

An economical support has sufficient high load capacity and sufficient capacity to yield.

Load capacity and economy of strate bolting ring and of support ring are explained on physical foundation.

In mining roadways three classes of support are to differentiate. The support classes differentiates by load capacity. By high load capacity of support sufficient hollow space of roadway can be saved, if rock pressure increases.

Relationships and diagrams of resistance, capacity to yield and economy of support systems accompanys the explains.

Gebirgsbeherrschung durch Ausbauwiderstand und Nachgiebigkeit des Ausbaus

Der Beitrag bringt wesentliche Ergebnisse einer Untersuchung der „Ausbausysteme zur Beherrschung großer Gebirgsdrücke — Nutzung von Tunnelbauerfahrungen im Bergbau", die mit Mitteln des Bundesministers für Forschung und Technologie der Bundesrepublik Deutschland gefördert wurde. Es werden nur die wesentlichen Ergebnisse der Gebirgsdrucktheorie und einige weitere Einzelheiten besprochen. Diese Theorie geht vom Gesteinselement an einer potentiellen oder tatsächlichen Bewegungsfläche im Gebirge aus und ist in den Veröffentlichungen des Verfassers (siehe Literaturverzeichnis auf der letzten Seite) seit 1967 entwickelt worden. Die Abbildungen zeigen den Einfluß der Gebirgsparameter und des Ausbauwiderstandes — als den gegebenen und gewählten Randbedingungen — auf Spannungen und Verformungen des Gebirges.

Gebirgsdrucktheorie

Abb. 1 zeigt Bereiche und Begriffe für den zentralsymmetrischen Fall hinsichtlich der geometrischen Form, der Gebirgsparameter und der Spannungen. An diesem Fall werden beispielhaft einige Erkenntnisse dargestellt.

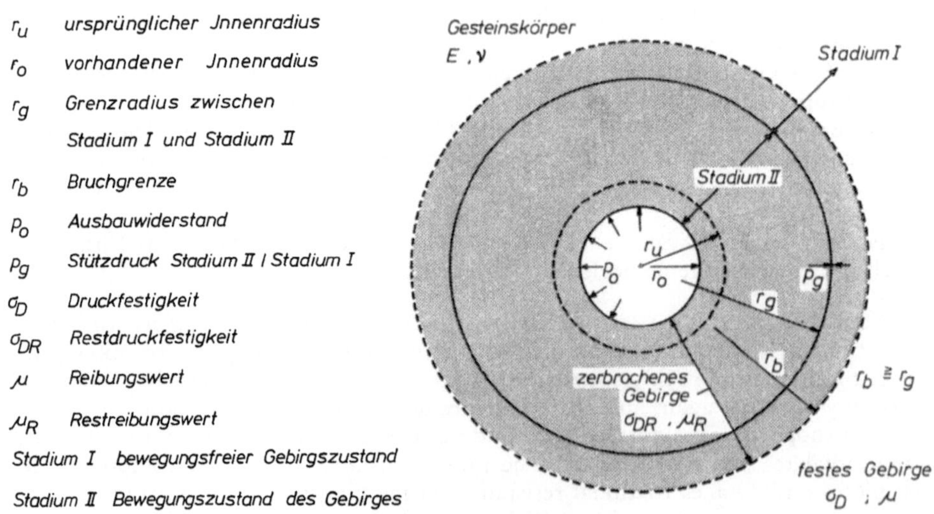

Abb. 1. Bereiche und Begriffe

Zones and terms

Bei den Gebirgsparametern ist zu unterscheiden zwischen unzerbrochenem und zerbrochenem Gebirge. Hinsichtlich des Spannungszustandes ist eine andere Unterscheidung von Gebirgsbereichen nötig. Im Ruhezustand — den ich als Stadium I bezeichne — wirkt das Gebirge wie ein Kontinuum, gleichgültig ob es fest oder zerbrochen ist.

Ist die Gebirgsfestigkeit am Gebirgsstoß kleiner als die Tangentialspannung, so entstehen Bruchbewegungen. Im Bewegungszustand des Gebirges — den ich als Stadium II bezeichne — wird der Spannungszustand

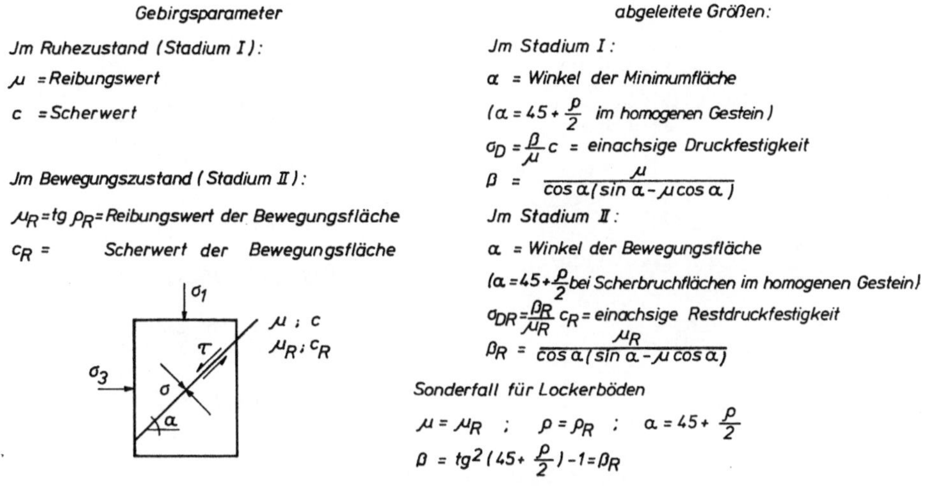

Abb. 2. Gebirgsparameter
Parameters of rock

durch Scherspannungen auf sich kreuzenden Scharen von Bewegungsflächen bestimmt. Der Spannungszustand im Stadium II wird vom Ausbauwiderstand und den Eigenschaften der Scherbewegungsflächen bestimmt.

Am Grenzradius zwischen Stadium I und Stadium II entsteht beim Vortrieb eine geringe nach innen gerichtete radiale Verformung entsprechend der Spannungsumlagerung im Stadium I. Der radiale Druck am Grenzradius entspricht den Bruch- oder Scherbedingungen des Stadiums I, die von der Tangentialspannung und den Gebirgsparametern abhängen.

Im Stadium II entspannt sich das Gebirge gegen den Scherwiderstand auf den Bewegungsflächen. Die Gesteinskörper nehmen an Volumen zu. Die Querschnittsabnahme des Untertagebaus ist die Folge der Verformung am Grenzradius und der Entspannung der Gesteinskörper.

Abb. 2 zeigt ein Gebirgselement an einer Scherbewegungsfläche. Aus den Gebirgsparametern Reibungswert und Scherwert werden die dimensionslose Größe β und die Druckfestigkeit abgeleitet. Beide Größen hängen vom Winkel α zwischen Hauptspannungsrichtung und Scherbewegungsfläche ab. Der Winkel α kann durch Schicht- oder Trennflächen im Gebirge vorgegeben sein. In homogenen Lockerböden und im homogenen Fels — zwei Sonderfälle — ist der Winkel α eine Funktion des Reibungswertes.

Abb. 3 zeigt Gleichungen für die Spannungen im Gebirge. Abb. 4, 5 und 6 zeigen Spannungskurven für verschiedene Gebirgsdrücke und Ausbauwiderstände. Bei zunehmendem Gebirgsdruck wächst der Bereich im Stadium II erheblich an, und zwar bei großem Elastizitätsmodul des Gesteins

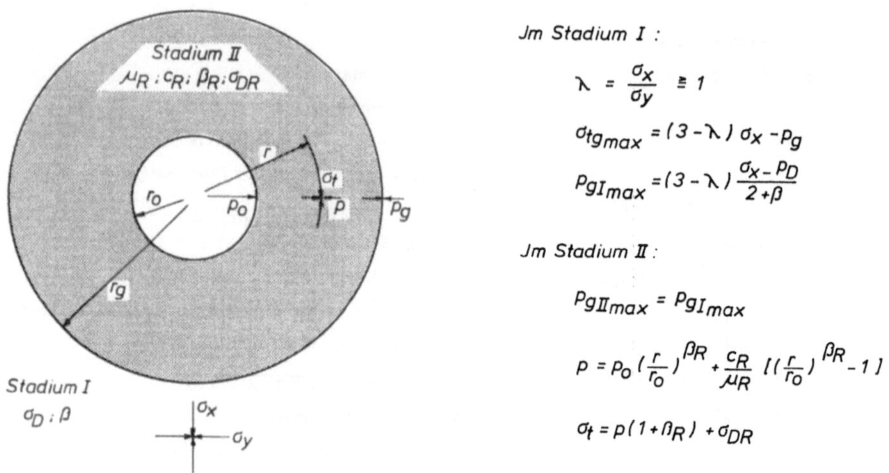

Abb. 3. Spannungen im Gebirge
Rock pressure

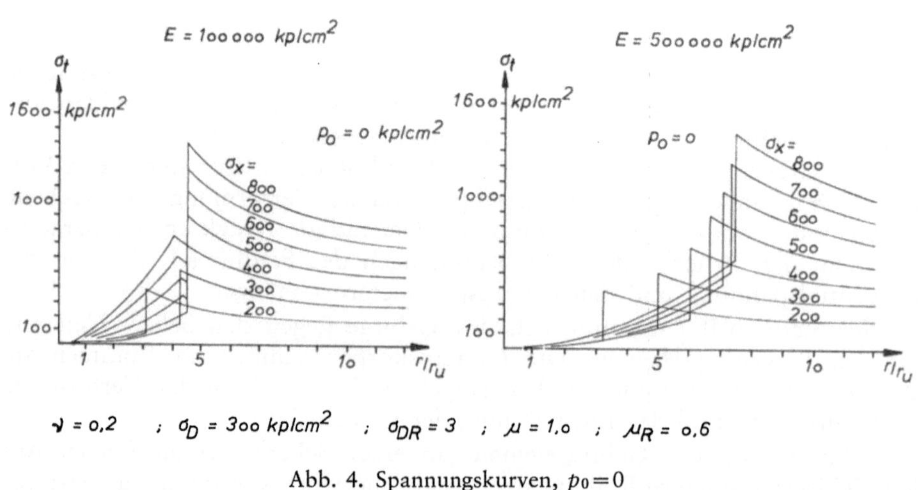

Abb. 4. Spannungskurven, $p_0 = 0$
Diagram of rock pressure

besonders stark. Ein großer Bereich im Stadium II hat einen großen Querschnittsverlust des Untertagebaus zur Folge, insbesondere bei kleinem Elastizitätsmodul des Gesteins.

Durch Ausbauwiderstand wird der Bereich im Stadium II klein gehalten und bleibt deshalb auch der Querschnittsverlust des Untertagebaus klein.

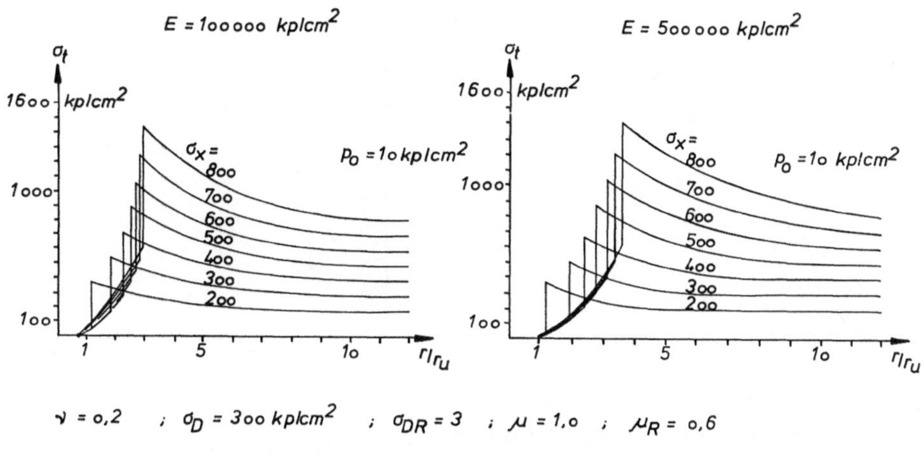

Abb. 5. Spannungskurven, $p_0 = 10$
Diagram of rock pressure

Die links in den Abb. 7 bis 9 dargestellten Diagramme zeigen, daß der Grenzradius zwischen Stadium I und II mit steigendem Gebirgsdruck zu-

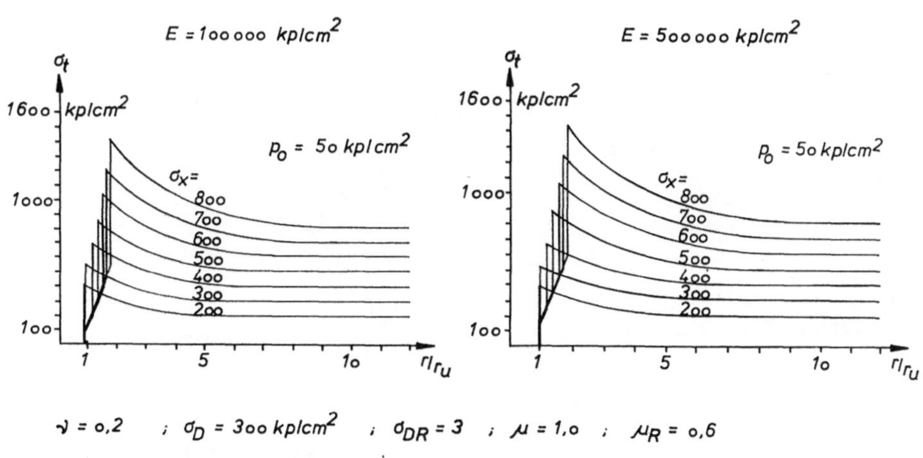

Abb. 6. Spannungskurven, $p_0 = 50$
Diagram of rock pressure

nimmt, ein Maximum erreicht und wieder abnimmt, weil der Innenradius mit zunehmendem Gebirgsdruck immer mehr abnimmt. Durch großen Ausbauwiderstand werden Grenzradius und Verformung des Innenradius klein gehalten.

Die rechts in den Abb. 7 bis 9 gezeigten Diagramme stellen den Querschnittsverlust des Untertagebaus bei verschiedenen Ausbauwiderständen und

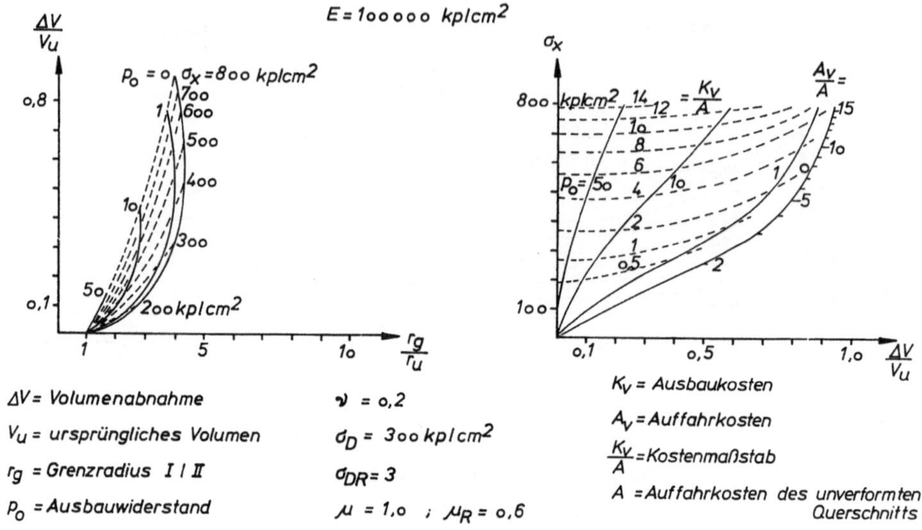

Abb. 7. Verformung und Kostenmaßstab
Deformation and scale of costs

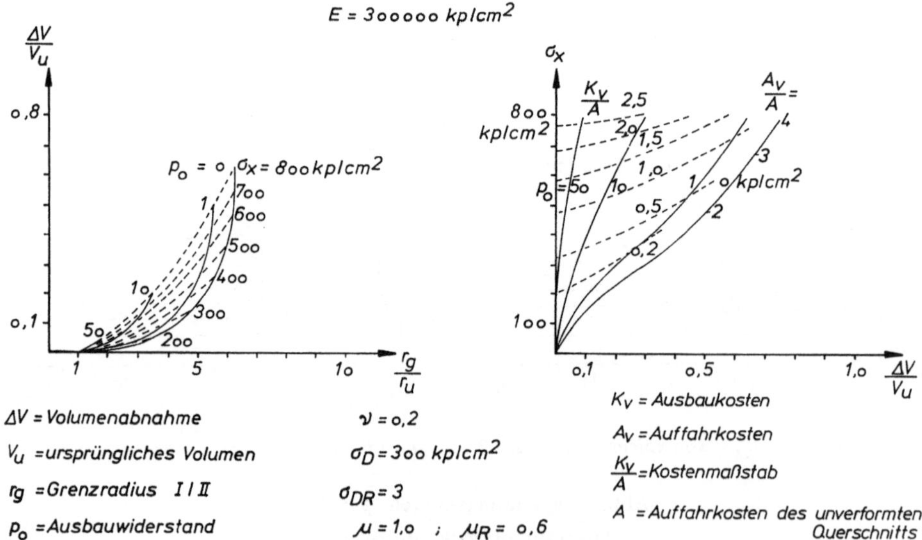

Abb. 8. Verformung und Kostenmaßstab
Deformation and scale of costs

Gebirgsdrücken dar. Die gestrichelten Kurven sind ein Kostenmaßstab für den Ausbau. Die Werte des Kostenmaßstabes steigen mit zunehmendem

Gebirgsdruck und erreichen bei kleinem Elastizitätsmodul des Gesteins besonders große Werte. Ein Wert von 1,5 beispielsweise zeigt an, daß der Kostenaufwand für den gewählten Ausbau so groß sein darf, wie die eineinhalbfachen Auffahrkosten des unverformten Querschnitts. Ist der tatsächliche Kostenaufwand niedriger als der Wert des Kostenmaßstabes, so ist der Ausbau wirtschaftlich, ist er höher, so ist er unwirtschaftlich. Der Kostenmaß-

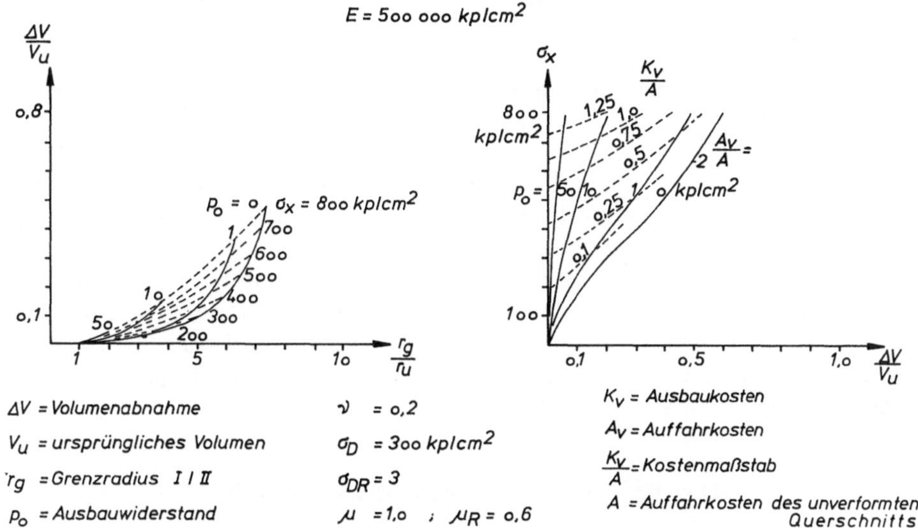

Abb. 9. Verformung und Kostenmaßstab
Deformation and scale of costs

stab zeigt hohe Werte, wenn ein Untertagebau ohne Ausbau bei den bestehenden Randbedingungen Querschnittsverluste von mehr als 50% erleiden würde.

Ein wirtschaftlicher Ausbauwiderstand vermindert den Querschnittsverlust des Untertagebaus auf 10% bis 20%. Es ist unwirtschaftlich, Querschnittsverluste vollständig verhindern zu wollen. Innerhalb des Systems Ausbau – Gebirge muß der für eine wirtschaftliche Lösung nötige Querschnittsverlust entweder im Gebirge oder durch konstruktive Maßnahmen im Ausbau erzielt werden.

Durch Bergbau ändert sich der äußere Gebirgsdruck

Das Ausbrechen eines unterirdischen Hohlraumes verursacht eine Spannungsänderung, die zu einem sekundären Spannungszustand führt. Ein großräumig geführter Abbau bewirkt einen entsprechend großräumigen sekundären Spannungszustand. Auf andere Hohlräume wirkt diese Spannungsänderung wie eine Änderung des äußeren Gebirgsdruckes. Durch den großflächigen Masseentzug beim Flözbergbau schwankt der bankrechte Druck in weiten Grenzen, und zwar zwischen einer Zunahme bis zum Mehrfachen des ursprünglichen bankrechten Drucks und einer Abnahme bis zur völligen Entspannung. Jeder neue Abbau erzeugt in seiner Umgebung neue Druckänderungen.

Wie beim Tunnelbau bildet sich außerhalb des Abbaueinwirkungsbereichs in einer bestimmten Zeit nach dem Vortrieb des Hohlraumes ein stabiler sekundärer Spannungszustand. Auf Hohlräume im Abbaubereich des Bergbaus wirkt lange Zeit eine Folge von Druckänderungen ein.

Im Bergbau treten über lange Zeit große Querschnittsänderungen der Hohlräume auf

Durch eine lange Zeit andauernde Folge großer Änderungen des äußeren Gebirgsdruckes erleiden bergmännische Hohlräume lange anhaltende Querschnittsverluste.

Nach einer bestimmten Zeit stellt sich ein neuer sekundärer Spannungszustand ein, der aber nur bis zur nächsten Gebirgsdruckänderung stabil ist.

Der Tunnelbau hat Lösungen für die Verformbarkeit des Ausbaues gfunden, die für den Bergbau abgewandelt werden müssen, um jederzeit Verformbarkeit des Ausbaus verfügbar zu haben.

Der Begriff „innerer Ausbauwiderstand“

Hoher radialer Ausbauwiderstand kann mit Ausbau aus unterschiedlichem Material in unterschiedlicher Dicke realisiert werden. In der Regel wirkt der Ausbauwiderstand am Außenrand des Ausbaus auf das Gebirge. Bei unterschiedlichen Außenradien ist ein Vergleich der Ausbauwirkungen schwierig.

Um einen Vergleich zu ermöglichen, habe ich den Begriff innerer Ausbauwiderstand eingeführt. Die Wirkung der Ausbaumittel wird durch einen radialen, am lichten Innenrand des Hohlraumes angreifend gedachten Ausbauwiderstand ersetzt, den inneren Ausbauwiderstand.

Der innere Ausbauwiderstand entsteht in einem Ersatzsystem, bei dem die Wanddicke des Ausbaus zu Null gesetzt und das tatsächliche Ausbaumaterial durch Gebirgsmaterial mit den Parametern des Stadium II ersetzt ist. Dieses Ersatzsystem erzeugt am tatsächlichen Außenrand des Ausbaus die gleiche radiale Druckspannung, nämlich den äußeren Ausbauwiderstand. Der fiktive innere Ausbauwiderstand erzielt nahezu die gleiche Wirkung im Gebirge, wie der tatsächliche äußere Ausbauwiderstand. In der Regel ist der Fehler, um den man dann den Querschnittsverlust zu groß errechnet, vernachlässigbar klein.

Mit Hilfe des inneren Ausbauwiderstandes wird es möglich, die Stützwirkung unterschiedlich dicker Ausbausysteme zu vergleichen. Auch läßt sich das Zusammenwirken verschiedener, gleich verformbarer Ausbaumittel einfach durch Addition ihrer inneren Ausbauwiderstände errechnen.

Tragwirkung und wirtschaftliche Bemessung des Ankerausbaus

Der Begriff innerer Ausbauwiderstand beschreibt in idealer Weise die Wirkung eines Tragringes aus radial angeordneten Ankern. Der Raum des geankerten Gebirgsringes ist durch die Parameter des Stadium II des Gebirges vollständig beschrieben.

Die radialen Ankerzugkräfte halten den radialen Druckspannungen im Gebirge das Gleichgewicht. In einem mittleren Bereich des Ankers zwischen

Ankerausbau im Stadium II

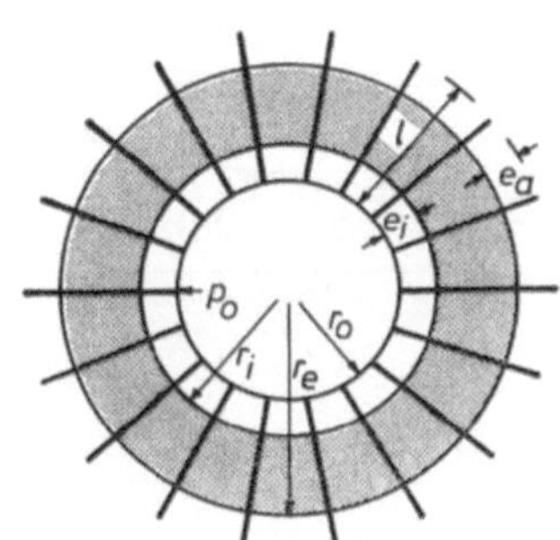

mittlere Ankerspannung am Hohlraumrand:

$A \cdot n$ *(bezogen auf* r_o*)*

Jnnerer Ausbauwiderstand p_o*:*

$$\frac{p_o}{A \cdot n} = \left(\frac{r_o}{r_i}\right)^{\beta_R+1} - \left(\frac{r_o}{r_e}\right)^{\beta_n+1} = \alpha$$

für $r_e < r_g$

$$\frac{p_o}{A \cdot n} = \left(\frac{r_o}{r_i}\right)^{\beta_R+1} = \alpha_o$$

für $r_i < r_g < r_e$

Kostenmaßstab:

$$\frac{K}{p_o} = \alpha_K \cdot r_o \cdot \frac{k}{A}$$

Optimale Ankerlänge: $\frac{l}{r_o} = \frac{e_i + e_a}{r_o} + 0{,}3$

A = *Ankerkraft*
n = *Anker je* m^2
K = *Ankerkosten je* m^2
k = *Kosten je Anker*

Abb. 10. Ankerausbau
Strate bolting

dem Abstand e_i vom inneren und dem Abstand e_a vom äußeren Ankerende dürfen die radialen Druckspannungen gleichmäßig im Gebirge verteilt angenommen werden (siehe Abb. 10).

Der innere Ausbauwiderstand des geankerten Gebirgsringes steigt mit zunehmender Ankerlänge zunächst stark, oberhalb der optimalen Ankerlänge aber nur noch wenig an (siehe Abb. 11 und 12).

Die Wirkungskennzahl α des Ankerausbaus ist kleiner als 1 und nähert sich bei der optimalen Ankerlänge ihrem Maximum. Die wirtschaftliche Ankerlänge ist bei kleiner relativer Verteilungstiefe e_i/r_0 und großem Reibungswert des Gebirges im Stadium II kleiner, in relativ schlechtem Gebirge und bei größerer relativer Verteilungstiefe größer.

Der innere Ausbauwiderstand ist immer kleiner als die auf die abgewickelte Innenfläche wirkende mittlere radiale Ankerspannung $A \cdot n$.

Mit Ankerausbau lassen sich innere Ausbauwiderstände zwischen 5 und 50 Mp/m^2 erreichen. Ein geankerter Gebirgsring ist nachgiebig, wenn die Anker in ganzer Länge im Bereich des Stadium II liegen und ein Schlupf zwischen Anker und Gebirge eintritt. Um einen Schlupf zu erreichen, müssen die einzelnen Anker so stark gewählt werden, daß das umgebende Ge-

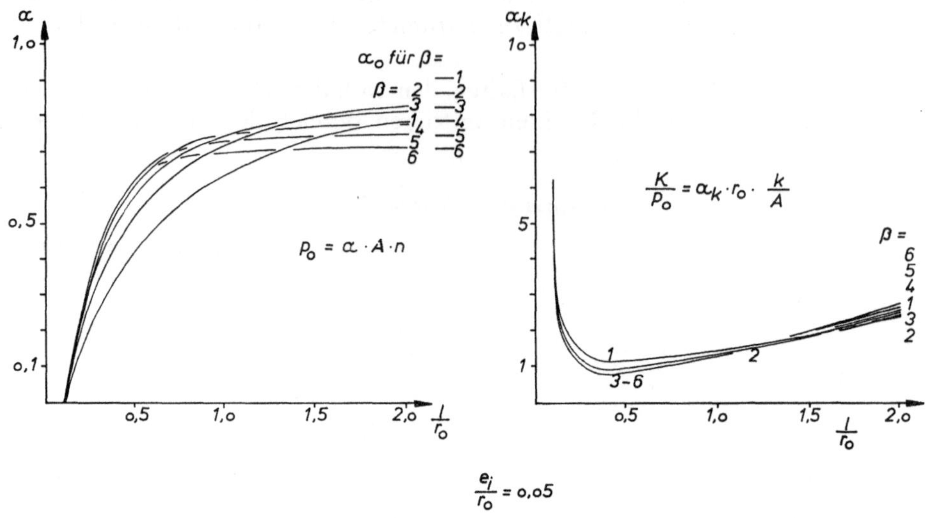

Abb. 11. Ankerausbau
Strate bolting

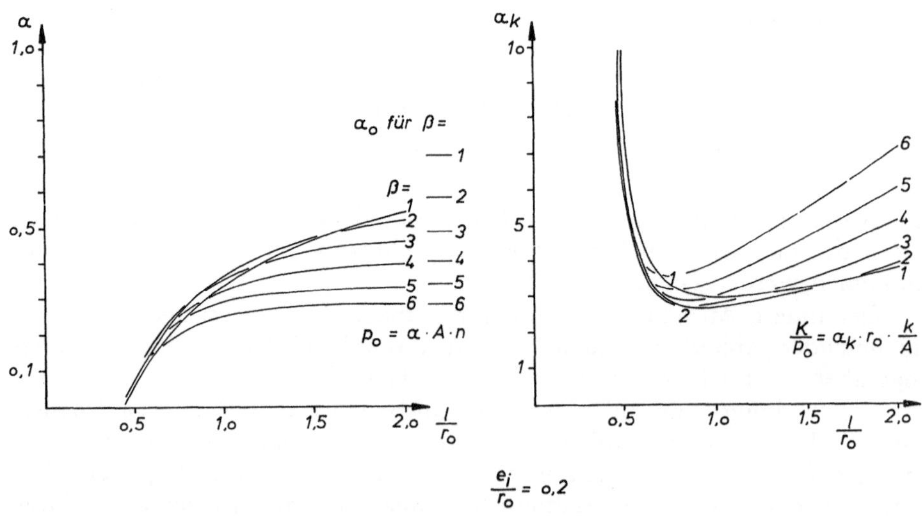

Abb. 12. Ankerausbau
Strate bolting

birge in Form von Scherbewegungen Relativbewegungen zum Anker ausführen kann.

Tragfähigkeit und wirtschaftliche Bemessung des geschlossenen Ausbauringes

Geschlossene Ausbauringe können als vorgesetzter Unterstützungsausbau oder als Injektionstragring im Gebirge ausgeführt sein. In beiden Fällen wird der radiale Ausbauwiderstand am Außenrand des Ausbaus (*ra*) auf das

Ausbaugewölbe im Stadium II

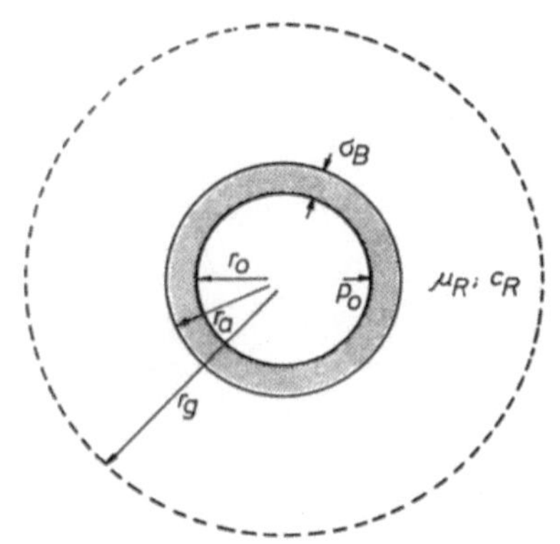

Jnnerer Ausbauwiderstand :

$$p_o = \sigma_B \left(\frac{r_o}{r_a}\right)^{\beta_R} \cdot \frac{r_a^2 - r_o^2}{2 r_a^2} - \frac{c_R}{\mu_R} \cdot \left[1 - \left(\frac{r_o}{r_a}\right)^{\beta_R}\right] = \delta \sigma_B$$

Kostenmaßstab :

$$\frac{K}{p_o} = \epsilon \cdot \frac{k\, r_o}{\sigma_B}$$

r_a = *Außenradius des Ausbaues*

σ_B = *Druckfestigkeit des Baustoffes*

K = *Ausbaukosten je* m^2 *Jnnenfläche (auf* r_o *bezogen)*

k = *Ausbaukosten je* m^3

Abb. 13. Ausbaugewölbe
Support ring

Gebirge übertragen. Unterstützungsausbau kann aus druckfesten Baustoffen hergestellt werden, insbesondere aus mineralischen Baustoffen oder aus Stahl (siehe Abb. 13).

Der innere Ausbauwiderstand geschlossener Ausbauringe, nach der Definition des inneren Ausbauwiderstandes am Innenradius r_0 angreifend gedacht, zeigt einen bemerkenswerten Verlauf hinsichtlich der Abhängigkeit von der Wanddicke des Ausbaus und von den Gebirgsparametern. Hoher Reibungswert und hoher Scherwert des Gebirges ergeben einen kleinen inneren Ausbauwiderstand p_0, ein scheinbar überraschendes Ergebnis, das aber bei kurzem Nachdenken einleuchtet (siehe Abb. 14).

Wird ein besseres Gebirge durch Ausbaumaterial verdrängt, tritt eben nur eine geringere Tragfähigkeitssteigerung des Gesamtsystems ein.

Schlechteres Gebirge erfordert einen wesentlich größeren inneren Ausbauwiderstand, um den Querschnittsverlust des Hohlraumes klein zu halten. Wenn auch insgesamt ein größerer Aufwand für den Ausbau nötig ist, so macht die Erkenntnis Mut, daß der Ausbauwerkstoff bei schlechterem Gebirge besser ausgenutzt und ein größerer innerer Ausbauwiderstand erzielt wird.

Eine wesentliche Erkenntnis ist insbesondere die technische begrenzte Dicke des geschlossenen Ausbaus mit 20% des Innenradius bei großem Reibungswert und 40% des Innenradius bei kleinem Reibungswert des Gebirges. Wird diese Grenzdicke überschritten, so sinkt die Stützwirkung des

Ausbaus, der innere Ausbauwiderstand p_0 nimmt ab! Es wäre ein grober technischer Fehler, diese Wanddicke zu überschreiten.

Nur ein dünner geschlossener Unterstützungsausbau ist wirtschaftlich. Ein Optimum wird erreicht, wenn ein Baustoff mit möglichst hoher Druckfestigkeit bei günstigem Verhältnis Druckfestigkeit zu Kosten gewählt wird.

Die spezifischen Kosten von Stahl liegen zu hoch, so daß Unterstützungsausbau aus mineralischen Stoffen wirtschaftlich vorteilhaft ist. Bei schlechtem Gebirge liegt die wirtschaftliche Wanddicke unter 10% des Innenradius. Mit mineralischen Baustoffen lassen sich Ausbauwiderstände zwischen 100 und 300 Mp/m² verwirklichen.

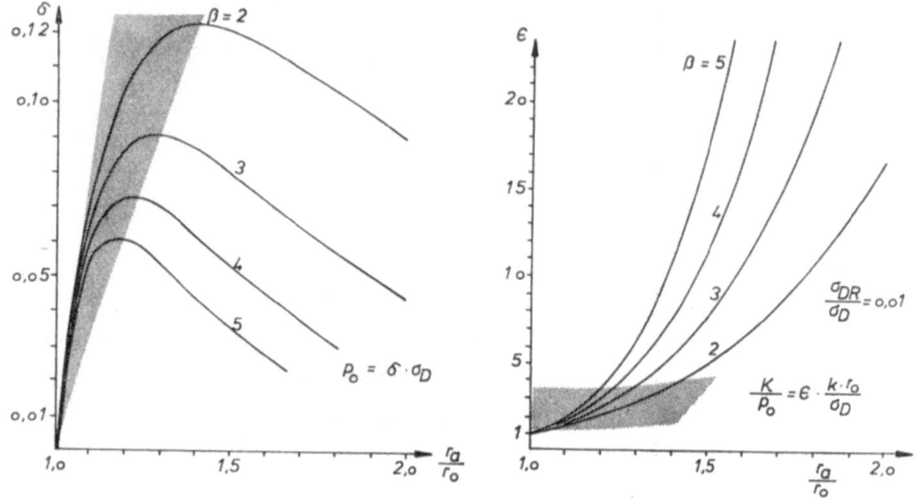

Abb. 14. Ausbaugewölbe
Support ring

Das System Gebirge/Unterstützungsausbau muß konstruktiv mit ausreichender Nachgiebigkeit ausgebildet werden. Im Bergbau wird diese Nachgiebigkeit über lange Zeit in mehreren Schritten in Anspruch genommen. Bei fehlender Nachgiebigkeit im Gebirge kann und muß der Unterstützungsausbau konstruktiv mit 10% bis 20% Nachgiebigkeit ausgestattet werden.

Ausbausysteme für Bergbaustrecken

Man unterscheidet zwischen leichtem Ausbau mit offener Sohle für relativ geringen Gebirgsdruck oder kurzlebige Grubenräume und verstärktem Ausbau für relativ großen Gebirgsdruck.

Leichter Ausbau

Beim leichten Ausbau herrscht noch der starre oder nachgiebige Stahlbogen vor, der mit einer Handverpackung hinter Drahtverzug mehr oder weniger gut an das Gebirge angeschlossen wird. Neuerdings wird auch maschinelles Hinterfüllen mit abbindendem mineralischem Material angewendet. Hohe Auffahrleistungen und geringe Ausbaukosten ergeben große Wirtschaftlichkeit, solange die Unterhaltungskosten klein bleiben.

Die Grenze der Wirtschaftlichkeit wird erreicht, wenn die Unterhaltungskosten die Auffahrkosten erreichen oder überschreiten. Völlig unwirtschaftlich ist leichter Ausbau, wenn die Strecke so stark verformt wird, daß sie neu aufgefahren oder sogar unter Aufrechterhalten des Betriebes durchgebaut werden muß. Ankerausbau mit offener Sohle wird zu einer wirtschaftlichen Verbesserung bei nahezu gleicher Querschnittsverminderung führen.

Verstärkter Ausbau — Geschlossener Ankerausbau

Gute Ergebnisse sind mit ringförmig geschlossenem Ankerausbau zu erwarten. Ein Ankerringausbau ist bei mittlerem Gebirgsdruck vorteilhaft, da bei hohen Auffahrleistungen und mittleren Ausbaukosten nur geringe Unterhaltungskosten entstehen.

Bei relativ großem Gebirgsdruck muß mit großem Querschnittsverlust gerechnet werden. Zum Ankerringausbau gehört ein flächiger Verzug, beispielsweise aus Maschendraht oder Baustahlgewebe, der das Gebirge zwischen den Ankern faßt.

Geschlossener Ausbau aus mineralischem Material

Die erste Stufe eines solchen Ausbaus ist bei mittlerem Gebirgsdruck ein Zusatzausbau aus Spritzbeton oder -mörtel in Verbindung mit Stahlbögen oder Ankerausbau und einem Sohlausbau aus Ankern. Die Nachgiebigkeit des Ausbaus muß deshalb in der Sohle erreicht werden.

Gute Erfahrungen liegen mit Spritzmörtel und offener Sohle vor, solange die Sohlenhebung durch Nachsenken beherrscht wird und eine ausreichende Streckenbreite erhalten bleibt. Die geankerte Sohle ist noch nicht genügend erprobt, um sichere Aussagen machen zu können. Versuche in dieser Richtung sind nötig.

Geschlossener, nachgiebiger Ausbau aus mineralischem Material ist seit Jahrzehnten in Form des Ausbaus aus Betonformsteinen mit Quetschlagen erprobt. Die Auffahrleistungen sind niedrig und die Ausbaukosten hoch. Die niedrigen Unterhaltungskosten wiegen diesen Nachteil nicht auf.

Eine erfolgreiche Entwicklung im tschechischen und belgischen Bergbau mit dem starren Paneelausbau ist auf den Steinkohlenbergbau der Bundesrepublik Deutschland nicht übertragbar.

Zwar sind die Auffahrleistungen ausreichend hoch und die Ausbaukosten vertretbar, aber die Unterhaltungskosten sind unsicher, weil bei häufigem Gebirgsdruckwechsel Brüche im starren Ausbau befürchtet werden müssen.

Bei relativ großem Gebirgsdruck ist ein nachgiebiger Ausbau mit hohem inneren Ausbauwiderstand nötig. Die Untersuchungen des Teilvorhabens

„Ausbausysteme zur Beherrschung großer Gebirgsdrücke — Nutzung von Tunnelbauerfahrungen im Bergbau" des mit Mitteln des Bundesministers für Forschung und Technologie der BRD geförderten Forschungsvorhabens „Steinkohlenbergwerk der Zukunft" haben ergeben, daß eine nachgiebig gestaltete Sohle aus Ortbeton mit einem nachgiebig gestalteten Oberbogen aus Stahlbetonfertigteilen wirtschaftlich ist.

In einem weiteren Forschungsvorhaben wurde bereits ein nachgiebiger geschlossener Ausbau aus Stahlbetonfertigteilen entwickelt, der im Untertageeinsatz erprobt werden soll.

Die Wirtschaftlichkeit dieses neuartigen hochtragfähigen nachgiebigen Ausbaus beruht auf den erwarteten geringen Unterhaltungskosten und auf den gegenüber dem Ausbau aus Betonformsteinen wesentlich größeren Auffahr- und Ausbauleistungen. Die Wirtschaftlichkeit gegenüber einem Ausbau mit mittlerem Ausbauwiderstand (beispielsweise Ankerringausbau) folgt aus dem wesentlich größeren nutzbaren Restquerschnitt des Hohlraumes.

Es ergeben sich nach der Höhe der Gebirgsbeanspruchung also drei Ausbauklassen für Bergbaustrecken.

1. Leichter Ausbau mit offener Sohle;
2. Mittelschwerer Ausbau als geschlossener Ankerringausbau oder Spritzbeton mit geankerter Sohle;
3. Schwerer Ausbau als geschlossener Ausbau aus Ortbetonsohlbogen und Oberbogen aus Stahlbetonfertigteilen.

Alle drei Ausbausysteme müssen wegen der ständigen Gebirgsdruckänderungen ausreichend verformbar gestaltet werden.

Mit wachsender Gebirgsbeanspruchung ist größerer Ausbauwiderstand wirtschaftlich, weil er einen ausreichend großen Nutzquerschnitt sichert.

Bei leichtem und mittelschwerem Ausbau kann die Sohle nicht unmittelbar befahren werden. Als Fahrbahn sind im Steinkohlenbergbau Gleise oder an der Firste aufgehängte Schienen erforderlich. Die feste Sohle des schweren Ausbaus ist voraussichtlich als eine gute Fahrbahn für gummibereifte Fahrzeuge nutzbar.

Zusammenfassung

Auf Hohlräume im Abbaubereich des Bergbaus wirkt eine Folge von Gebirgsdruckänderungen ein. Die Größe des Gebirgsdrucks schwankt über Jahre hinaus mehrmals zwischen einem Mehrfachen des ursprünglichen bankrechten Drucks und völliger Entspannung.

Bergmännische Hohlräume erleiden lange anhaltende Querschnittsverluste. Jederzeit muß Verformbarkeit des Ausbaus verfügbar sein.

Ursache des Querschnittsverlustes bergmännischer Hohlräume im Abbaubereich ist eine Überbeanspruchung des Gesteins und eine Zone zerbrochenen Gesteins, die sich unter Scherbewegungen entspannt.

Physikalische Eigenschaften des Gebirges sind die Grundlage für die gebirgsmechanischen Gesetze über das Zusammenwirken von Ausbau und zerbrochenem Gebirge. Der radiale Ausbauwiderstand erzeugt einen Anpreßdruck auf den Scherbewegungsflächen des Gebirges und erhöht den Reibungswiderstand entsprechend dem Reibungswert. Die radiale Traglast des Systems Ausbau — zerbrochener Gebirgsring beträgt ein Mehrfaches des radialen Ausbauwiderstandes. Die Gesetzmäßigkeiten werden erläutert; sie zeigen dem Ingenieur den Weg zur Lösung der Aufgabe, den Querschnittsverlust des Hohlraumes mit wirtschaftlichen Mitteln klein zu halten.

Hoher Ausbauwiderstand begrenzt die Querschnittsabnahme des Hohlraumes. Aus wirtschaftlichen Gründen muß ein Kompromiß gewählt werden, bei dem eine bestimmte Querschnittsabnahme des Hohlraumes in Kauf genommen wird. Ausreichende Nachgiebigkeit des Ausbaus ist wirtschaftlich.

Maßstab für die Wahl der Ausbaumittel ist der als *neuer* Begriff eingeführte innere Ausbauwiderstand. Die Ergebnisse der Untersuchung der Gebirgsankerung werden dargestellt. Der innere Ausbauwiderstand des Ankerringausbaus ist kleiner als die auf die Hohlraumwandung bezogene mittlere Ankerspannung. Es wird eine optimale Ankerlänge herausgestellt. Zu kurze und zu lange Anker sind unwirtschaftlich.

Die Untersuchung des Unterstützungsausbaus in Form des geschlossenen Ausbauringes hat eine technisch und wirtschaftlich begrenzte Dicke ergeben. Je dünner der Ausbau gewählt wird, um so höher wird der Werkstoff ausgenutzt. Es sollten Baustoffe mit hoher Druckfestigkeit bei einem günstigen Verhältnis Druckfestigkeit zu Kosten gewählt werden.

Für Bergbaustrecken sind drei Ausbauklassen zu unterscheiden, die mit steigenden Kosten einen steigenden radialen Ausbauwiderstand leisten. Alle drei Systeme sind ausreichend nachgiebig zu gestalten, um jederzeit Verformbarkeit verfügbar zu haben.

Die Ausführungen werden durch Formeln und Diagramme über Tragfähigkeit und Wirtschaftlichkeit der Ausbausysteme untermauert.

Literatur

Lütgendorf, H. O.: Spannungen und Verformungen im festen Gebirge um kreiszylindrische Grubenräume. Glückauf-Forschungshefte *28*, H. 2 (1967).

Lütgendorf, H. O.: Spannungen und Verformungen im lockeren Gebirge um kreiszylindrische Grubenräume. Glückauf-Forschungshefte *28*, H. 4 (1967).

Lütgendorf, H. O.: Die Wirkung des Grubenausbaus auf die Spannungen in einer durch Gleitlösen begrenzten, angeschnittenen Gebirgsschicht. Glückauf-Forschungshefte *29*, H. 4 (1968).

Lütgendorf, H. O.: Der Mindestausbauwiderstand des Grubenausbaus in Strecken und Schächten. Glückauf-Forschungshefte *29*, H. 5 (1968).

Lütgendorf, H. O.: Quantitative Gebirgsmechanik der Untertagebauten im geklüfteten Gebirge. Essen: Verlag Glückauf 1971.

Lütgendorf, H. O.: Konvergenz der Grubenräume im elastischen Gebirge bei Gebirgsdruckschwankungen. Glückauf-Forschungshefte *35*, H. 5 (1974).

Lütgendorf, H. O.: Tragfähigkeit bogenförmiger Grubenausbauarten aus Stahl, Beton und Stahlbetonfertigteilen. Glückauf-Forschungshefte *36*, H. 1 (1975).

Lütgendorf, H. O., Schuermann, F.: Beanspruchung von Grubenbauen durch Gebirgsdruckänderungen beim Flözbergbau und ihre Überwachung durch Langmeßanker. Glückauf-Forschungshefte *35*, H. 4 (1974).

Anschrift des Verfassers: Prof. Dr.-Ing. Hans Otto Lütgendorf, Baudirektor der Gewerkschaft Auguste Victoria (BASF), Viktoriastraße 47, D-4370 Marl, (Kr. Recklinghausen), Bundesrepublik Deutschland.

Rock Mechanics, Suppl. 7, 103—128 (1978)

Rock Mechanics
Felsmechanik
Mécanique des Roches

Bergbau und Tunnelbau — Anregungen und Ergänzungen

Von

G. Feder und A. Olsacher

Mit 25 Abbildungen

Zusammenfassung — Summary

Bergbau und Tunnelbau — Anregungen und Ergänzungen. Zum Vergleich mit den folgenden Ausführungen wurden in Abschnitt 1 die Bemessungsgrundsätze für Tunnel dargelegt, bei denen keine Brucherscheinungen des Gebirges zugelassen sind. Anschließend werden Tunnel mit planmäßigen Bruchzonen erörtert und die Zweckmäßigkeit solcher Bruchzonen aufgezeigt. Im Abschnitt 3 werden 2 Beispiele aus dem Bereich der Fernrohrleitungen und U-Bahntunnel in Bergsenkungsgebieten behandelt.

Mining and Tunnelling — Interaction and Cross-Fertilization. Demands made upon deformation-behaviour and life-expectancy of man-made underground excavations in mining and tunnelling differ significantly. Rock-mechanics, however, remains the same. Whenever both fields come into contact mutual stimulation and complementation results. This applies particularly in the case of major rock-deformations (see Fig. 1c). To illustrate the above the following examples are presented.

In the Radenthein magnesite mine arc and slope effects on the periphery of the subsiding fracture-zone linked to top-slicing, result in particularly intensive concentrations of rock pressures. It is here where the main haulage-ways had to be placed. Support-measures of roof and wall had to be separated quantitatively in order to achieve rational control of tunnel-construction. It was found that a marked wall-convergence was feasible without impairing roofsupport. SN anchors prevented wall collaps during convergence. Model tests and measurements in situ assist in elucidating fracture mechanism.

A tunnel with a cross section 21 m² revealed a further advantage of SN anchors as warning indicator for unexpectedly worthening rock conditions. The fracture patterns which did in fact develop confirm the statements made in Fig. 1.

Subsidence troughs in the hanging wall of mined-out areas result in the deformation of pipelines and shallow tunnels. Deformation of pipelines result in anchor point effects. Data obtained during the investigation of anticipated fracture-mechanism are presented (Lehner 1977, Leoben). Underground railway tunnels below the watertable in subsidence and swamp areas require a watertight and flexible tubeshell. A recently developed method to achieve this by using the leave spring principle is presented.

Die Anforderungen hinsichtlich geringer Verformung und hoher Lebensdauer sind im Tunnelbau meist wesentlich schärfer als im Bergbau. Bergmännische Hohlraumbauten können daher mit bedeutend kleinerem finanziellem Aufwand erstellt werden als Tunnelbauwerke gleicher Abmessung. Die im Bergbau durchaus zulässigen Bruchvorgänge können aber dem Tunnelbauer wertvolle Einblicke bieten, die es ihm leichter ermöglichen, die im Tunnelbau tatsächlich vorhandenen Sicherheiten zu erfassen.

Vom Standpunkt des vom Bergbau her mit eingeplanten Brucherscheinungen vertrauten Montanisten, zeichnet sich die Gebirgsmechanik des Tunnelbaues wie folgt ab:

Vorbemerkung: Zur einfacheren Darlegung wird hier angenommen, daß die größte Hauptspannung des primären Gebirgsdruckes vertikal wirkt. Die Ergebnisse gelten aber in gleicher Weise auch für schräge Primärhauptspannungen, doch sind dann die First-, Ulm- und Sohlpunkte so versetzt anzunehmen, daß die Verbindungslinie von First- und Sohlpunkt mit der Richtung der größten Hauptspannung übereinstimmt.

Die mathematische Formulierung der folgenden Ergebnisse wurde von Feder 1977 veröffentlicht.

1. Tunnel ohne Druck- oder Scherbrucherscheinungen des Gebirges

Abgesehen von Zugbrüchen und Entspannungsschwellen im First- und Sohlbereich (im Falle kleiner primärer Seitendruckbeiwerte) treten keine

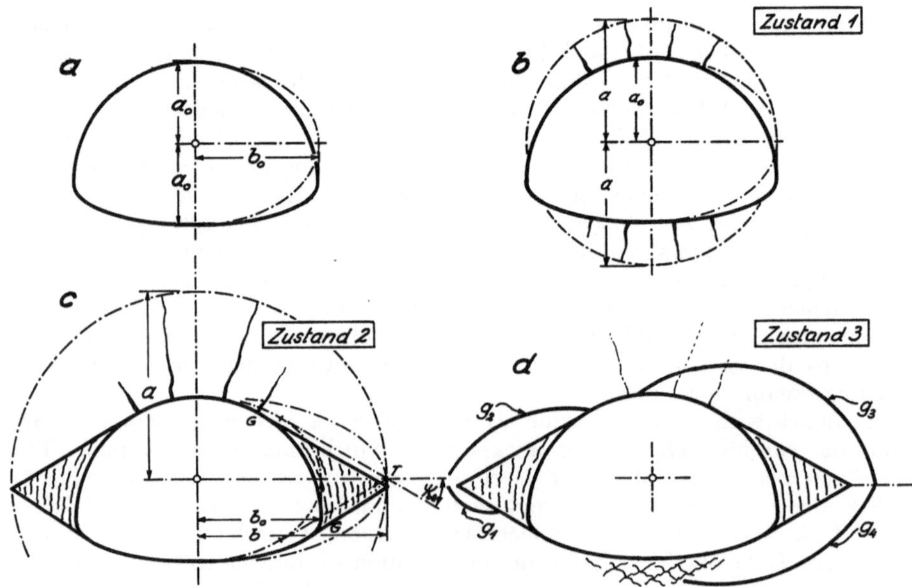

Abb. 1. Bruchzustände in festem Gebirge bei vertikal richtungsbetontem Primärdruck nach Feder 1977

Failure stages in solid rock mass as a result of primary stresses predominantly vertically acting

Brucherscheinungen auf. Das Bauwerk wird also bereits im Zustand 1 (Abb. 1) stabil. Die Gebirgsdeformationen sind entsprechend klein, was wohl besonders im städtischen Tiefbau und U-Bahnbau von Bedeutung ist. Ein weiterer Vorteil liegt darin, daß die ungebrochene Tragkraft des Gebirges am ganzen Umfang erhalten bleibt, oder mit anderen Worten: daß kein Bruch- oder Plastizierungsvorgang auftritt, der einen Verlust an Kohäsion zur Folge hätte.

Ist die größte Druckspannung „max σ" (sie tritt am Ulm auf) mit Sicherheit kleiner als die einachsige Druckfestigkeit des Gebirges, dann sind im Ulmbereich — außer der Deckung örtlicher Anisotropie und einer eventuellen Oberflächenversiegelung — keine Stützmaßnahmen erforderlich. Eine einfache Ermittlung von max σ wurde in der erwähnten Veröffentlichung vorgelegt.

Ist max σ größer als die durch einen Sicherheitsfaktor abgeminderte einachsige Druckfestigkeit, dann sind zum Stabilisieren des Zustandes 1 eine oder mehrere der folgenden Maßnahmen im Ulmbereich erforderlich:

— Vergrößern der einachsigen Gebirgsdruckfestigkeit durch Injizieren (falls möglich).

— Vergrößern der einachsigen Gebirgsdruckfestigkeit durch Abbau des Porenwasserdruckes z. B. mit Vakuumlanzen, Vorstollen etc. (falls möglich).

— Aufbringen einer radialen Querpressung auf das Gebirge, so daß die einachsige Druckfestigkeit bedeutungslos und eine höhere Triaxialfestigkeit maßgebend wird. Die Querpressung wird im allgemeinen durch eine *vorgespannte* Systemankerung mit eventueller Sekundärkonstruktion (z. B. Spritzbeton mit Baustahlgitter) zur Verteilung der Ankerkräfte auf den Ausbruchrand erreicht. Nicht vorgespannte Anker wirken nur im Sinne einer Querdehnungsverminderung (nächster Absatz).

— Verhinderung der Querdehnung des Gebirges, so daß die einachsige Druckfestigkeit bedeutungslos und die höhere Druckfestigkeit bei Querdehnungsbehinderung maßgebend wird. Zur Querdehnungsbehinderung sind Konstruktionen, die allein durch ihren Biegungswiderstand wirken, unbrauchbar (z. B. hufeisenförmige Rahmen ohne Ringschluß), weil sie zu weich reagieren, was Konvergenzwege erfordert, bei denen das Gebirge bereits zerbricht. Geeignet sind geschlossene oder unnachgiebig (z. B. auf dem Verbau von Ulmstollen) gelagerte Gewölbe, die kompakt am Ausbruchrand anliegen. Letzteres ist beim Spritzbeton von vorneherein erfüllt, bei Tunnelbögen (unnachgiebige Schlösser!) erfordert es ein sorgfältiges Hinterspritzen mit Beton.

Jede dieser Maßnahmen muß zu einem Zeitpunkt wirksam werden, zu dem noch die stützende Wirkung der Ortsbrust ein Zerbrechen der Ulmbereiche verhindert (Anker und Ringschluß in Ortsbrustnähe; kurze Ringschlußzeit bei Absinken des stützenden Orstbrusteinflusses durch progressive Bruchvorgänge oder durch viskoses Gebirgsverhalten).

Bei Anwendung der Neuen Österreichischen Tunnelbauweise (NATM) wird bekanntlich die Gebirgsdeformation laufend gemessen und der Einsatz der Stützmittel so gesteuert, daß in dem, in diesem Kapitel behandelten Fall nur Deformationen auftreten, bei denen noch keine schädlichen Brucherscheinungen des Gebirges („Auflockerungen" nach Pacher 1964) zu erwarten sind (Abb. 2). Dadurch werden Wirtschaftlichkeit und Sicherheit erhöht.

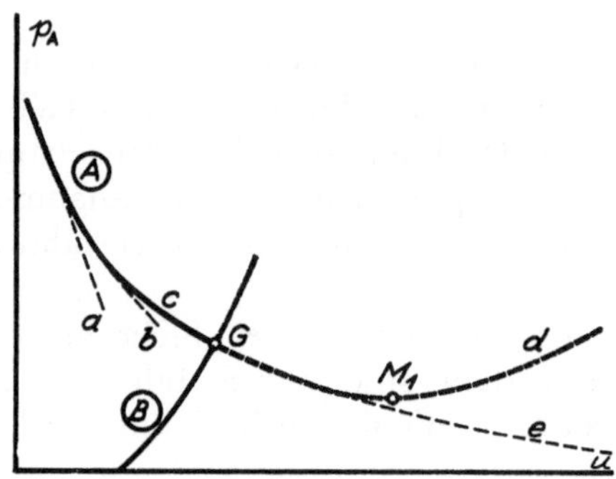

Abb. 2. Zusammenhang zwischen Ausbauwiderstand p_A und Konvergenz u unter vorwiegend zentralsymmetrischen Gegebenheiten nach Pacher 1964

A — erforderlicher Ausbauwiderstand um das Gebirge bei der Konvergenz u im Gleichgewicht zu halten; *B* — vom Ausbau tatsächlich aktivierbarer Widerstand; *G* — Gleichgewichtszustand; M_1 — Minimalstelle, sie stellt den kritischen Zustand bei der Ausbaubemessung dar

a) Gebirge mit linear-elastischem Verhalten und ohne Bruch; b) Gebirge mit einem bei steigender Belastung absinkenden Verformungsmodul; c) Gebirge mit zunehmenden Bruch- oder Plastizierungsvorgängen; d) Gebirge mit merklichen Auflockerungszonen als Folge der Brucherscheinungen („Pacherlinie", aus Messungen festgestellt); e) Theoretische Gebirgsverhalten ohne Auflockerungszonen („Fennerlinie")

Relation between lining resistance p_A and convergency u in predominantly isotropic shape- and stress condition (Pacher 1964)

A — required lining resistance for rock mass stabilizing after reaching convergency u; *B* — possible lining resistance after reaching convergency u; *G* — equilibrium; M_1 — minimum point. Critical stage for lining design

a) rock mass in linear elastic condition without fracture; b) rock mass with degressive variable deformation modulus; c) rock mass with progressive increasing failure mechanisms; d) rock mass with significant depressed and loosened areas (found via measurements by Pacher); e) theoretical rock mass behaviour without loosened areas (Fenner)

2. Tunnel mit planmäßigen Quetschbrucherscheinungen in den Ulmbereichen

2.1. Allgemeines

Im Falle richtungsbetonten Primärdruckes sind am Ulm die Druckspannungsspitzen (Umfangsrichtung) meist um ein Vielfaches größer als an First oder Sohle. Vor allem bei großer Überlagerung oder bei Gebirge geringer Festigkeit gelingt es dann nicht, mit vertretbarem Aufwand ein Brechen bzw. Plastizieren des Gebirges im Ulmbereich zu verhindern. Dies ist auch der im Bergbau übliche Fall, den wir als Zustand 2 (Abb. 1) bezeichnen. Erfreulicherweise tritt bei diesem Bruchvorgang im festen Gebirge ein Effekt auf, der die Bruchzone zwingt, einen eng begrenzten Bereich GTG nicht zu überschreiten (Feder 1977). Die Ursache liegt darin, daß das zerbrochene Gebirgsmaterial hohlraumwärts herausgedrückt werden muß, was nur bis zu einer bestimmten Bruchzonentiefe möglich ist, da sonst die Gleitflächen mit einem zu spitzen Winkel $2\psi_{bg}$ gegeneinander anlaufen, so daß eine Selbstsperrung eintritt. Ein einfacher Versuch mit einer sandgefüllten Klappe zeigt dies deutlich (Brunner 1978).

Ist diese Endphase des Zustandes 2 erreicht, dann hat sich folgendes Tragsystem ausgebildet:

— Das natürliche Gebirgsgewölbe (der sogenannte „Tragring") hat sich vom Ausbruchrand weg bergwärts verschoben. Die Stützlinie verläuft nun knapp bergwärts des Punktes *T*.

— Die Druckspannungsspitze des natürlichen Gebirgsgewölbes ist infolge des Nachgebens der Bruchzone vom Ausbruchrand zum Punkt *T* verlagert worden, wobei ihre Größe noch etwas zugenommen hat. Für die Aufnahme dieser Spannungsspitze ist nun aber nicht mehr die einachsige Druckfestigkeit maßgebend, sondern eine Druckfestigkeit bei nahezu absoluter räumlicher Querdehnungsbehinderung. Letztere wird durch den Stauchwiderstand der bergwärts der Linie *GT* liegenden Gebirgszonen in einem Ausmaß wachgerufen (Abb. 1), das durch künstliche Stützmaßnahmen nie erreicht werden könnte.

— Der Stauchwiderstand der bergwärts *GT* liegenden Zonen wird noch wesentlich dadurch vergrößert, daß das zerbrochene Material hohlraumwärts *GT* nicht frei ausfließen kann, sondern so verspannt bleibt, daß es vertikale Druckkräfte auf die Flächen *GT* überträgt. Ein solches Verspannen erfolgt durch Stützmittel, die auch bei großen Verformungen ihre Tragkraft nicht verlieren, wie nicht vorgespannte Ulmenanker aus Stahl mit großer Gleichmaßdehnung oder — falls deren Ringschluß möglich ist — gut hinterfüllte Streckenbögen mit Reibungsschlössern. Die hinterfüllte Türstock- bzw. Polygonzimmerung hat in beschränktem Maße ähnliche Wirkung.
Da bei Gebirgsverhältnissen, die ein planmäßiges Zerquetschen der Ulmbereiche *GTG* erfordern, im Tunnelbau meist ein voreilender Kalottenvortrieb notwendig wird, ist ein rascher Ringschluß nur selten möglich. Es kommt daher im Tunnelbau als primäres Stützmittel meist nur die Systemankerung in Frage. Sie muß eingebracht werden, solange noch die nahe Ortsbrust entlastend einwirkt. Anderenfalls kann es zu Scherbrüchen bergwärts *GT* kommen, eine als Zustand 3 (Abb. 1) bezeichnete Situation, die unbedingt vermieden werden soll.
Dieser Zustand 3 kann auch eintreten, wenn sich die Gebirgseigenschaften unerkannt so verschlechtert haben, daß die Ankerlänge zu kurz (Abb. 5) oder die Ankertragkraft zu gering (Abschnitt 2.4) geworden ist, um den Zustand 2 stabil zu halten.

— Die Systemankerung ist im allgemeinen nicht in der Lage allein das zerbrochene Gebirgsmaterial am Ulm zu halten, da dieses an den Ankern vorbeizufließen vermag. Es ist daher eine Sekundärkonstruktion erforderlich, welche am Ulm die vom hereindrängenden Bruchmaterial ausgeübten Flächenlasten aufnimmt und an die örtlichen Ankerplatten abgibt. Baustahlgitter, Verzugleitern, Spritzbeton und Streckenbögen dienen dazu. Da sie aber in diesem Falle nur der Lastverteilung dienen und trotz großer Konvergenz tragfähig bleiben müssen, ist für diese Stützmittel auch kein Ringschluß erforderlich. Es sind daher auch Stauchschlitze im Spritzbeton zulässig bzw. bei großen Konvergenzen notwendig. Der „Tragring“ selbst liegt wie erwähnt tiefer im Gebirge; er ist geschlossen und seine Stützlinie verläuft im Ulmbereich nahe *T*.

— Der Zustand 2 (Abb. 1) wird meist schon vor seiner Endphase stabil (Abb. 3). Ulmanker, die bis in das kompakt verbliebene Gebirge reichen, vermindern die Tiefe der Bruchzone (Feder 1977). Der Bruchtyp innerhalb der Quetschzone *GTG* ist bei festem Gebirge der Spaltbruch (Abb. 3) bei pseudofestem Gebirge oder bindigem Boden plastische

Abb. 3. Spaltbrucherscheinungen an den Ulmen. Typisch für festes Gebirge unter vertikal richtungsbetontem Primärdruck (Bild von Herrn o. Prof. Dr. G. Horninger, Wien, freundlicherweise zur Verfügung gestellt)

Squashing of tunnel wall by cleavage failures. Typical for solid rock mass as a result of predominantly vertically acting primary stresses (Photo: o. Prof. Dr. G. Horninger)

Dilatanz, begünstigt durch Porenwasserdruck oder Zerquetschen des die Poren umgebenden Traggerüstes. Die den Spaltbruch begleitende große Auflockerung hat meist zur Folge, daß die horizontale Konvergenz wesentlich größer wird als die in vertikaler Richtung.

— Soferne es sich nicht um stark hochovale Querschnitte handelt, sind bei richtungsbetontem Primärdruck die Druckspannungen in First und Sohle wesentlich kleiner als am Ulm, vielfach ändern sie sich sogar zu Zugspannungen. Wegen letzterer und der damit verbundenen Gefahr von Niederbrüchen in Ortsbrustnähe hat sich der Einsatz von geankerten Streckenbögen und Baustahlgitter wegen der sofort vorhandenen Tragfähigkeit bewährt.

2.2. Anwendung der NATM bei planmäßigem Quetschbruch der Ulme

Der Zusammenhang zwischen erforderlichem Ausbauwiderstand p_A und Konvergenz u ist für das Ulmverhalten in Abb. 4 dargestellt, wobei richtungsbetonter Primärdruck (Regelfall) vorausgesetzt wird. Entsprechend den 3 unterschiedlichen Bruchzuständen (Abb. 1) treten auch 3 unterschiedliche Kurvenzweige auf.

Der Zweig 1 entspricht dem Zustand 1 (Abb. 1) und damit auch der Abb. 2.

Der Zweig 2 wird durch die Spaltbrüche im Ulmbereich (Zustand 2, Abb. 1) geprägt. Zunächst steigt mit zunehmender Konvergenz die Größe jenes Ausbauwiderstandes an, welcher in der Lage ist, den Verformungsvorgang zum Stillstand zu bringen, da der zerbrochene Ulmbereich an Tragkraft verloren hat. Dies stimmt auch mit den Messungen von Pacher (1964) überein. Mit zunehmender Tiefe der Bruchzone kommt es aber zur beschriebenen Selbstsperrung des Ausquetschvorganges und damit zum raschen Absinken des erforderlichen Ausbauwiderstandes.

Der Zweig 3 entspricht dem Zustand 3 (Abb. 1). Er wird maßgebend, wenn der Ausbauwiderstand am Ulm so klein wird, daß die vertikale Querverspannung der Bruchzone nicht mehr ausreicht den Scherbruch des Bereiches bergwärts *GTG* (Abb. 1) zu verhindern. Mit dem progressiven Anwachsen eines solchen Scherbruches steigt p_A rasch an. Dieser Zustand ist unbedingt zu vermeiden (Abb. 5).

Im gleichen Sinne wie in Abb. 2 gezeigt, kann man nun gemäß Abb. 4 das Prinzip der NATM sowohl für den Zweig 1 (kleine Verformung, großer Ausbauwiderstand) als auch in dafür geeigneten Fällen für den Zweig 2 (wesentlich kleinerer Ausbauwiderstand, große Verformung) anwenden. Die Erfahrungen am Arlberg haben dies bestätigt. Das folgende Beispiel zeigt ferner, daß die hier getroffene unterschiedliche Behandlung von First und Ulm zweckmäßig ist.

2.3. Ausführungsbeispiel aus dem Bergbau hinsichtlich planmäßiger Quetschbrucherscheinungen in Ulmbereichen

Der abbauwürdige Magnesitkörper auf der Millstätter Alpe ist plattenförmig entlang einer ausgeprägten Störungszone in altkristalline Schiefer eingelagert. Diese steilstehende Platte wird scheibenförmig von oben nach unten abgebaut (Abb. 6). Die schematische Darstellung der Verhältnisse zeigt, daß die Abbaustrecken der einzelnen Scheiben entlang des bergseitigen Lagerstättenkontaktes geführt werden, wo es aufgrund einer Reihe von Ursachen Spannungskonzentrationen gibt, die zu einer erhöhten asymmetrischen Belastung der Strecken führen (Abb. 7).

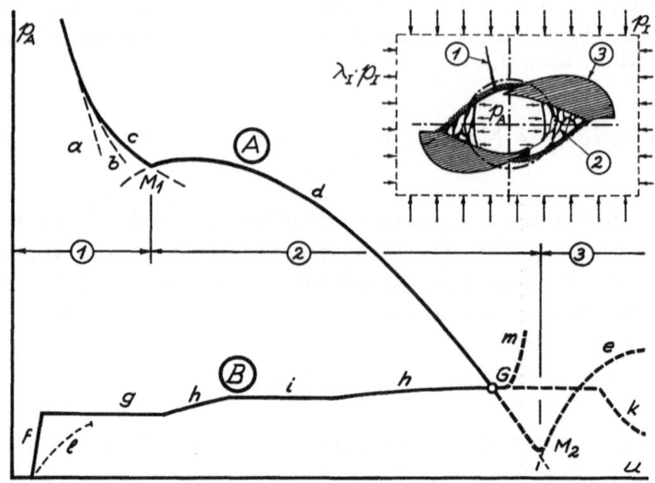

Abb. 4. Zusammenhang zwischen Ausbauwiderstand am Ulm p_A und Ulmkonvergenz u unter vorwiegend vertikal richtungsbetontem Primärdruck (Feder 1977) am Beispiel des Einsatzes der NATM am Arlbergtunnel

A, B, G, M_1, a) und b) — wie in Abb. 2

(1) — Zustand 1 nach Abb. 1 (Ulme ungebrochen, Firste und Sohle mit versagender Zugzone); (2) — Zustand 2 nach Abb. 1 (Ulme in begrenzter Zone zerquetscht); (3) Zustand 3 nach Abb. 1 (Verbruch durch Felsgrundbruch); c) Gebirgsverhalten bei fortschreitendem Versagen der Zugzone in First und Sohle; d) Gebirgsverhalten bei fortschreitendem Zerquetschen der Ulmzonen (Zustand 2), Abb. 3. Anfängliches Ansteigen, da die durch das Zerquetschen verlorene Kohäsion durch eine vom zusätzlichen Ausbaudruck wachzurufende Reibung ersetzt werden muß. (Widerspricht nicht den Meßergebnissen von Pacher, Abb. 2, Kurve d.) Späteres Abfallen infolge der Selbstsperrung des Ausquetschvorganges nach Erreichen einer bestimmten Bruchzonentiefe; e — Gebirgsverhalten beim Verbruch (Zustand 3) Abb. 5; M_2 — Zweiter Minimalwert. Er stellt für die Bemessung auf Zustand 2 (bei planmäßig zerquetschter Ulmzone, Abb. 3) den „kritischen Zustand" dar. M_2 kann auch unter der Linie $p_A = 0$ liegen (Natürliche Höhle mit zerquetschten Ulmen); f — Widerstand der Systemankerung vor Erreichen der Streckgrenze; g — Streckgrenzendehnung der Anker; h) Ankerdehnung im Verfestigungsstadium; i) Deformation der Ankerplatten; k) Ankerbruch oder Abscheren der Muttern; f) bis k) Verhalten von Ankern aus verformungsfreundlichem Stahl bei richtiger Abstimmung mit Platte und Mutter; l) Verhalten von Ankern aus glasfaserverstärktem Kunststoff (gegenwärtiger Stand der Technik); m) Steigerung des Ausbauwiderstandes durch Schließen der Kontraktionsschlitze im Spritzbeton und/oder Einbau der Innenschale nach Abklingen der Konvergenzgeschwindigkeit

Relation between the support resistance p_A on tunnel wall and the wall convergency u in a rock mass loaded by predominantly vertical acting primary stress

A, B, G, M_1, a) and b) — as described in Fig. 2

(1) — stage 1 (walls sound; on top and bottom failing tensile zones) see Fig. 1b; (2) — stage 2 (walls squashed in limited zone) see Fig. 1c; (3) — stage 3 (tunnel collaps by rockslide) see Fig. 1d

c) rockmass behaviour during progressive tensile failure in top and bottom zone; d) rockmass behaviour during progressive squashing failure of walls (stage 2) see Fig. 3. Initial curve rising as a result of required additional wall supporting forces for compensating lost cohesion (squashing failure) by friction. Later curve decreasing as a result of self-blocking of squash failure mechanism as soon as a certain deepnes of failure zone is reached; e) rock mass behaviour during collaps by rockslide (stage 3), see Fig. 5; M_2 — second minimum point. Critical stage for design of wall support if stage 2 (squashed wall zones, Fig. 3) are taken into consideration. M_2 may also be located below $p_A = 0$ as it can be found in natural caves with squashed wall zones; f) resistance of steel anchor systems before yielding; g) yield elongation of anchors; h) anchor elongation in strain hardening condition; i) deformation of anchor-plates; k) total failure of anchor or nut; f) up to k) be-

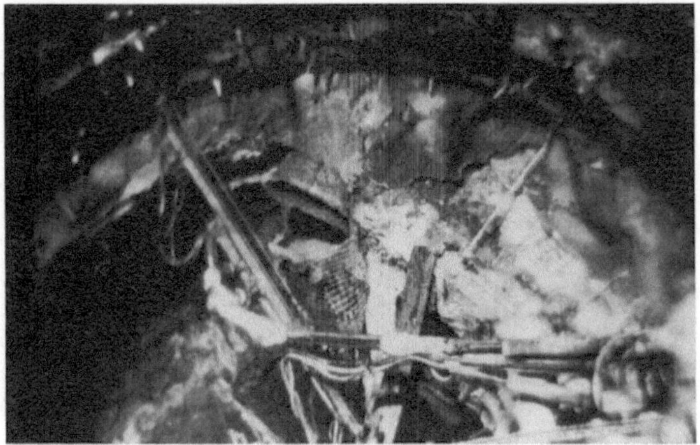

Abb. 5. Tunnelverbruch (Zustand 3) vgl. Abb. 1
Tunnel collaps by rock slide (stage 3) see Fig. 1

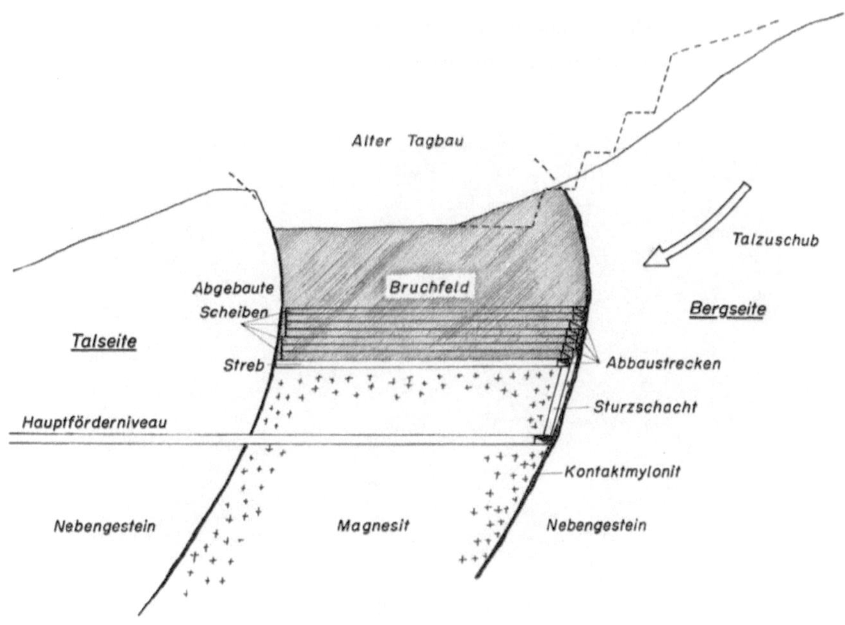

Abb. 6. Schematisches Lagerstättenprofil
Schematic deposit profile

haviour of anchors made of ductile steel, well fitted to anchor-plate and -nut design; l) behaviour of anchors made of glass fibre reinforced polyester (present known state of art); m) increasing of wall support resistance as a result of closing the deformation gaps in shotcrete shell and/or of building in an inner concrete shell as a second supporting system after calming down rock deformation velocity

Der zunächst eingesetzte nachgiebige Stahlausbau versagte, weil er hinsichtlich seiner Nachgiebigkeit der Belastungssituation nicht gerecht wurde. Die Untersuchungen, welche eine Anpassung des Ausbaues an die gegebene

Abb. 7. Typische Verschiebung des Streckenprofiles gegen die Talseite
Typical deformation of heading shape towards valley

Abb. 8. Verformte Ausbauelemente
Deformated lining elements

Situation ermöglichen sollten, zeigten, daß hauptsächlich der bergseitige Ulm die Tendenz hat, gegen den Streckenhohlraum zu wandern (Abb. 8).

Diesbezügliche Modellversuche an der Montanuniversität Leoben zeigten dazu folgende Bruchmechanismen. Die Skizze *a* der Abb. 9 zeigt die Belastungsanordnung am Modell, welches seitlich und unten unverschieblich gelagert war. Die Skizze *b* zeigt das gefundene Bruchbild bei nicht geankertem Ulm. Im Vergleich zum geankerten Ulm in Skizze *c* sind die durch Spaltbrüche (2) gelösten Blöcke des Ulmbereiches bei nicht geankertem Ulm wesentlich größer.

Zwischen dem Bruchbereich und dem zunächst kompakt verbleibenden Bereich bildet sich eine Gleitfläche (2a) deutlich aus. Nach einer wesentlichen Laststeigerung zeichnet sich schließlich ein Scherbruch (3) ab, welcher

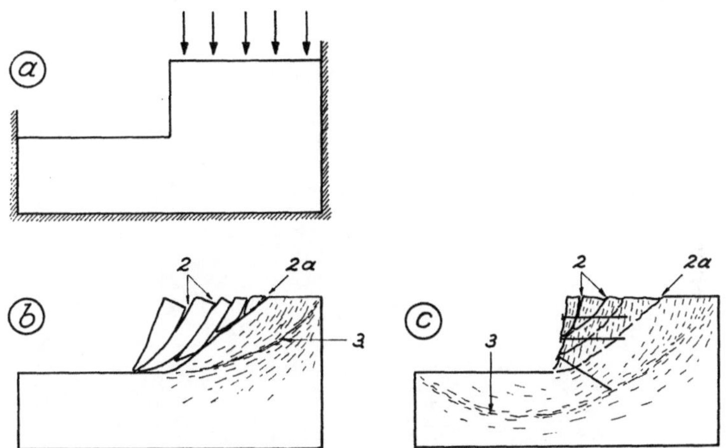

Abb. 9. Am Modell ermittelter Bruchmechanismus

a) Modell, Belastung und Lagerung des Ulm- und Sohlenbereiches; b) Bruchmechanismus bei ungeankerter Ausführung; c) Bruchmechanismus bei Ulmverbau mit Systemankerung 2 — Spaltbrüche des Zustandes 2; 2 a — Grenzfläche der Spaltbruchzone; 3 — beginnende Ausbildung des Scherverbruches (Zustand 3)

Failure mechanism found out by model tests

a) model test specimen, load and support of wall- and bottom-area; b) failure mechanism with not anchored walls; c) failure mechanism with anchor system in wall 2 — cleavage failures (stage 2, Fig. 1); 2 a — border of squashing zone; 3 — first signs of collaps by rock slide (stage 3)

sich zunächst durch eine Anhäufung kleiner Risse andeutet. Die Skizze *c* zeigt, daß bei geankertem Ulm die Spaltbrüche kleinere Blöcke mit besserem Zusammenhalt hervorrufen. Sie treten außerdem erst bei höherer Belastung auf. Die Blöcke selbst sind wiederum von kleineren Spaltbrüchen mit geringem Durchtrennungsgrad durchsetzt. Die Gleitfläche 2a deutet sich auch oberhalb einzelner Ankerreihen an und die Scherfläche 3, welche dem *Zustand 3* (Abb. 1), der einleitend beschriebenen Bruchzustände, entspricht, wird wesentlich länger als bei nicht geankerter Ausführung.

Abb. 10 zeigt ein Modell mit geankertem Ulm, oben im Zustand 2, unten nach Entfernen des Spaltbruchmaterials, wobei auch die Anrisse für den späteren Scherbruch des Zustandes 3 zu erkennen sind. Die oberen Ankerreihen wurden bereits mit dem Spaltbruchmaterial abgeräumt; die unterste Ankerreihe (zu tief) konnte hier das Vorbeifließen des Bruchmaterials nicht verhindern.

Aufgrund dieser Versuche kann man sich den Bruchvorgang in der Natur vereinfacht gemäß Abb. 11 vorstellen. Der linke Teil des Bildes stellt oben das Bruchfeld, unten den noch nicht abgebauten Magnesitstock dar; der rechte Teil das Kontaktgebirge mit seinen wahrscheinlichen Bruchformen bei nicht geankertem und im untersten Horizont bei geankertem Ulm.

Die unterste Ankerreihe hat die Aufgabe gegen Sohlverbruch zu sichern, die oberen Reihen fassen eine ca. 3 m mächtige Bruchzone so zusammen, daß sie trotz der Spaltbrüche eine erhöhte Resttragkraft bietet, wobei das

Abb. 10. Modell mit Systemankerung im Ulm beim Übergang vom Zustand 2 auf Zustand 3
Oben: Bruchzustand; unten: wie zuvor, jedoch nach Entfernen des durch Spaltbrüche zerstörten Materials

Test specimen with wall anchor system in transition from stage 2 to stage 3
Above: failure situation; below: same situation after removing squashed material

begrenzte Hereinwandern des ganzen Bruchblockes den Ulm in Ausbruchsnähe auf Kosten weiter bergwärts liegender, tragfähigerer Bereiche entlastet.

Für die vertikale Entlastung des Ausbruchrandes sind entsprechend dem Prinzip des *Zustandes* 2 (Abb. 1) entsprechend große horizontale Verformungen erforderlich.

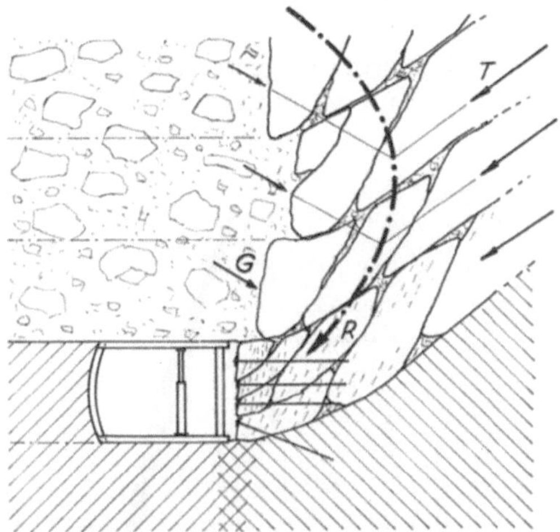

Abb. 11. Vermutliches Bruchverhalten des bergseitigen Ulms im Zuge des scheibenförmigen Abbaues von oben nach unten

Presumable failure behaviour of upstream wall during top slicing

Der ursprünglich gewählte geschlossene Stahlausbau war aber gerade in horizontaler Richtung nicht nachgiebig, wie Abb. 12a zeigt, so daß ein Spannungsabbau durch Verformung nicht erfolgen konnte. Die Folge davon war das in Abb. 8 gezeigte Ausbauversagen.

Der gegebenen lagerstätten- und abbaumethodenmäßig bedingten asymmetrischen Belastungssituation wurde nun durch Trennung der Ausbaufunktionen Rechnung getragen. Der bergseitige Streckenulm wurde durch Gebirgsanker so gesichert, daß er sich kontrolliert gegen den Streckenhohlraum bewegen konnte, ohne daß dabei Kräfte auf den übrigen Ausbau übertragen wurden. Dies ist in Abb. 12b schematisch dargestellt.

Durch diese Maßnahme konnte das Problem in der Praxis gelöst werden, wie die Abb. 13 beweist, welche die Abbaustrecke vor dem Strebdurchgang, oben vor und unten nach der Ausbauumstellung zeigt.

Mit den Ergebnissen der Modellversuche wonach zum Spannungsabbau gerade im Ulmbereich eine Bewegungsmöglichkeit gegeben sein muß — konnten die Ursachen des praktischen Erfolges sichtbar gemacht und systematisch begründet werden.

Bewegungsmessungen im geankerten Streckenulm und Messungen der Ankerlastaufnahme, welche für die Optimierung des Ankereinsatzes durchgeführt wurden, erbrachten Ergebnisse, welche sich ebenfalls mit den Modellversuchen decken. Abb. 14 zeigt die in situ ermittelte Trennfläche zwischen bewegtem Gebirge und Gebirge in relativer Ruhe (Olsacher 1976).

2.4. Beispiel eines Totalverbruches im Sinne des Zustandes 3 (Abb. 1 und 5) infolge Überlastung der Systemankerung

Beim Vortrieb einer Förderwendel im Magnesitbergbau Radenthein wurden im Lagerstättennahbereich tektonisch hoch beanspruchte Schichten durchfahren.

Die Ortsbrust zeigte dort trotz extremer Kleinbrüchigkeit des anstehenden Granatglimmerschiefers ein gutmütiges Verhalten. Die Sicherung des

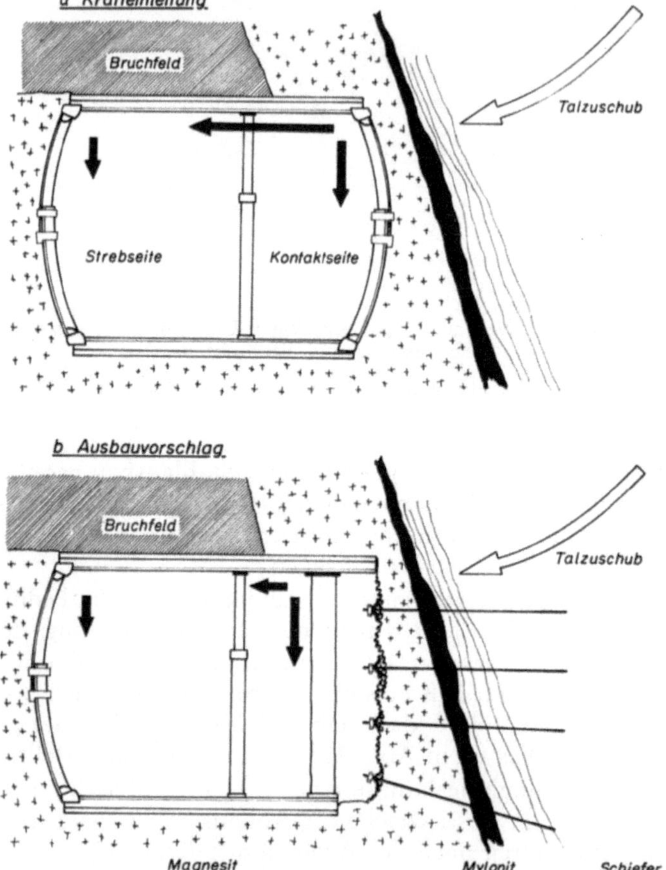

Abb. 12. Trennung der Ausbaufunktionen durch Einsatz von Gebirgsankern am bergseitigen Ulm

Separation of supporting functions by installation of an anchor system in upstream heading wall

Hohlraumes erfolgte mittels Gebirgsankern, Baustahlgitter und Spritzbeton entsprechend der Neuen Österr. Tunnelbauweise. Erst 5 bis 10 m hinter der Brust traten in diesem Bereich nennenswerte Horizontalkonvergenzen auf, welche sich während eines organisatorisch bedingten Vortriebsstillstandes beruhigten. Nach Wiederbeginn des Vortriebes stiegen die Horizontalkon-

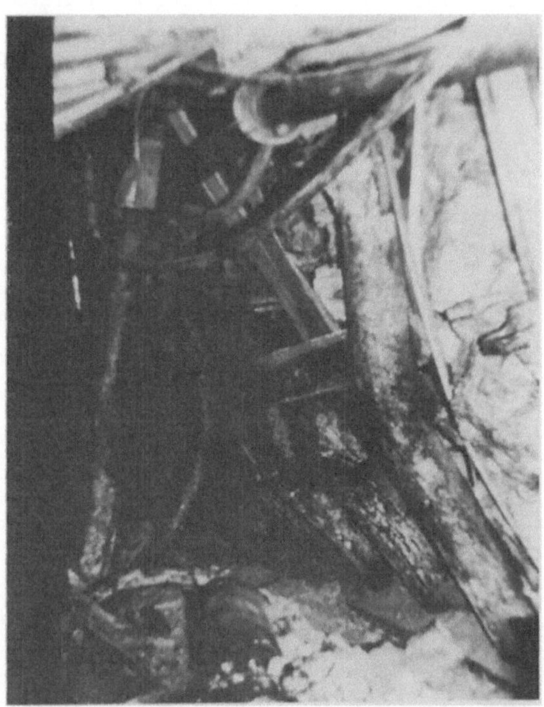

Abb. 13. Restquerschnitt der Strecke unmittelbar vor Strebdurchgang
Oben: bei alter Ausbauart; unten: bei neuer Ausbauart

Residual cross section of heading immediately before passage of a face
Above: with ancient kind of support; below: with improved kind of support

vergenzen sprunghaft auf täglich ca. 5% des Stollendurchmessers an. Die Vertikalkonvergenzen waren unbedeutend, wie aus der Abb. 15 ersichtlich ist.

Im Gefolge der großen Horizontalkonvergenzen traten Verformungen und Zerstörungen des Ausbaues auf. Es kam sehr rasch zu Ankerrissen, so daß aus Sicherheitsgründen eine Verstärkung des Ausbaues nicht mehr möglich war, weil die abspringenden Ankerplatten die Mannschaft gefährdeten.

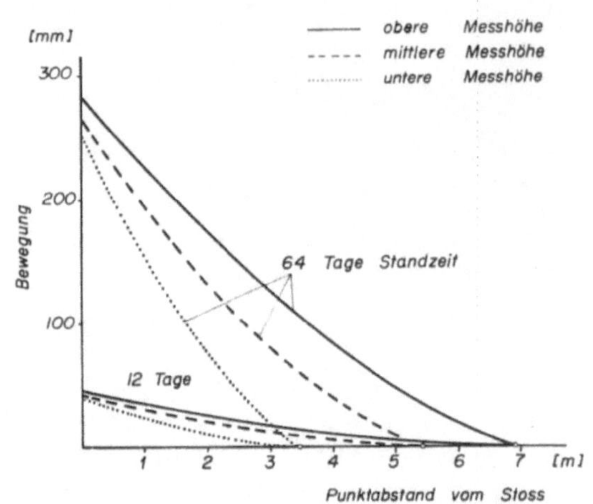

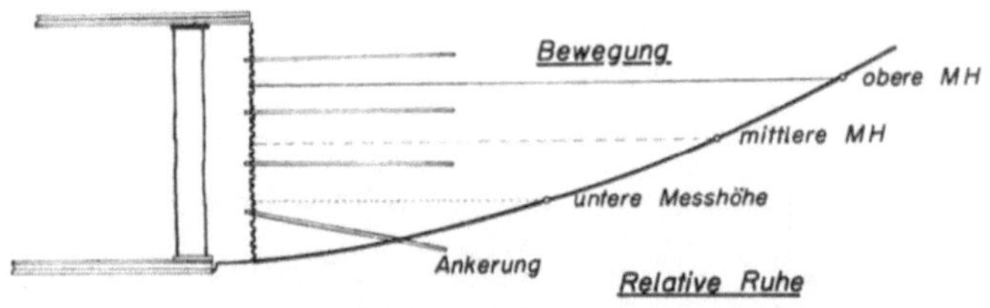

Abb. 14. Auswertung der Meßankerdaten
Results of extensometer measurements

Es wurde deshalb am Beginn des gefährdeten Bereiches mit Nachrißarbeiten und Verstärkung des Ausbaues begonnen. Leider war der dabei erzielte Arbeitsfortschritt zu gering, so daß ein Verbruch von ca. 15 m Stollenröhre nicht mehr verhindert werden konnte.

Im Zuge der Nachrißarbeiten und bei der anschließenden Gewältigung des Verbruches konnten die Bruchvorgänge um den Streckenhohlraum studiert und fotografiert werden. Die Auflockerungen in den Ulmen, wie sie aus der Natur (Abb. 3) und aus Modellversuchen (z. B. Feder 1977) bekannt sind, waren auch hier zu beobachten (Abb. 16). Im Verbruchbereich kam es zusätzlich zur Scherbruchbildung entsprechend dem *Zustand 3* der Abb. 1. Dies zeigt auch der Bruch und die Überschiebung des Sohlgewölbes deutlich (Abb. 17).

Abb. 15. Spaltbruch der Ulmen mit anschließendem Hereinschieben der Bruchzone
Fracture cleavage of heading wall followed by pushing squashed rockmass into cave

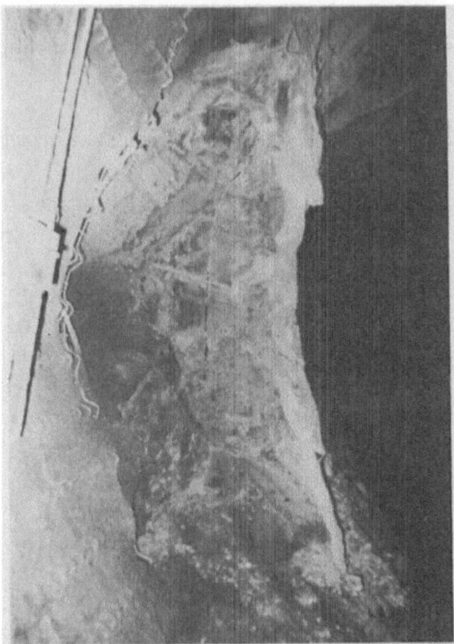

Abb. 16. Spaltbruchzone am Ulm, wie sie sich während der späteren Nachrißarbeiten zeigte
Fracture cleavage an heading wall, visible during ripping

Diese Beobachtungen und Erfahrungen zeigen, daß in erster Linie von der Sicherung der Ulme das weitere Verformungsverhalten des Hohlraumes abhängt.

In der Praxis ist es gerade bei Extremsituationen notwendig, möglichst rasch umfangreiche Auskünfte über dieses Verformungsverhalten zu bekommen. Neben der Anwendung der heute üblichen Meßmethoden, könnte man sich der alten Bergmannsweisheit „das Holz warnt bevor es bricht" erinnern und auf die heutigen Gegebenheiten umlegen. Auch die Anker können eine

Abb. 17. Ansatz eines Scherbruches im Sohlenbereich. Ein ähnlicher Scherbruch (Zustand 3) an der Firste leitete den Verbruch von 15 m Stollenlänge ein

Starting shear failure in bottom. The 15 m long collaps was introduced by a similar shear failure (stage 3) in top zone

warnende Funktion, ja sogar die Funktion eines zugegebenermaßen groben Meßinstrumentes haben, welches aber den großen Vorteil hat, über das ganze Profil verteilt zu sein. Etwa dann, wenn man statt einer zwei Ankerplatten verwendet, von denen sich eine bei einer bestimmten Belastung verformt und die zweite als eigentlich tragendes Element ausgebildet ist.

Eine derartige Messung der Ankerbelastungen hätte trotz der geringen Genauigkeit der Einzelmessung, wegen der Häufigkeit der „Meßpunkte" großen statistischen Wert.

3. Städtische Tunnelbauwerke im Einflußbereich von Abbaueinwirkungen aus Bergbaubetrieben

3.1. Mögliche Grenzwerte der Querbelastung von Tunnelkrümmern

Der Senkungstrog (Abb. 18), der sich über Abbauzonen an der Tagoberfläche abzeichnet, verursacht bekanntlich nicht nur Vertikalbewegungen, sondern auch Stauchungen und Dehnungen des Erdkörpers — in der Sprache des Markscheiders: „Pressungen“ und „Zerrungen“. Im unteren Teil des Bildes sind diese Zonen im Grundriß aufgetragen, wobei zu beachten ist, daß die ganze Erscheinung weiterwandert.

Ein, in solche Abbaueinwirkungen geratener Tunnelstrang wird in Längs- und Querrichtung mit dem Erdkörper deformiert, wobei er es versteht, sich den Längsdeformationen durch wurmartiges Kriechen teilweise zu entziehen (Meissner 1976).

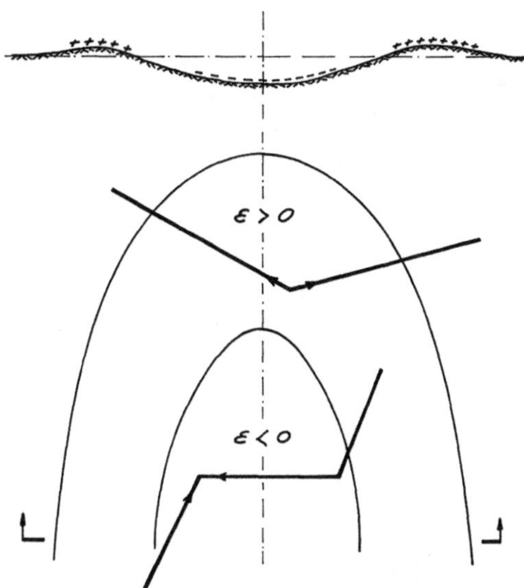

Abb. 18. Schematische Darstellung des Senkungstroges über Abbauzonen
Oben: Querschnitt; unten: Draufsicht
$+\varepsilon$ = Dehnung (Zerrung) der Tagesoberfläche

Schematic sketch of subsidence trough
Above: cross section; below: layout
$+\varepsilon$ = tearing on day surface

Gerät eine Krümmerstrecke einer Tunnelröhre in den Bergsenkungsbereich, dann verursachen die durch die Bodendeformation eingeleiteten Tunnellängskräfte, quer dazu gerichtete Abtriebskräfte. Sie sind in Zerrungszonen zum Krümmungsmittelpunkt des Tunnelstranges gerichtet, in Pressungszonen im Gegensinn.

Für uns hat sich nun die Frage nach dem Bruchmechanismus gestellt: Ist der Erdkörper der schwächere Teil, kommt es also zu einem Grundbruch — oder wird die Tunnelröhre abgerissen oder gestaucht, soferne keine Längenausgleicher vorhanden sind?

Mathematische Ansätze zu den folgenden Ausführungen finden sich in einer Arbeit von Lehner 1977.

Im Bereich eines Krümmers wird bekanntlich der Erdkörper quer zum Rohr schräg nach oben weggeschoben (Abb. 19). Der Vergleich mit einer

Stützmauer oder Ankerwand, die den passiven Grenzzustand des Erdkörpers wachruft, liegt nahe. Die Untersuchung zeigte auf Grund von Grenzwertanalysen und Modellversuchen folgende Einzelheiten:

A. Im ideellen Grenzfall des geraden, quer zu seiner Richtung verschobenen Rohrstranges ergibt sich ein vor dem Rohr hergeschobener Gleitkörper mit folgenden Eigenschaften:

— Die hintere Begrenzung des Gleitkörpers besteht aus 2 gegenläufigen Gleitflächen, zwischen denen ein Erdkeil absackt (Abb. 19).

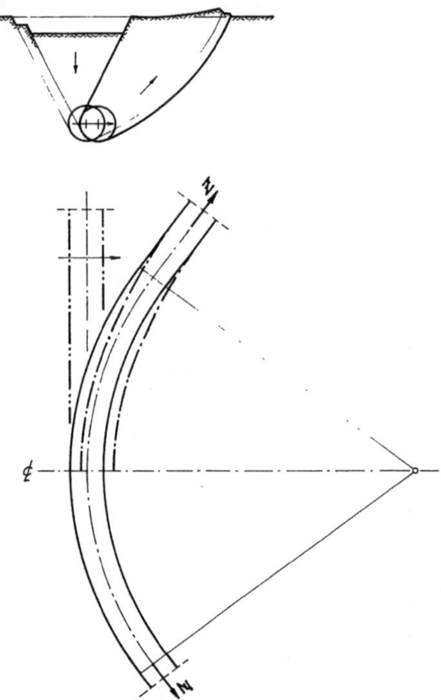

Abb. 19. Grundbruch infolge eines nicht dehnbaren Tunnel- oder Rohrleitungskrümmers in einer Zerrungszone

Ground failure in consequence of a bend of a not extensible tunnel- or pipeline situated in a tearing zone

— Die vordere Gleitfläche ist umso stärker gekrümmt und umso steiler je kleiner der Durchmesser im Vergleich zur Verlegungstiefe des Rohres ist (Abb. 20). Wie aber Abb. 21 zeigt, wird der Verschiebungswiderstand P von diesem Verhältnis nur gering beeinflußt.

Modellversuche und durch Variation von Gleitkreisen rechnerisch ermittelte Bruchbilder stimmen gut überein und finden fast nahtlosen Anschluß an die Ergebnisse von Großversuchen (Buchholz 1930).

B. Der *gekrümmte* Rohrstrang (Abb. 22) unter Längs*druck*kräften preßt sich wie ein Schiffsbug mit seiner konvexen Seite in den Erdkörper (Abb. 23). Bezieht man den Gleitbruchwiderstand des Bodens auf die

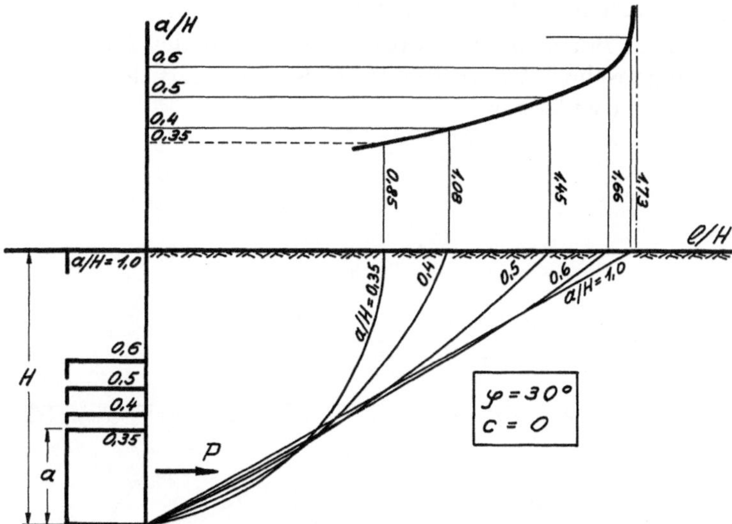

Abb. 20. Grundbruchform in Abhängigkeit vom Verhältnis a/H
Ground failure shape as a function of a/H-relation

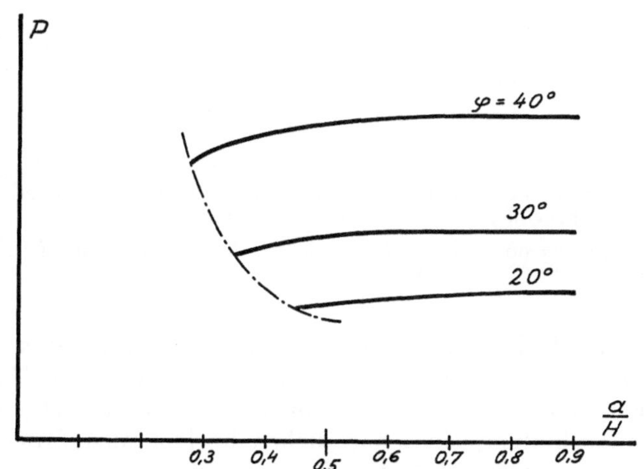

Abb. 21. Für ein Auslösen des Grundbruches erforderliche, von einem geraden Rohrstrang auszuwirkende Querbelastung P in Abhängigkeit von a/H
Cross load P required for ground failure (see Fig. 20) represented as a function of a/H

Bogensehne, welche eine Querbewegung durchführt, so zeigen sich im Vergleich zum geraden querverschobenen Strang folgende 2 zusätzliche Effekte:

— Im Bereich, in welchem der halbe Zentriwinkel β größer als der Wandreibungswinkel δ ist, sinkt der Erdwiderstand mit zunehmendem β ab. Er würde bei $\beta = 90^0$ den Wert Null erreichen. Dieser Effekt vermindert den Erdwiderstand, obwohl natürlich im allgemeinen $\max \beta = \alpha$ wesentlich kleiner als 90^0 ist.

— Der vor dem Bogen hergeschobene Gleitkörper des Bodens ist länger als die Bogensehne. Dieser Randeinfluß vergrößert den auf die Sehne bezogenen Erdwiderstand.

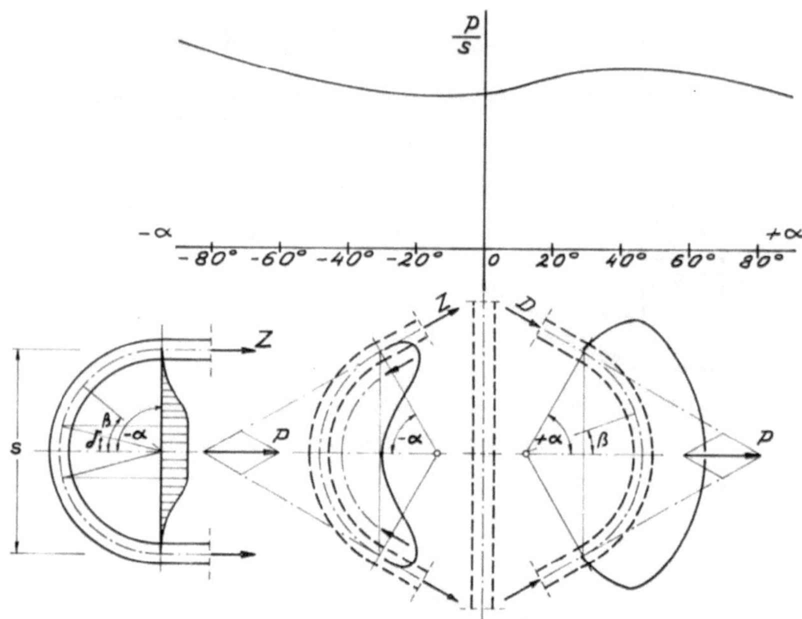

Abb. 22. Einfluß des Krümmerwinkels 2 α auf die den Grundbruch auslösende und auf die Längeneinheit der Sehne bezogene Querbelastung P/s

Influence of bend-angle 2 α on crossload P/s (related to longitudinal unit of chord) required for ground failure

C. Der gekrümmte Rohrstrang unter Längs*zug*kräften preßt sich wie ein Zugband mit seiner konkaven Seite gegen den Erdkörper (Abb. 24). In diesem Falle tritt zu den zuvor angeführten 2 Effekten noch ein 3. hinzu, nämlich die Ausbildung eines natürlichen horizontalen Erdgewölbes im vor dem Krümmer hergeschobenen Grundbruchkörper.

3.2. Vorschlag für einen Tunnelverbau unter Grundwasserdruck und unter Abbaueinwirkung

Die Bergsenkungsvorgänge verlaufen meist so langsam, daß es durchaus möglich ist, zunächst unter Grundwasserabsenkung den Tunnel mit den üblichen Stützmaßnahmen nach den Grundsätzen der NATM vorzutreiben und damit einen provisorischen Ausbau zum Erstellen einer besonderen Innenschale zu schaffen.

Diese muß aber dann trotz 10 bis 100 mal wiederholter Ovalverformungen mit Tunnelbreitenveränderungen bis zu 10% und trotz Längsdehnung und -Stauchung in gleicher Größenordnung, wasserdicht und voll

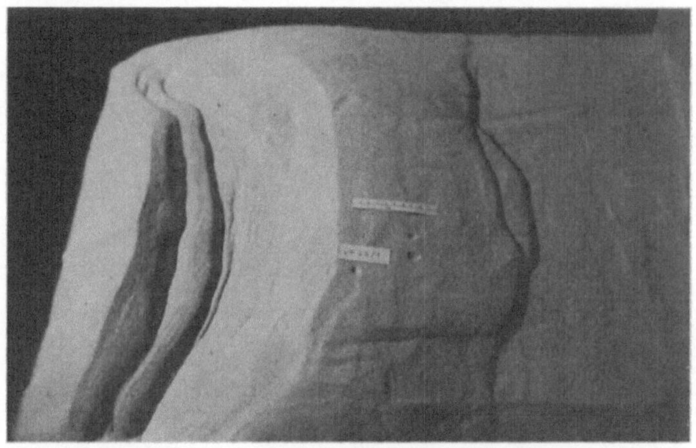

Abb. 23. Grundbruch über einem durch Längs-Druckkräfte beanspruchten Tunnel- oder Rohrkrümmer in kohäsionslosem Sand (Modellversuch)

Ground failure in not cohesive sand as a result of a tunnel- or pipeline-bend loaded by axial compression (model test)

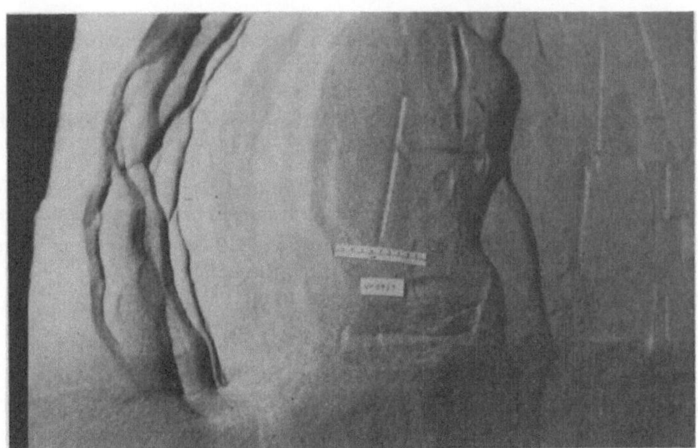

Abb. 24. Wie Abb. 23, jedoch Längs-Zugkräfte im Krümmer

Equal as Fig. 23 but tensile axial forces in bend (model test)

tragfähig bleiben, wobei sich außerdem der Vertikaldurchmesser der Tunnelröhre trotz der Tunnelbreitenänderung nicht verändern darf.

Ein Konstruktionsprinzip, das insbesondere bei gleichzeitigem Grundwasserandrang in einfacher Weise diesen Forderungen gewachsen ist, nützt (aus dem Wunsche nach der Fähigkeit zu großen, oft wiederholbaren Verformungen) das Prinzip des Blattfederpaketes (Abb. 25 a). Durch ähnliches Aufspalten der erforderlichen Wanddicke eines gewellten Stahlrohrverbaues wird bei gleicher Querschnittsfläche das Trägheits- und Widerstandsmoment mit Absicht auf ein erforderliches Mindestmaß verkleinert (Abb. 25 b). Da-

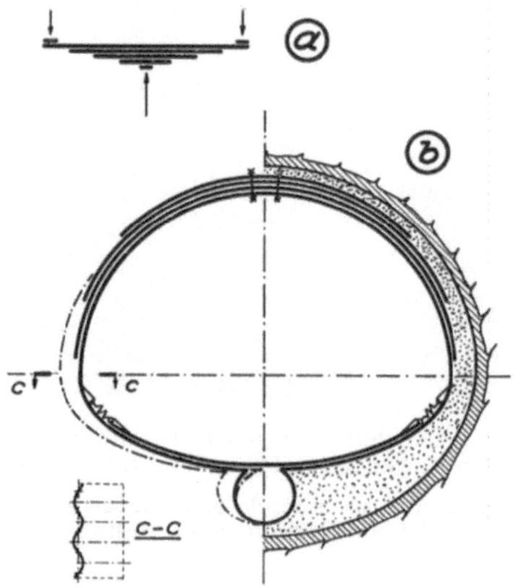

Abb. 25. Nutzung des Blattfeder-Effektes zur bruchlosen Aufnahme großer und wiederholt aufgezwungener Änderungen der Breite tragender, tagnaher Tunnelauskleidungen in Bergsenkungsgebieten

a) Blattfeder; b) Anwendung des Prinzipes im Tunnelbau (Dicke der Einzellagen übertrieben gezeichnet); c) Wellung in Längsrichtung

Use of leaf-feather effect for shallow tunnels in damage zones due to subsidence. The bearing lining becomes fit for multireversible large changes of tunnel width

a) leaf-feather; b) application of leaf-feather principle in tunnelling; c) corrugation in longitudinal direction

durch werden wesentlich größere Deformationen möglich, ohne dabei die Plastizitätsgrenze zu erreichen, was die bruchlose Wiederholbarkeit der Verformungsvorgänge um Zehnerpotenzen vergrößert. Selbstverständlich ist auch dem Aufspalten der Wanddicke eine Grenze gesetzt, denn ein Mindestwert an Steifigkeit muß aus Gründen der Beul-Stabilität erhalten bleiben.

Die dargestellte Konstruktion bietet aber nun die Möglichkeit, das Trägheitsmoment am Umfang so zu verteilen, daß es sowohl der Forderung nach möglichst großer elastischer Verformbarkeit (Blattfederprinzip), als auch der nach ausreichender Beulsteifigkeit entspricht.

Die gesamte Konstruktion besteht wegen der geforderten großen Dehnbarkeit aus Wellrohrelementen, deren Wellenprofil weitgehend stapelbar ist, also lagenweise ineinanderpaßt (Abb. 25c).

Nur die im Firstbereich innerste Lage dieser Tunnelschale umschließt den ganzen Umfang als „Innenrohr". Sie übernimmt die Abdichtung und die Aufnahme der Umfangskräfte. In dem in Abb. 25 gezeigten Beispiel wird im Sohlenbereich (wegen des Mechanismus zum Ausgleichen der vom Gebirge aufgezwungenen Tunnelbreitenänderungen) die Umfangskraft von einer zusätzlichen Innenlage aufgenommen.

Die äußeren Lagen der oberen Tunnelhälfte sind kürzer und dienen nur zur Erhöhung der Biegesteifigkeit, wobei eine dichte Schraubenverbindung im Scheitel genügt.

In der Längsrichtung des Rohres betrachtet, liegen diese äußeren Lagen ohne besondere Verbindung nebeneinander am „Innenrohr" auf. Es sind also für die Außenschalen keine Stoßverbindungen erforderlich, was neben der Serienfertigung der Hauptgrund für die geringen Kosten einer solchen Bauweise ist.

Eventuelle weitere Ausstattungen zur Umfangsveränderung können in den unteren Quadranten angeordnet werden, so daß trotz gleichzeitiger Abschirmung des Grundwassers die Wirkungsweise eines „Offenen Verbaues" oder auch die eines horizontal „Nachgiebigen Verbaues" mit einstellbarem Reibungswiderstand erzielt werden kann.

Literatur

Brunner, W.: Das Spannungsfeld der Bruchzonen im Bereich von Hohlraumbauten unter richtungsbetontem Primärdruck. Diplomarbeit am Institut für Konstruktiven Tiefbau der Montanuniversität Leoben 1978.

Buchholz, W.: Erdwiderstand auf Platten. Jahrbuch der Hafenbautechnischen Gesellschaft (Hannover), Bd. *12,* (1930).

Buchholz, W., Petermann, H.: Berechnungsverfahren für Ankerplatten und Wände. Der Bauingenieur H. 19/20 (1935).

Feder, G.: Zum Stabilitätsnachweis für Hohlräume in festem Gebirge bei richtungsbetontem Primärdruck. BHM H. 4 (1977).

Feder, G.: Versuchsergebnisse und analytische Ansätze zum Scherbruchmechanismus im Bereich tiefliegender Tunnel. Rock Mech., Suppl. 6. Wien – New York: Springer 1978.

Lehner, A.: Das Bruchverhalten von Rohrfernleitungen in Bergsenkungsgebieten. Diplomarbeit am Institut für Konstruktiven Tiefbau der Montanuniversität Leoben 1977.

Meissner, H.: Zum Spannungsverhalten eingeerdeter Rohrleitungen in Bergsenkungsgebieten, Diss. Montanuniversität Leoben 1976.

Olsacher, A.: Optimierung des Abbaustreckenausbaues durch den Einsatz von mörtelgebetteten Ankerbolzen im Magnesitbergbau Radenthein. Diss. Montanuniversität Leoben 1976.

Pacher, F.: Deformationsmessungen im Versuchsstollen als Mittel zur Erforschung des Gebirgsverhaltens und zur Bemessung des Ausbaues. Felsmechanik und Ingenieurgeologie, Suppl. 1, Wien – New York: Springer 1964.

Anschrift der Verfasser: o. Prof. Dr. techn. Georg Feder, Institut f. Konstruktiven Tiefbau der Montanuniversität Leoben, Franz-Josef-Straße 18, A-8700 Leoben; Dr. A. Olsacher, Österr.-Amerik. Magnesit AG, A-9545 Radenthein, Österreich.

Rock Mechanics, Suppl. 7, 129—138 (1978)

Rock Mechanics
Felsmechanik
Mécanique des Roches

Bemessung von Druckstollenauskleidungen unter Berücksichtigung des Einflusses der Gebirgsverankerung

Von

Z. Gergowicz

Mit 5 Abbildungen

Zusammenfassung — Summary — Résumé

Bemessung von Druckstollenauskleidungen unter Berücksichtigung des Einflusses der Gebirgsverankerung. Der Einfluß des Ankerausbaues auf die Verformungen von Stollenauskleidungen wird untersucht, welche den Druck des Triebwassers aufnehmen. Es wird nachgewiesen, daß die Anker zwei Aufgaben zu erfüllen haben: die Erhöhung der Tragfähigkeit des den Ausbruchraum umgebenden Gebirges sowie die Verminderung der Zugspannungen in der Betonauskleidung des Triebwasserstollens. Unter Benutzung der allgemein angewandten Berechnungsverfahren werden entsprechende Formeln abgeleitet, welche es ermöglichen, die Wirkungsweise der Ánker zu berechnen. Es wird nachgewiesen, daß der Einfluß der Ankerung vor allem in wenig tragfähigen Gebirgsarten und in Fels von sehr geringer Tragfähigkeit, der zu plastischen Verformungen neigt, zur Wirkung kommt. Es wird ferner festgestellt, daß die Anker stärker auszubilden sind, als dies einer Dimensionierung auf reinen Gebirgsdruck entsprechen würde, und daß nicht die Länge der Anker, sondern deren Tragfähigkeit entscheidend ist.

Dimensioning of the Lining of Pressure Shafts With Regard to the Influences of Rock Anchoring. Strata bolting is frequently utilized in pressure tunnels and in pressure shafts owing to its inquestionably proven technical advantages. However it is solely considered to act as a mean of rock support without utilizing its fitness to contribute to the bearing capacity of the tunnel lining which has to withstand the pressure of the transported water. A preliminary investigation shows that a cooperation of the strata bolting with the concrete lining always exists, although being sometimes quite moderate. As its magnitude depends on the ratio of stresses induced to the rock by the bolts to those induced by the water pressure it is shown how to determine the limit of the influence which has to be taken into account. A water pressure in the tunnel of about 20 at may be this limit and therefore the theory applies chiefly to pressure tunnels and to the upper regions of pressure shafts.

The above mentioned deliberations lead to the conclusion that the rock bolts should be able to withstand a stronger stress than that derived from rock pressure, in order to utilize their advantageous influence upon the concrete lining. Therefore the forces of the rock bolts were added up to withstand the inner water pressure of the tunnel jointly with the rock pressure.

It is assumed that the tunnel lining transmits one component of the water pressure indirectly into the rock mass. As the rock strata are prestressed by the anchor bolts, they will be disburdened in consequence of the dilatation of the tunnel lining which will push back the rock mass, taking over the forces which were prevailing within the bolts that will diminish their anchor strains to zero. The outer surface of the tunnel lining will therefore be affected by the passive rock deformation resistance and additionally the remaining forces within the bolts will act from the outside of the tunnel.

Starting from the formulae published by Kastner and using the classic solution of the Lamé problem which is their base, a formula is derived determining the ratio of pressures acting towards the tunnels center versus the inner pressure against the tunnel lining. Finally some calculated examples show the possibility of reducing the tensile stresses within the inner regions of the concrete lining up to 30 percent provided that the cooperation of the rock bolt forces is taken into account.

Le dimensionnement des supports de puits en charge, en considération de l'influence du boulonnage de la roche. Le boulonnage est utilisé pour la construction économique des galéries forcées et des puits en charge, grâce aux avantages technologiques rendus possibles par une coopération du rocher et des moyens de soutien. En général le boulonnage est utilisé en type de support temporaire et son potentiel de participer à la décharge du revêtement d'un tunnel n'est pas mis en compte, quand celui-ci est chargé par la pression de l'eau à turbiner. Des essais préliminaires ont démontrés qu'une interaction du boulonnage et du revêtement d'un tunnel a lieu toujours, bien qu'étant parfois de moindre importance. Son valeur dépend du rapport entre les contraintes du rocher excitées par les forces des boulons et les contraintes provoquées par la déformation du revêtement en béton du tunnel due à la pression de l'eau à turbiner.

Une tolérance limite a pu être définie à partir duquel cette interaction doit être prise en considération. Il semble que cette limite soit égale à 20 at de pression de l'eau, ce que veut dire que la théorie s'applique surtout aux galéries forcées et aux régions superficielles des puits en charge.

La conséquence des réflexions présentées concernant l'influence propice du boulonnage serait qu'il fallait dimensionner les moyens de soutien d'être plus forts qu'il était nécessaire pour résister à la poussée du rocher. Ainsi la construction de soutien du rocher est intégrée au système qui prend charge de la pression de l'eau à l'intérieur du tunnel. Pour y arriver il s'avérait nécessaire de fixer les conditions de travail du système et de préciser la méthode d'estimer les données de base nécessaires pour le calcul statique-élastique.

La réflexion fondamentale est la hypothèse que le revêtement du tunnel déformé par la pression interne de l'eau à turbiner, s'appuye au rocher et en le déformant prend charge des forces du boulonnage qui a comprimé préalablement le rocher. Ainsi les tensions des boulons sont diminuées jusqu'à être zéro. Le revêtement est affecté à son extérieur par la poussée du rocher et par la résistance à compression des boulons.

En utilisant la formule de Kastner et la solution classique de Lamé qui est sa base, un terme est décrit qui représente le rapport des contraintes exercées de l'extérieur au revêtement et la pression de l'eau qui agit contre sa surface intérieure. En calculant des cas exemplaires on peut démontrer que la quote-part du boulonnage à la diminution des tensions du béton de l'intérieur du revêtement s'amonte vers 30%. On a constaté que la réduction la plus importante des tensions se réalise en rochers souples étant caractérisés par déformations plastiques considérables.

Der Ankerausbau wird in den Druckstollen und -schächten infolge seiner technologischen Vorteile recht gerne angewendet. Man betrachtet ihn aber nur als einen Vorausbau, dessen Einfluß auf die Bemessung der Stollen-

oder Schachtauskleidung, welche den Druck des Triebwassers übernimmt, nicht berücksichtigt wird.

Unabhängig von seinen allgemein bekannten Vorteilen im technologischen Sinne wirkt sich der Ankerausbau ebenfalls günstig auf einige andere Faktoren aus, die bei den Druckstollen oder -schächten eine ausschlaggebende Rolle spielen. Wichtig ist vor allem die Tatsache, daß die Stollenauskleidung nach vorheriger Verdichtung des Gebirges ausgeführt wird, was sich zweifellos vorteilhaft auf das dichte Anliegen des Betons an das Gebirge auswirkt. Ein weiterer bedeutender Effekt der Ankerung ist deren Einfluß auf die mechanischen Eigenschaften des Gebirges, was insbesondere in bezug auf die Verformung der Betonauskleidung nicht gleichgültig ist. Die Ankerung verbessert ferner das geophysikalische Verhalten des Gesteinsmassivs und spielt hinsichtlich seiner Unregelmäßigkeit, Ungleichartigkeit, Anisotropie usw., eine positive Rolle.

Aus durchgeführten Untersuchungen [1, 4] ergab sich aber auch, daß sich der positive Einfluß der Ankerung nicht nur auf die oben angeführten Faktoren beschränkt, die sich ohnehin zahlenmäßig schwer erfassen lassen, sondern daß er sich auch direkt in konkret bestimmbarer Weise auf die Arbeit der Stollenauskleidung auswirkt. Ein meßbarer Effekt dieses Einflusses ist die Verminderung der Zugspannungen im Betonring der Auskleidung. In manchen Fällen ist dieser Einfluß zwar unbedeutend, nichtsdestoweniger ist er immer vorhanden. Seine Größe hängt vom Verhältnis der im Gebirge durch die Anker hervorgerufenen Spannungen zur Pressung des Triebwassers auf die Stollenauskleidung ab. Aus der Praxis ist bekannt, daß die Größenordnung der von den Ankern herrührenden Spannungen um den Wert von 100 kN/m^2 (ca. 1 kp/cm^2) schwankt. Unter Zugrundelegung einer Reihe unterschiedlicher Wasserdrücke war es nach entsprechenden Durchrechnungen möglich, die Grenze zu bestimmen, bis zu der bei der Bemessung der Stollenauskleidung der Einfluß der Ankerung noch zu berücksichtigen ist. Als diese Grenze erwies sich ein Triebwasserdruck von ca. 20 at. Da in den Druckschächten im allgemeinen bedeutend höhere Drücke auftreten, ergibt sich die Schlußfolgerung, daß der Einfluß der Ankerung hauptsächlich für Druckstollen zu berücksichtigen ist.

Eine weitere Schlußfolgerung, die aus den Untersuchungen konkreter Fälle gezogen werden konnte, betraf die Art der Ankerung. Es wurde festgestellt, daß in denjenigen Fällen, in denen der Einfluß der Ankerung wichtig ist, der Ankerausbau stärker zu dimensionieren ist, als dies der Gebirgsdruck erfordern würde. Dem Ankerausbau werden demnach zwei Aufgaben zugewiesen: Die Übernahme des Gebirgsdruckes und die Absteifung der Betonauskleidung des Stollens.

Zur Bemessung der Stollenauskleidung können wir uns verschiedener Formeln bedienen, die von den Autoren Mühlhofer, Galerkin [5], Jaeger [2] und anderen aufgestellt wurden. Alle diese Formeln basieren auf der Lösung des klassischen Problems durch Lamé über die Spannungen in einem dickwandigen Ring, der von innen und außen durch radial wirkende Kräfte belastet ist. Wenn wir also für ein solches Berechnungsschema annehmen, daß die von innen wirkende Belastung p_i gleich dem Druck des

Triebwassers ist (Abb. 1), so wird vom Gebirge her ein Widerstand p_a entgegenwirken, dessen Wert wir allgemein anhand der Formel

$$p_a = \lambda\, p_i \tag{1}$$

bestimmen können, worin der Koeffizient λ denjenigen Teil der Belastung charakterisiert, der auf das Gebirge übertragen wird, also auch den Druckabfall im Betonring bestimmt.

Zum Beispiel kann nach Kastner [3] der Koeffizient λ mittels der Formel (2) ausgedrückt werden:

$$p_a = \frac{2\, m_b}{(a^2+1)\, m_b - (a^2-1) + (1+\beta)\, \alpha\, (m_g+1)\, (a^2-1)} \cdot p_i$$
$$\alpha = \frac{E_b\, m_b}{E_g\, m_g}\,; \quad a = \frac{r_a}{r_i} \tag{2}$$

worin r_i, r_a = Radien der Betonauskleidung, E_b, m_b = Elastizitätsmodul und Poisson-Zahl des Betons, E_g, m_g = Deformationsmodul und Poisson-Zahl des Gebirges, β = ein Koeffizient ist, der die plastischen Deformationen des Gebirges charakterisiert.

Wenn wir p_i und p_a kennen, können wir nach der Formel von Lamé die gesuchten Spannungen im Betonring berechnen. So würde sich die Sachlage gestalten, wenn die Betonauskleidung direkt an einem unverankerten Gebirge anläge. Da der Betonring jedoch mit dem Ring des vorverdichteten Gebirges in Berührung steht, kompliziert sich das Problem ganz wesentlich.

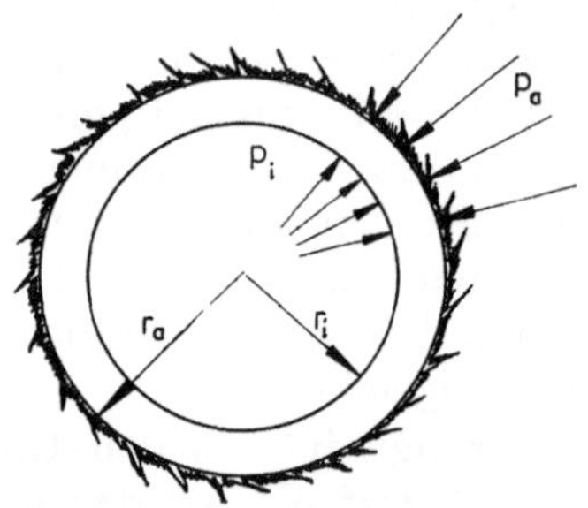

Abb. 1. Berechnungsschema der Druckstollenauskleidung
Calculation diagram for pressure tunnel lining
Schéma de calcul du soutènement d'une conduite forcée

Diese Komplikationen ergeben sich aus der Tatsache, daß der Rand des Ausbruchraumes unter dem Einfluß der Ankerkräfte Verschiebungen erfuhr und daß dieser Zustand durch die kontinuierliche Anspannung der Anker aufrechterhalten wird. Würden wir also nach der Herstellung der am verankerten Ring anliegenden Betonauskleidung die Anker entspannen, so würde das sich entspannende Gebirge auf diese Auskleidung einen Druck auszuüben beginnen.

In analoger Weise kommen wir zu den Schlußfolgerungen, daß im entgegengesetzten Fall, nämlich dann, wenn der Wasserdruck aus dem Stolleninneren eine Verschiebung der Auskleidung nach außen verursacht, diese

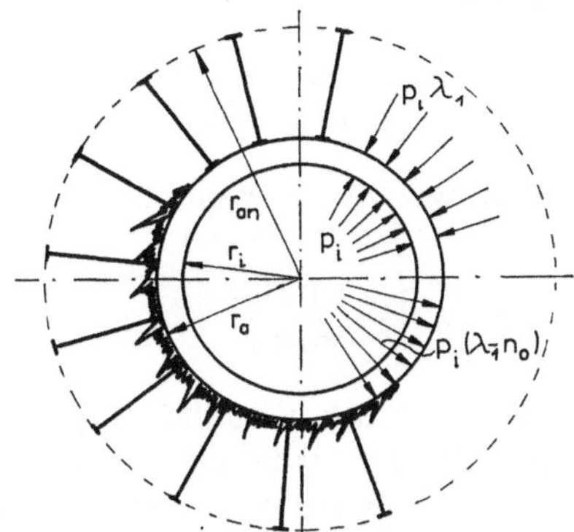

Abb. 2. Berechnungsschema der Druckstollenauskleidung unter Berücksichtigung des Ankerungseinflusses

Calculation diagram for pressure tunnel lining with regard to the influence of bolting

Schéma de calcul du soutènement d'une conduite forcée avec la prise en considération de l'influence du boulonnage

Auskleidung die Rolle eines das Gebirge verdichtenden Stützmittels übernimmt und die Spannungen in den Ankern auf Null herabsetzt. Auf die Auskleidung wird demzufolge entgegengesetzt zur Richtung des Wasserdruckes nicht nur der zuvor als p_a bezeichnete Widerstand des Gebirges wirken, sondern auch Drücke, die von den Kräften in den Ankern herrühren. Wenn wir diese Spannungen mit p_{an} bezeichnen, so wird

$$p_a = \lambda\, p_i + p_{an} = p_i\, (\lambda + n_0) = \lambda_1\, p_i; \qquad n_0 = p_{an} : p_i \tag{3}$$

Könnten wir von vornherein annehmen, daß sich die Situation tatsächlich so gestaltet, so würden wir hier auf keine speziellen Schwierigkeiten stoßen, da die Spannungen p_{an} bekannt sind. Die Ursache der sich in diesem Falle erhebenden Zweifel liegt in der Tatsache, daß die Stollenauskleidung nur die Folgen der in den Ankern wirkenden Kräfte übernimmt, nicht aber die Kräfte selbst. Notwendig ist also eine erneute Untersuchung des gesamten Systems zur Bestimmung von λ_1 als Funktion λ und n_0, also

$$\lambda_1 = F\, (\lambda_1\, n_0) \tag{4}$$

Wir wollen von der Annahme ausgehen, daß der Stollen kreisförmig ausgebildet ist mit einem Radius des Ausbruchraumes r_a und einem Radius der Betonauskleidung r_i (Abb. 2). Der Ankerausbau ist stärker dimensioniert, als dies der Gebirgsdruck erfordern würde und bildet den Ring des verankerten Mediums mit einem Radius r_{an}. Auf die Betonauskleidung wirkt von innen der Wasserdruck p_i, von außen die radiale, gleichförmig verteilte Belastung $p_a = p_i \cdot \lambda_1$, auf das Gebirge dagegen die nach außen gerichtete radiale Belastung mit einem Wert $p_a - p_{an} = p_i(\lambda_1 - n_0)$. Wenn wir die Lösung von Lamé heranziehen, dann ergibt sich die Verschiebung des äußeren Randes der Betonauskleidung

$$u_{ba} = \frac{m_b\,[\lambda_1\,(a^2+1)-2]-\lambda_1\,(a^2-1)}{m_b\,E_b\,(a^2-1)} \cdot r_a\, p_i \tag{5}$$

Auf ähnliche Weise können wir die Verschiebung des Ausbruchrandes u_g wie folgt bestimmen:

$$u_g = -z\,\frac{(m_g+1)\,(\lambda_1 - n_0)}{E_g\, m_g} \cdot r_a\, p_i \tag{6}$$

Aus der Bedingung der Verschiebungsgleichheit an der Berührungsstelle der Betonauskleidung mit dem Gebirge erhalten wir eine Gleichung, die nach der Durchführung einfacher rechnerischer Operationen die folgende Form annimmt:

$$p_a = \lambda_1\, p_i = \frac{2\, m_b + z\,\alpha\, n_0\,(m_g+1)\,(a^2-1)}{(a^2+1)\, m_b - (a^2-1) + z\,\alpha\,(m_g+1)\,(a^2-1)} \cdot p_i \tag{7}$$

Alle Benennungen in dieser Formel sind die gleichen wie zuvor, mit Ausnahme des Faktors $(1+\beta)$, an dessen Stelle der Koeffizient z eingeführt wurde, der ebenfalls, nur auf etwas andere Weise ausgedrückt, den Einfluß der plastischen Verformungen des Gebirges charakterisiert.

Bei der Bestimmung dieses Koeffizienten wurde von einer Annahme ausgegangen, welche die spezifische Wirkungsweise der Ankerarbeit berücksichtigt. Diese Wirkungsweise läßt sich so zusammenfassen, daß sie sich ausschließlich als Veränderung des sekundären (statischen) Gebirgsdruckes äußern kann, bzw. eventuell dann, wenn infolge der Verlagerungen des Mediums — was hauptsächlich bei nachgiebigen und sehr nachgiebigen Medien der Fall sein wird — der primäre (dynamische) Gebirgsdruck auf hinlänglich kleine Werte reduziert wurde.

Es ist bei den Druckstollen nicht angebracht, die Bildung einer Zone zuzulassen, die mit einem sekundären Gebirgsdruck entsteht, also einer Zone, in welcher Zerreißungen, Verschiebungen und größere Deformationen auftreten. Es ist aber auch bekannt, daß der Primärdruck hinlänglich klein sein muß, wenn die Anker irgendeinen realen Nutzen bringen sollen. Wenn wir also die beiden vorstehend angeführten Grenzbedingungen in Betracht ziehen, dann ergibt sich zwangsläufig der Einsatzbereich der Ankerung. Die Überlegungen zeigen, daß die Anker nur dann nützen und ihr Einsatz ausschließlich in demjenigen Intervall von Spannungen und Deformationen sinnvoll ist, die im Medium im Augenblick seines Überganges vom elastischen zum plastischen Zustand auftreten. Die durch die Anker hervorgerufenen Spannungen müssen mit anderen Worten in der Endphase der sogenannten elastischen Deformationen liegen und den Deformationsprozeß des Mediums so hemmen, daß es weder eine Verbandsauflösung noch größere bleibende Veränderungen erleidet.

Zur Veranschaulichung des vorstehend Gesagten wollen wir uns noch der Hypothese von Sałustowicz [6] über den im elastisch-plastischen Medium verlaufenden, deformierenden Druck um einen kreisförmigen Ausbruchraum bedienen. Von der Voraussetzung ausgehend, daß in dem unverritzten Medium ein hydrostatischer Spannungszustand herrschte, nimmt die Formel von Sałustowicz, die den Druck am Rand des Abbauraumes bestimmt, folgende Form an:

$$p_0 = p - \frac{2k}{\sqrt{3}} - \frac{2k}{\sqrt{3}} \ln \frac{u}{u_e} \tag{8}$$

worin: p = Primärdruck, $2k$ = Plastizitätsgrenze der Gesteine bei Druck, u_e = radiale elastische Verschiebung, u = radiale Gesamtverschiebung. Die Beziehung (8) kann graphisch wie in Abb. 3 dargestellt werden. Drücken

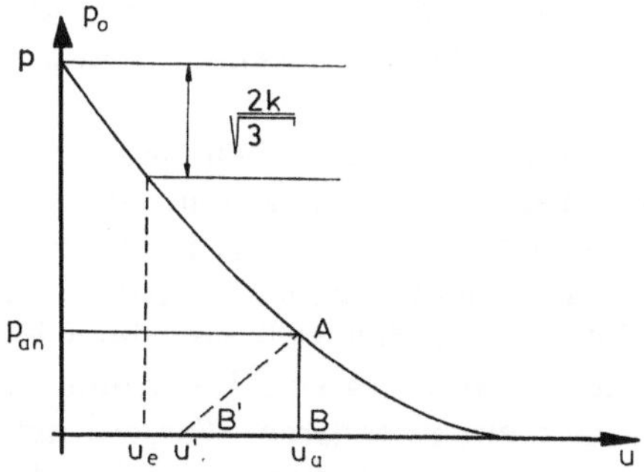

Abb. 3. Zusammenhang zwischen den Radialspannungen und den Radialverschiebungen
Dependence of the radial stresses on the displacements of excavation edge
Relation entre les contraintes radiales et les déplacements du contour de la voie

wir die radialen Spannungen, die durch die Anker am Rande des Abbauraumes hervorgerufen werden, durch p_{an} aus, so wird sich die Verschiebung des Mediums nach dem Erreichen eines bestimmten Wertes u_a nicht weiter vergrößern, wodurch der vorher genannte Übergangszustand zwischen den dynamischen und statischen Drücken entsteht.

Die vorstehenden Feststellungen gelten jedoch ausschließlich für Spreizhülsenanker. Bei Klebeankern gestaltet sich die Sachlage etwas anders, da zum Hervorrufen der zuvor erwähnten Spannungen p_{an} in den Ankerbolzen entsprechend große Deformationen auftreten müssen. Wenn also die Gerade *AB* die Arbeit des Ausbaus mittels Spreizhülsenankern charakterisierte, so bezieht sich die Gerade *AB'* auf den Ausbau mittels Klebeankern und der Abschnitt $u'\, u_a$ stellt die Deformationen der Ankerbolzen dar.

Aus den oben angeführten Erwägungen geht also hervor, daß sich die durch die Anker hervorgerufene Verdichtung des Gebirges in erster Linie auf die plastischen Verformungen auswirken wird. Damit zugleich vermindert sich ihr Einfluß auf die Verformung des Gebirges infolge des Druckes der Betonauskleidung. Von dieser Annahme ausgehend wurde die folgende Beziehung abgeleitet:

$$\frac{z p_i (\lambda - n_0)(m_g + 1) r_a}{E_g} = \frac{(1+\beta) p_i (\lambda - n_0)(m_g + 1) r_a}{E_g} - \frac{\beta n_0 p_i (m_g + 1) r_a}{E_g} \quad (9)$$

was nach einfachen Transformationen zu folgender Beziehung führte:

$$z = (1+\beta) \frac{\lambda - 2 n_0}{\lambda - n_0} + \frac{n_0}{\lambda - n_0} \quad (10)$$

worin λ ein Reduktionskoeffizient ist, der nach der Formel von Kastner berechnet wird.

Aus einer Analyse der Formel (10) geht hervor, daß nach Maßgabe des Größerwerdens von β auch der Unterschied zwischen z und $(1+\beta)$ ansteigt, womit der Einfluß der Ankerung auf die Verformung der Stollenauskleidung zunimmt. Diese Feststellung ist sehr wichtig, da sie den positiven Einfluß der Ankerung insbesondere bei wenig standfestem Gebirge augenscheinlich macht.

Wie sich die Unterschiede zwischen den Werten λ_1, die den Einfluß der Ankerung berücksichtigen, und λ gestalten, ist aus Abb. 4 ersichtlich, auf welcher diese beiden Koeffizienten für konkrete Fälle verschiedener Druckstollen dargestellt sind. Herangezogen wurden hierbei die von Kastner angegebenen Beispiele. Es handelt sich um einen kreisförmigen Stollen mit einem Radius des Ausbruchraumes von 2,0 m und um eine 0,35 m dicke Betonauskleidung, die unter einem inneren Wasserdruck von 11,5 at steht. Für diese als konstant angenommenen Parameter wurden abwechselnd der Elastizitätsmodul E_g des Gebirges sowie der Koeffizient β variiert. Aus den Diagrammen ist ersichtlich, daß für größere Kennzahlen α, d. h. wenn das Gebirge gering tragfähig ist, sowie für $\beta=2$, also wenn es zu plastischen Verformungen neigt, λ_1 um rund 60% größer wird. Für $\beta=0$ ist dieser Unterschied relativ klein, aber man darf hierbei nicht außer acht lassen, daß in diesem Falle nur elastische Verformungen des Gebirges in Frage kommen, auf die die Anker im Hinblick auf die beschränkte Spannkraft nur einen relativ kleinen Einfluß ausüben.

Die in Abb. 4 dargestellten Diagramme illustrieren jedoch den Einfluß der Ankerung nicht ausreichend. Es geht darum, daß besonders die Zugspannungen in der inneren Randschicht des Betonringes wichtig sind. Die Werte dieser Spannungen wurden für die in Abb. 4 dargestellten Beispiele λ_1 und λ sowie für einen Wasserdruck von 11,5 at berechnet und in Abb. 5 dargestellt. Aus diesen Diagrammen geht hervor, daß für $\beta=0$ die Abminderung der Betonspannungen relativ klein ist und 12% beträgt, während der Unterschied für $\beta=2$, also für ein wenig tragfähiges und zu plastischen Verformungen neigendes Gebirge, bereits über 25% beträgt.

Von diesen Ergebnissen ausgehend kann man sich die Frage stellen, ob es sich bei einer Verminderung der Zugspannungen in der Betonauskleidung in Grenzen von 20 bis 30% lohnt, eine stärkere Ankerung anzuwenden, als diese wegen des Gebirgsdruckes erforderlich wäre. Diese Frage muß man bejahen, und die Maßnahme sollte für gering tragfähiges Gebirge sowie unter schwierigen und komplizierten geotechnischen Verhältnissen geradezu selbstverständlich sein. Hierfür sprechen mehrere Ursachen.

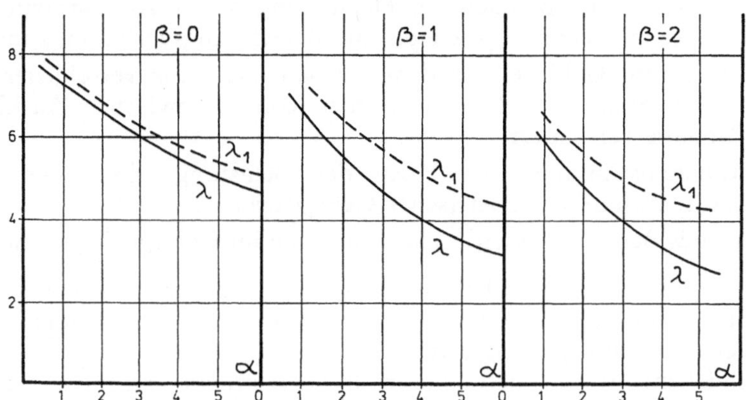

Abb. 4. Die Werte des Koeffizienten λ nach Kastner und λ_1 nach Gl. (7)
Coefficient λ acc. to Kastner and λ_1 acc. to Eq. (7)
Valeurs du coefficient λ calculé à l'aide de la formule de Kastner et du coefficient λ_1, calculé à l'aide de la formule (7)

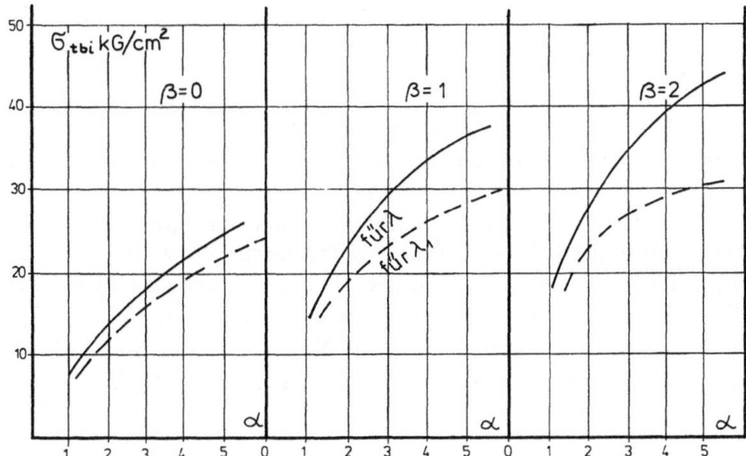

Abb. 5. Randzugspannungen in der Betonauskleidung, berechnet: nach Kastner — kontinuierliche Linie; unter Berücksichtigung des Ankereinflusses — gestrichelte Linie
Tensile stresses within the concrete lining acc. to Kastner — continuous line, and with regard to the influence of bolting — dotted line
Contraintes de traction dans le soutènement en beton, calculées: d'après Kastner — ligne continue, avec la prise en considération de la influence du boulonnage — ligne en pointillés

Den konkreten Effekt selbst, nämlich die Verminderung der Zugspannungen in der Betonauskleidung, außer acht lassend, muß festgestellt werden, daß die Ankerung des den Ausbruchraum umgebenden Mediums den ungünstigen Einfluß einer Reihe von zusätzlichen Faktoren vermindern oder sogar völlig aufheben kann. Zu diesen Faktoren gehören nach Kastner

— die Ungleichartigkeit des Gebirges und seine Anisotropie;
— das Gebirge verbleibt nach der Herstellung des Ausbruchraumes nicht im elastischen Zustand, so wie das die Berechnungsverfahren annehmen;
— zwischen dem Gebirge und dem Ausbau kann aus verschiedenen Ursachen ein Spalt entstehen, der einen Kontakt zwischen Auskleidung und Gebirge verhindert;
— Nichtübereinstimmung der Form der Belastung (drehsymmetrischer Wasserdruck und asymmetrischer Gebirgsdruck);
— infolge des Betriebswasserdruckes entstehen im Gebirge Zugspannungen.

In bezug auf alle diese Faktoren wirkt sich die zusätzlich eingeführte Vorverdichtung des Gebirges in Form eines symmetrischen Systems radialer Spannungen zweifellos günstig aus.

Abschließend wäre noch eine recht wesentliche Tatsache hervorzuheben, und zwar die, daß in keiner der dargestellten Formeln die Länge der Anker als Parameter vorkommt. Daraus kann die Schlußfolgerung gezogen werden, daß es möglich ist, relativ kurze Anker größter Tragfähigkeit zu verwenden. Diese Tatsache ist sowohl in bezug auf die Ankerungsarbeiten als auch auf die Kosten von Bedeutung.

Literatur

[1] Gergowicz, Z.: Ankerausbau hydrotechnischer Druckstollen. Komun. Inst. Geotechn. Pol. Wr. Wrocław, Nr. 182 (1977).

[2] Jaeger, C.: Rock Mechanics and Engineering. Cambridge: University Press 1972.

[3] Kastner, H.: Statik des Tunnel- und Stollenbaues auf der Grundlage geomechanischer Erkenntnisse. 2. Aufl. Berlin – Heidelberg – New York: Springer 1971.

[4] Pham Cong Sang: Analiza pracy obudowy betonowo-kotwiowej hydrotechnicznych sztolni cieśnieniowych. Kom. Inst. Geol. Pol. Wr. 1976.

[5] Széchy, K.: Tunnelbau. Wien – New York: Springer 1969.

[6] Salustowicz, A.: Zarys mechaniki górotworu. Wyd. “Sląsk”, Katowice, 1965.

Anschrift des Verfassers: Prof. dr. hab. inż. Zdzisław Gergowicz, Instytut Geotechniki, Politechniki Wrocławskiej, Pl. Grunwaldzki 9, 50-370 Wrocław, Polen.

Rock Mechanics, Suppl. 7, 139—155 (1978)

Rock Mechanics
Felsmechanik
Mécanique des Roches

Wechselwirkung zwischen Anker und Gebirge

Von

Peter Jirovec

Mit 19 Abbildungen

Zusammenfassung — Summary — Résumé

Wechselwirkung zwischen Anker und Gebirge. Aufgrund der unterschiedlichen konstruktiven Gestaltung der Haftstrecken moderner Daueranker werden drei Ankertypen unterschieden. Mechanische Vorgänge in der Haftstrecke eines Felsankers hängen von der Art der Krafteinteilung ab. Das Zusammenwirken von Anker und Gebirge kann mit Versuchen an Meßankern erforscht werden. Die Ergebnisse liefern nicht nur Aussagen über die Krafteintragung entlang der Haftstrecke, sondern auch über deren veränderliches Partizipieren an der Kraftübernahme während der Laststeigerung. Dadurch kann die Sicherheit eines optimal dimensionierten Ankers als die Mobilisierung der Verbundschubspannung des zur Kraftübertragung herangezogenen Haftstreckenteiles erläutert werden.

Die Messung des radialen Druckes in der Haftstrecke der verschiedenen Anker-Typen, läßt Schlüsse auf eine mögliche Rißausbildung zu. Anhand von Meßergebnissen werden einerseits die Krafteintragung, andererseits die Größe der radialen Spannung in der Haftstrecke dargelegt.

Behaviour of Anchors in Rock Masses. Modern permanent rock anchors can be divided into three types depending upon the nature of construction of the fixed anchor length. Mechanical processes in the fixed anchor length depend upon the manner in which force is transmitted from the anchor to the surrounding rock mass. The interaction between the anchor and the rock mass can be investigated using measurement anchors. The results permit the prediction not only of the stress distribution along the field anchor length but also of the variable portion of the fixed anchor length effective in load transmission. The safety of an optimally designed anchor can this be defined as the mobilization of the bond stress in that portion of the fixed anchor length which is effective in load transmission. Measurement of radial stress in grout at point of high stress concentration is very important for fracture analysis. On the basis of results of measurements the manner of load transmission in the fixed anchor length and the magnitude of radial pressures developed are discussed.

Comportement des ancrages en roches. Sur la base des différentes formes de construction de la section d'adhérence on distingue trois types d'ancrages modernes. Le comportement mécanique de la section d'adhérence d'un ancrage dépend de la façon avec laquelle se fait la mise en charge. L'interaction entre l'ancrage et la roche peut être étudié à l'aide d'essais. Les résultats indiquent non seulement la répartition des efforts le long de la section d'adhérence mais aussi les changement qui y ont lieu durant l'augmentation de l'effort total. Ceci permet de définir le coefficient de sécurité d'un ancrage optimal comme étant la fraction de la contrainte de cisaillement qui est mobilisée dans la section d'adhérence effective.

Les mesures de la contrainte radiale dans la section d'adhérence des différents types d'ancrage, permettent d'avoir une idée de la formation de fissures éventuelles.

La répartition des efforts et la grandeur des contraintes radiales dans la section d'adhérence sont illustrés à l'aide de résultats de mesure.

1. Einleitung

Die Ankerung im Bauwesen, in Böden und im Fels, ist eine neue, wirtschaftliche Methode, die sich in den letzten 20 Jahren sehr verbreitete.

Bei den, seit den fünfziger Jahren benutzten Verpreßankern, wurden durch etwa 400 Ausgrabungen zahlreiche Erkenntnisse über ihr Tragverhalten, die Konstruktion, die Rißbildung, das Korrosionsverhalten usw. gewonnen (Ostermayer 1975), während bei Felsankern eine Untersuchung der Haftstrecke noch nicht vorgenommen wurde. Diese bestehende Erkenntnislücke, die Erforschung der Sicherheit des einzelnen Felsankers, muß noch gefüllt werden, da von ihr die Gesamtstabilität eines verankerten Systems abhängt (Jirovec 1976).

Die Belastbarkeit eines vorgespannten Freispielankers hängt in erster Linie von dem Verbund an der Grenze zwischen Spanngliedern und Verpreßmörtel der Haftstrecke ab.

In diesem Beitrag werden Ergebnisse der Messung der Krafteintragung entlang der Haftstrecke und die dadurch entstehenden Spannungen im Verpreßkörper dargelegt.

Die Untersuchungen wurden sowohl an Ankern in situ — in Sandstein — als auch in einem Betonzylinder mit einem Durchmesser von 130 cm und einer Höhe von 300 cm durchgeführt. Es wurden aus meßtechnischen Gründen und wegen der direkten Interpretation der Ergebnisse für den projektierenden Ingenieur keine Versuche an Modell-Ankern in Angriff genommen. Außer den Druckrohrankern wurden keine Anker oder Ankerteile von Ankerherstellern oder Baufirmen herangezogen. Zur Verwendung kam allgemein käuflicher Spannstahl; in eigener Werkstatt wurden Abstandhalter hergestellt, welche jeweils dem Bohrloch von 76 mm bzw. 82 mm angepaßt waren. Korrosionsschutzmaßnahmen im Bereich der Freispiel- oder Haftstrecke wurden nicht vorgenommen.

2. Anker-Typen

Aus der Sicht des mechanischen Verhaltens moderner Daueranker in der Verpreßstrecke können wir heute folgende drei Anker-Typen unterscheiden:

Anker-Typ A (Abb. 1) bildet gerade, mit der Ankerachse parallel laufende Einzelstäbe, die durch Distanzscheiben im Bereich der Verpreßstrecke auseinandergehalten werden. Auf dem Bild ist die Haftstrecke durch einen

Packer begrenzt. Zu dieser Gruppe gehören Bündelanker, auf deren bergseitigem Ende die einzelnen Spanndrähte durch eine Stahlplatte gezogen

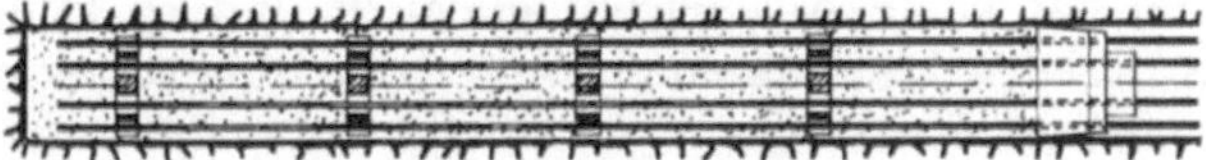

Abb. 1. Anker-Typ A
Anchor-type A
Ancrage du type A

sind, oder sie sind mit einer anderen konstruktiven Maßnahme versehen, so daß die Krafteinleitung am bergseitigen Ankerende erfolgt (Abraham, Porzig 1973). Zu dieser Ankergruppe gehört auch der Mono-Anker.

Anker-Typ B (Abb. 2) besteht aus Einzelstäben, die in der Verankerungsstrecke durch abwechselnde Spreizung und Einschnürung mit dem umgebenden Verpreßmörtel innig verbunden sind.

Abb. 2. Anker-Typ B
Anchor-type B
Ancrage du type B

Bei den Ankern dieser beiden Gruppen sind die einzelnen glatten oder gerippten Stäbe in ein Bündel gefaßt. So können sie je nach Drahtanzahl und Bohrlochdurchmesser eine Vorspannkraft von mehreren 1000 kN übertragen. Durch die Wirkungsweise dieser Anker-Typen ist der Verpreßmörtel im allgemeinen auf dem luftseitigen Anfang der Haftstrecke auf Zug und Abscherung stark beansprucht.

Anker-Typ C (Abb. 3) besteht aus einem auf der Oberfläche glatten oder mit flachem Gewinde bestückten Rohr, in das ein Ankerzugglied ein-

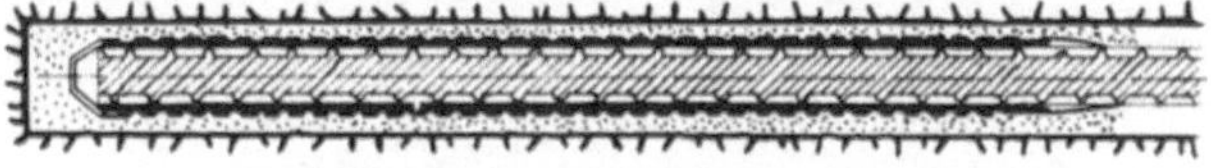

Abb. 3. Anker-Typ C
Anchor-type C
Ancrage du type C

geschraubt ist. Nach der Art der Beanspruchung des Rohres wird der Anker unter der Bezeichnung Druckrohranker geführt (Stump 1976).

Der Mechanismus der Krafteinleitung unterscheidet sich grundsätzlich von dem der zwei obengenannten Anker-Typen. Die Krafteinleitung vom Zugglied in den Verpreßkörper erfolgt nämlich an der Bergseite. Der in das

Druckrohr eingeschraubte Stab ist stetig auf Zug beansprucht, von ihm aus wird die Kraft auf das Druckrohr übertragen. Die Haftstrecke wird im Bereich der Krafteinleitung auf Druck und Abscheren beansprucht. Der stark gedrückte Mörtel hat keine Ausweichmöglichkeit und so kommt es durch den Anstieg des radialen Druckes zu einer Verbesserung seiner mechanischen Beschaffenheit. Die zu übertragende maximale Vorspannkraft ist durch die Fläche des Zuggliedes begrenzt.

Kennt man die Verbundschubspannung an der Grenze zwischen Anker und Injiziergut, so erhält man daraus Aufschluß über die Wechselwirkung zwischen Anker und Gebirge. Vom Verbundschubspannungsverlauf hängen ab: die optimale Ankerlänge, die Lage der maximalen Beanspruchung, die zur Kraftübertragung herangezogene Länge und schließlich das Versagen des Ankers. An der Grenze zwischen Verpreßmörtel und Bohrlochwandung ist die Verbundschubspannung jedoch — mit Ausnahme des unter Anker-Typ A zitierten Ankers — nicht maßgebend, da das Verpreßgut selbst in die feinsten Poren des Gesteins eindringt. Nach Erhärten ist es mit dem Fels verwachsen, so daß es sich um keine gefährdete Fläche mehr handelt.

Eine mikroskopische Aufnahme des Verbundes an der Grenze Sandstein bzw. Spanndraht mit Verpreßmörtel zeigen die Abb. 4 und 5.

Abb. 4. Verbund zwischen Sandstein und Verpreßmörtel
Bond between sandstone and grout
Soudure entre le grès et le coulis d'injection

3. Verbund

Die Untersuchung der Verbundschubspannungen wurde an drei verschiedenen Meßankern vorgenommen:

1. an einem von uns selbst entworfenen Anker des Types A,
2. an einem gerippten Spannstab des Types A,
3. an einem Druckrohranker des Types C.

Bei den zwei erstgenannten Ankern ist jede Laststufe durch eine bestimmte Krafteintragungskurve innerhalb der Haftstrecke gekennzeichnet — wie in der oberen Hälfte des Bildes (Abb. 6) gezeigt. Dem jeweiligen Krafteintragungsverlauf entspricht eine Kurve der Verbundschubspannungsver-

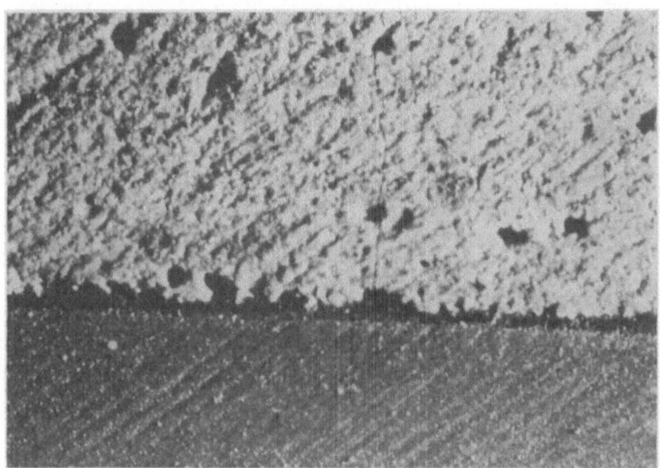

Abb. 5. Verbund zwischen Verpreßmörtel und Spanndraht
Bond between grout and tendon
Soudure entre le coulis d'injection et le tiran

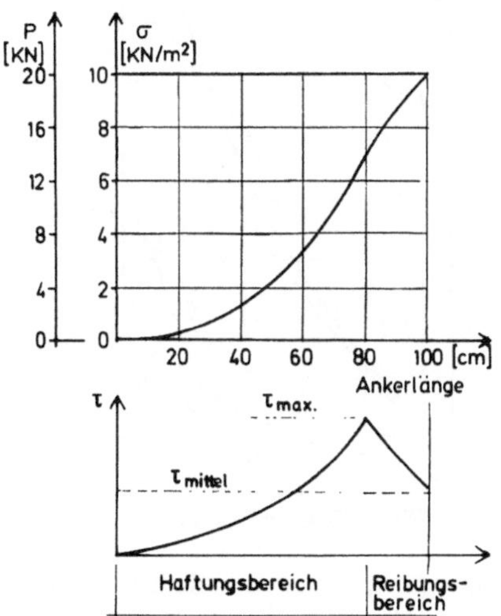

Abb. 6. Zusammenhang zwischen Krafteintragung und Verbundschubspannung in der Haftstrecke. Verteilung der Haftstrecke auf Haftungs- und Reibungsbereich

Relation between force transmission and bond stress in the fixed anchor length. Distribution of the fixed anchor length in the adhesion part and friction-shear part

Relation entre la force transmise et la contrainte de cisaillement dans la section d'adhérence. Répartition de la section d'adhérence en section d'adhérence et en section de frottement

teilung (untere Hälfte in Abb. 6), welche als erster Differentialquotient der Zugspannungen berechnet wird.

Die Verbundschubspannungskurve hat einen steigenden und einen fallenden Ast. Diese beiden bestimmten die Aufteilung der Haftstrecke in einen Haftungs- und einen Reibungsbereich. Die Grenze zwischen beiden Bereichen ist durch den Wendepunkt der Krafteintragungskurve gegeben.

Ein anderes Kriterium für die Gliederung der Haftstrecke ist das Verhalten des Schlupfes. Mit seiner Zunahme wird sie in folgende Bereiche unterteilt: elastisch,
elastoplastisch,
plastisch,
gerissen.

4. Selbst entworfener Anker

Der von uns entworfene Anker besteht aus einem dickwandigen, glatten Rohr, das bei einer gewählten Haftstreckenlänge von 1,0 m auf 160 cm Länge geschlitzt wurde (Egger 1971, Jirovec 1974). Nach dem Aufbringen

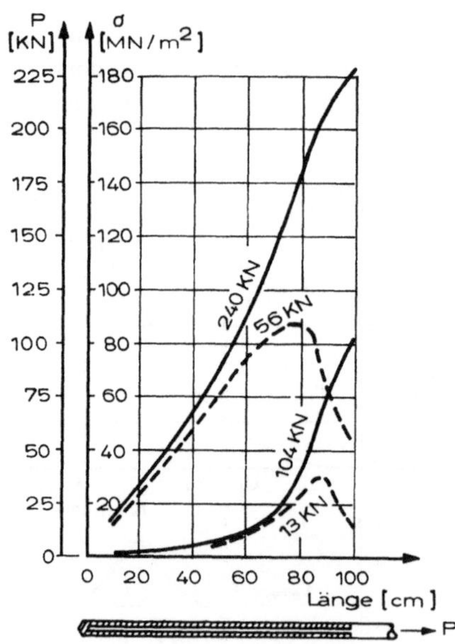

Abb. 7. Krafteintragung in der Ankerhaftstrecke des Meßankers
Force transmission in the fixed anchor length of the measurement anchor
Efforts dans la section d'adhérence

von Dehnungsmeßstreifen an der Rohrinnenseite wurden die beiden Rohrhälften wieder mit Agomet-U3 verklebt. Um die Haftstrecke abgrenzen zu können, wurde auf das Rohr ein aufblasbarer Gummipacker aufgezogen.

Der Krafteintragungs- und Verbundschubspannungsverlauf in der Haftstrecke des Meßankers ist in den Abb. 7 und 8 dargestellt. Der Verbund dieses Ankers ist nur durch Haftung und Reibung gegeben.

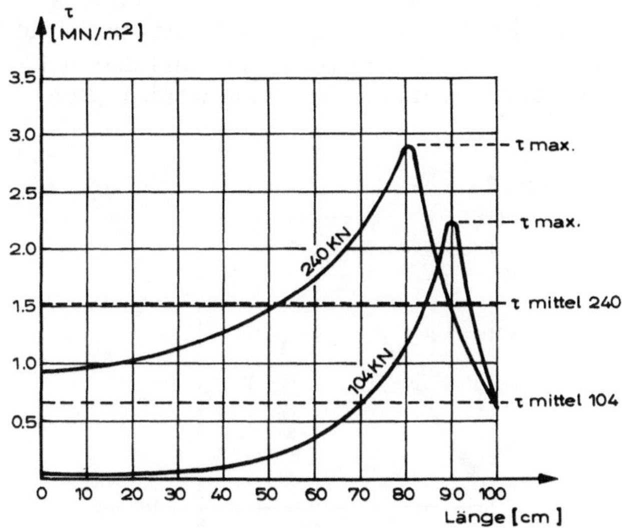

Abb. 8. Verbundschubspannungsverlauf in der Ankerhaftstrecke des Meßankers
Distribution of bond stress in the fixed anchor length
Distribution des contraintes de cisaillement dans la section d'adhérence de l'ancrage d'essai

Auf der Abszisse ist die Ankerlänge eingetragen, auf der Ordinate die Kraft P bzw. Verbundschubspannung τ. Der steile Abbau der Kraft von 104 kN bedeutet steilen Zuwachs der Verbundschubspannung. In den ersten 40 cm werden 85% der Kraft eingeleitet, was bedeutet, daß nur die luftseitige Haftstreckenhälfte zur Kraftübernahme herangezogen wird. Dies zeigt deutlich die entsprechende Verbundschubspannungskurve. Mit wachsender Belastung wird Schritt für Schritt auch die bergseitige Haftstreckenhälfte zur Kraftübernahme herangezogen. Die Kraftkurve wird flacher, was einen flacheren Verbundschubspannungsverlauf zur Folge hat. Die Spannungsspitze wandert so lange ins Berginnere, bis der Schlupf des Ankers gleich mit dem im Verpreßmörtel ist. In diesem Fall wird auch die Haftung am bergseitigen Ankerende mobilisiert. Wegen der Dilatation im Reibungsbereich kann die Verbundschubspannungsgröße am luftseitigen Haftstreckenanfang nicht den Nullwert erreichen.

Betrachten wir nun die Verbundschubspannungskurve der Kraft von 104 kN und ihren Mittelwert: Er ist das arithmetische Mittel der Verbundschubspannung im Verlauf der Haftstrecke des Ankers. Man sieht, daß somit Verbundschubspannung an Stellen angenommen wird, wo keine oder ein Mehrfaches davon vorhanden ist. Anders gesagt, die an der luftseitigen Haftstreckenhälfte wirkende Spannung bleibt unberücksichtigt, in der anderen Haftstreckenhälfte wird sie jedoch geltend gemacht. Das Verhältnis zwischen dem Wert τ_{max} zu τ_{mittel} wird mit steigender Belastung kleiner.

Die gestrichelten Linien in der Abb. 7 stellen Entlastungsstufen dar. Es ist interessant, daß die Spannung im Haftungsbereich gespeichert bleibt. Dieser Effekt ist auf die Reibung, die der Ankerverkürzung entgegenwirkt, und auf die irreversiblen Verschiebungen im Mörtel zurückzuführen.

Das Vorhandensein des Reibungsbereiches und dessen Länge kann aus dem Last-Verschiebungs-Diagramm bestimmt werden (Abb. 9). Es werden

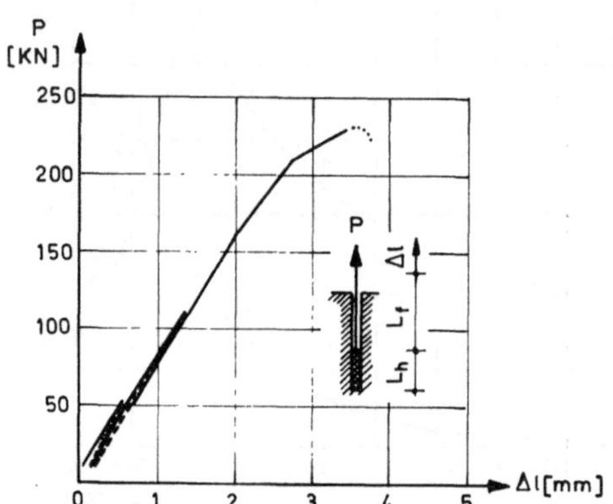

Abb. 9. Last-Verschiebungs-Diagramm
Load-displacement-diagram
Diagramme charge-déplacements

die gemessenen Dehnungen den theoretischen gegenübergestellt. Bei der Voraussetzung der Kenntnis der Deformation am Ankerkopf bzw. bei Kenntnis der eventuellen Verbindungen kann die Länge des neugebildeten Reibungsbereiches errechnet werden.

Andere Ergebnisse des Ausziehversuches zeigen die Abb. 10 und 11.

Die Krafteintragungskurve der Kraft von 110 kN ist flacher, was als beginnende Belastung der bergseitigen Haftstreckenhälfte zu interpretieren ist. Der Verbund an der Grenzfläche Anker-Verpreßmörtel war demnach kleiner, was somit das Erreichen des Ankerversagens (bei höherer Last) zur Folge hatte. Den Beginn dieses Versagens stellt die Kraft von 215 kN dar. Ihr Verlauf entlang der Haftstrecke ist durch den konstanten Differentialquotienten gekennzeichnet. Aufgrund dessen bekommt man gleichmäßigen Verbundschubspannungsverlauf im Haftungsbereich, der im Reibungsbereich zum Nullwert führt. Dasselbe gilt auch für die Kraft von 230 kN.

Nach dem ersten Schlupf konnte die Belastung weiter gesteigert werden. Mit dieser Kraft wurde das Maximum erreicht, was mit dem Anwachsen der Verschiebungen verbunden war (Abb. 12).

Erst nach Erreichen des Ankerversagens ist der Verbundschubspannungsverlauf der Annahme am nächsten, mit der man allgemein die Ankerverbundschubspannungen errechnet. Nun aber herrscht Verbundschubspannungsmangel am Krafteinleitungspunkt vor. Dieses Versuchsergebnis stellt graphisch dar, daß bei Laststeigerung Schritt für Schritt die Haftung in der luftseitigen Haftstreckenhälfte verloren geht. Dadurch wird mehr und mehr

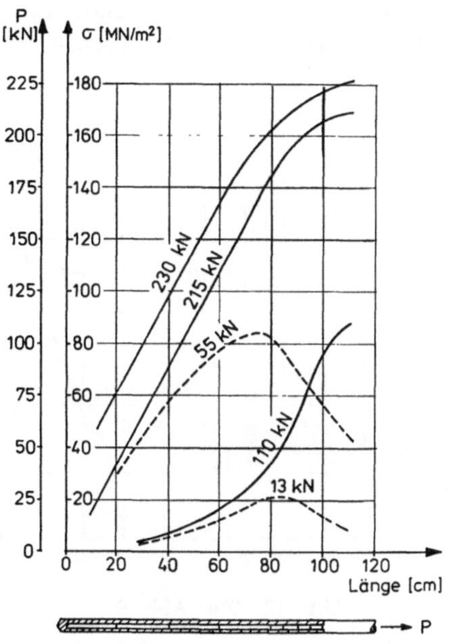

Abb. 10. Wie Abb. 7

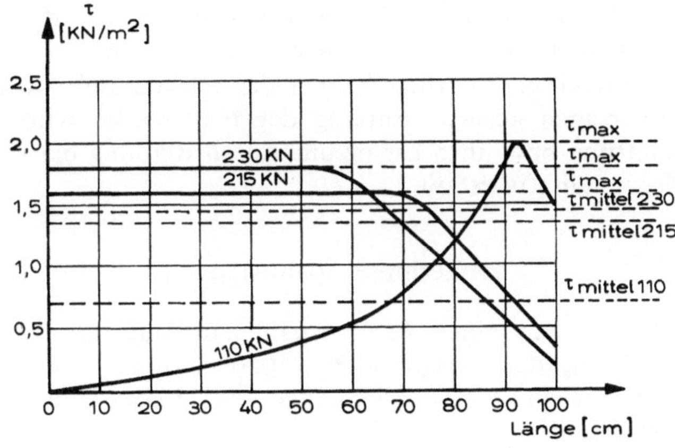

Abb. 11. Wie Abb. 8

die bergseitige Haftstreckenhälfte zur Kraftübernahme herangezogen. Dies bedeutet eine Reserve im Tragverhalten eines Ankers. Sie ist schließlich durch das Ankerversagen erschöpft.

Dieses Versagen des Ankers hat für die Praxis elementare Bedeutung, jedoch mit dem ursprünglichen Sinn des Wortes „Versagen“ wenig zu tun.

Der Anker gibt bei einer optimalen Haftstreckenlänge lediglich nach in dem Sinne, daß er sich bei Überschreiten der theoretisch berechneten Ankerkraft allmählich aus der Haftstrecke abzulösen beginnt. Das heißt, bei Überlastung gibt er langsam nach, was rechtzeitig aufgedeckt und durch Einbau von Zusatzankern behoben werden kann. Die Zusatzanker ersetzen in diesem Fall nicht den überlasteten Anker, sie unterstützen in lediglich, indem sie

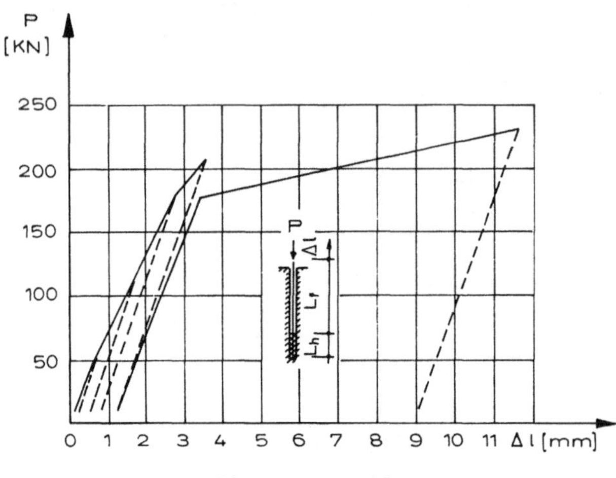

Abb. 12. Wie Abb. 9

seine tatsächliche Last auf das zulässige Maß reduzieren. Ist die Haftstrecke jedoch so lang, daß das Zugglied beim Überlasten des Ankers in der Freispielstrecke abreißen kann, tritt plötzliches Versagen ein und die Sicherheit des Bauwerkes ist mehr gefährdet als im vorherigen Falle.

An diesem Meßanker wurden Folgen des ganzen Belastungsvorganges bis zum Ankerversagen gezeigt. Entlang der Haftstrecke wurde die Überwindung der Haftung bzw. ihre Ergänzung durch Reibung bei verschiedenen Belastungsstufen qualitativ erfaßt.

5. Gerippte Spannstäbe

Es handelt sich um einen Rippenspannstahl-Stab mit einem Durchmesser von 26,5 mm, der Festigkeit 850/1050 kN/mm². Es wurden Dehnungsmeßstreifen auf eine Haftstreckenlänge von 110 cm aufgeklebt und mit Schutzstreifen versehen.

Das Ergebnis eines Zugversuches zeigen die Abb. 13 und 14.

An diesem, wie auch an jedem anderen profilierten Stab spielt bei der Kraftübertragung neben Haftung und Reibung der Formverbund eine große Rolle. Es sind dies die drei Komponenten, die den physikalischen Vorgang des Verbundes bestimmen. Durch die Profilierung von hochwertigen Stahlstäben ergibt sich ein hoher Verbund, durch welchen die großen Dehnungen kompensiert werden. Sie wirken der Zugkraft entgegen. Der Verbund zeigt sich in höheren Verbundschubspannungsspitzen, kleineren Längen des Reibungsbereiches und höherer Belastbarkeit, als es beim Meßanker der Fall war.

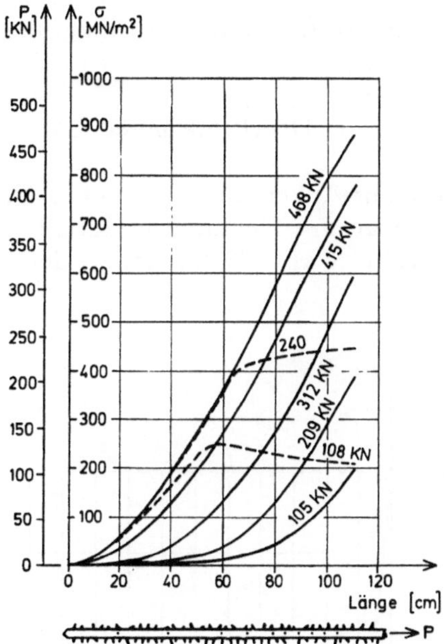

Abb. 13. Krafteintragung in der Ankerhaftstrecke eines gerippten Spannstabes
Force transmission in the fixed anchor length of a ribbed bar
Efforts dans la section d'adhérence d'un tiran en acier tor

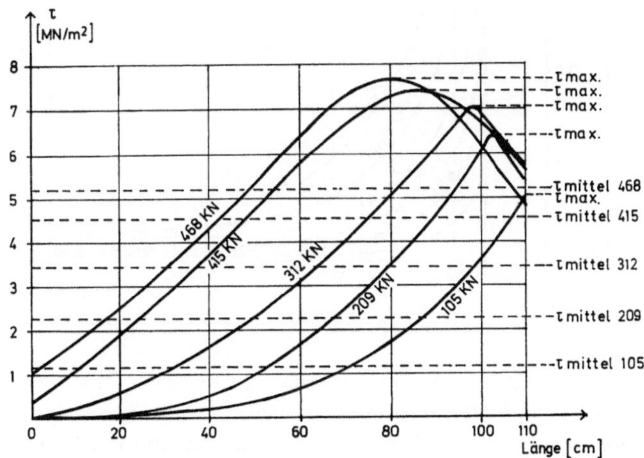

Abb. 14. Verbundschubspannungsverlauf in der Ankerhaftstrecke eines gerippten Spannstabes
Bond stress distribution in the fixed anchor length of a ribbed bar
Distribution des contraintes de cisaillement dans la section d'adhérence d'un tiran en acier tor

Die schon erwähnte optimale Haftstreckenlänge dieses Ankers liegt um 110 cm. Die nach DIN bestimmte Gebrauchslast beträgt dann 270 kN und die Länge des Reibungsbereiches ungefähr 10 cm. Da zur Kraftübertragung nur die erste Haftstreckenhälfte herangezogen wird, bietet die andere Hälfte immer noch genügend Reserve für eventuelle Ankerüberspannung.

6. Druckrohranker

Ein Original-Druckrohr-Anker, der Anker-Typ C, wurde auf ähnliche Weise wie der Dywidag-Stab mit Dehnungsmeßstreifen versehen und in-situ eingebaut. Das Druckrohr war 143 cm lang, der Durchmesser betrug

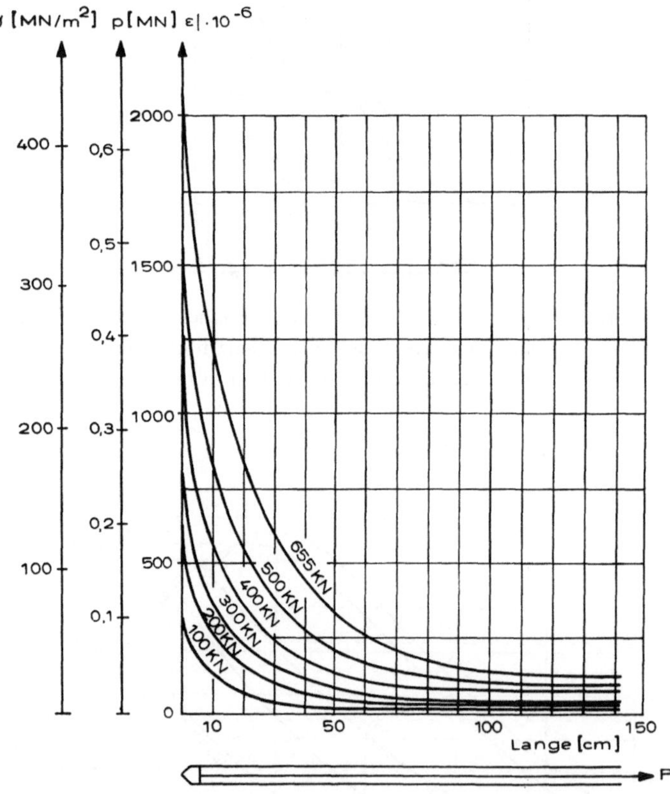

Abb. 15. Krafteintragung in der Ankerhaftstrecke eines Druckrohrankers
Force transmission in the fixed anchor length of a pressure pipe anchor
Efforts dans la section d'adhérence d'un ancrage en forme de tube

61/11 mm und das Rohr war mit Gewinde versehen. Die Festigkeiten des Druckrohres betrugen $\sigma_{Str}/\sigma_{Zug}$ 480 N/mm²/660 N/mm². Als Zugglied wurde ein Stab des Durchmessers 26,5 mm der Festigkeit 1,10/1,25 kN/mm² verwendet. Er war für eine Gebrauchslast von 300 kN zugelassen.

Die Belastungsstufen und die ermittelten Verbundschubspannungen stellen die Abb. 15 und 16 vor.

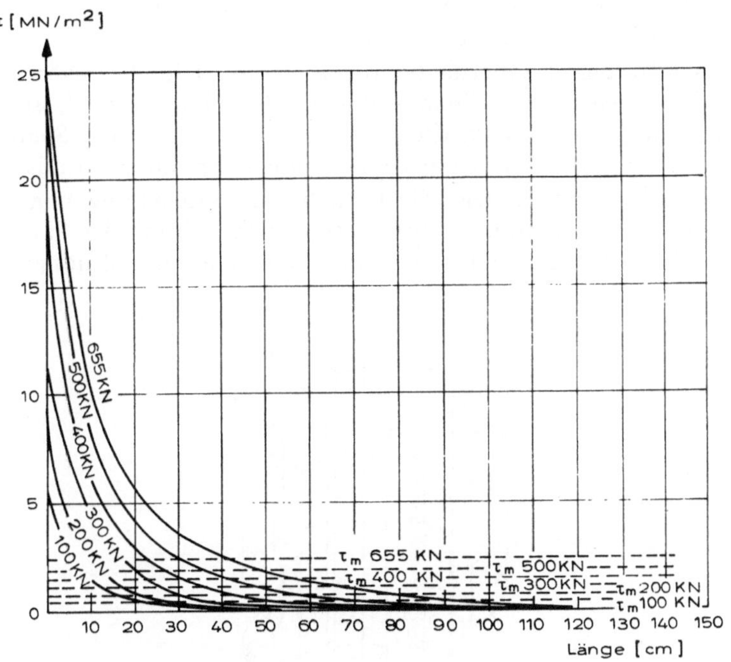

Abb. 16. Verbundspannungsverlauf in der Ankerhaftstrecke eines Druckrohrankers
Distribution of bond stress in the fixed anchor
Distribution des contraintes de cisaillement dans la section d'adhérence d'un ancrage en forme de tube

Der erste Dehnungsmeßstreifen war auf dem Druckrohr 5 cm von der Gewindemuffe angeklebt. Der gemessene Kraftabfall deutet an, daß in den ersten fünf Zentimetern 40% bis 60% der zu übertragenden Kraft eingeleitet wird. Das luftseitige Drittel des Rohres ist nur mit 10% bis 6% der jeweiligen Kraft belastet. Durch die hohen Anfangs-Verbundschubspannungen wird die Belastung entlang des Druckrohres so abgebaut, d. h. in den Mörtel der Haftstrecke eingeleitet, daß am Rohrende nur 4% der Kraft bleiben.

Die luftseitige Rohrhälfte kann als Widerlager angesehen werden, das der Stauchung des Rohres entgegenwirkt. Es gewährleistet die Stabilität gegen Abscheren an der Grenzfläche Druckrohr/Mörtel. Das Druckrohr darf nur im elastischen Bereich beansprucht werden.

Wegen seiner Wirkungsweise braucht dieser Anker-Typ nicht unter Druck verpreßt zu werden.

7. Radiale Spannungen

Bei den Ankern, in welchen Haftung und Reibung die Komponenten des Verbundes darstellen, ist die Reibung als Ursache der radialen Spannung zu erklären. Die Reibung hängt vom Manteldruck der Beschaffenheit der Stahloberfläche und von der Schlupfgröße ab. Sie ist nahezu unabhängig von der Größe der Zugspannung. Als Phänomene dieser Komponente können die Dilatation und die Keilwirkung angesehen werden.

Bei Ankern, deren Oberfläche profiliert ist, tritt Formverbund auf. Größere Dehnungen solcher Spannstähle in der Haftstrecke lassen, durch die Rippen bedingt, Spaltzugkräfte auftreten. Die unter dem Spaltzug stehenden Verpreßkörperteile üben zusätzlich radiale Spannungen aus.

Beim Druckrohranker liegt die Ursache der Entstehung von radialem Druck im Zuwachs des Rohrdurchmessers, welcher durch das Stauchen des Druckrohres hervorgerufen wird. Der in Längsrichtung gedrückte Verpreß-

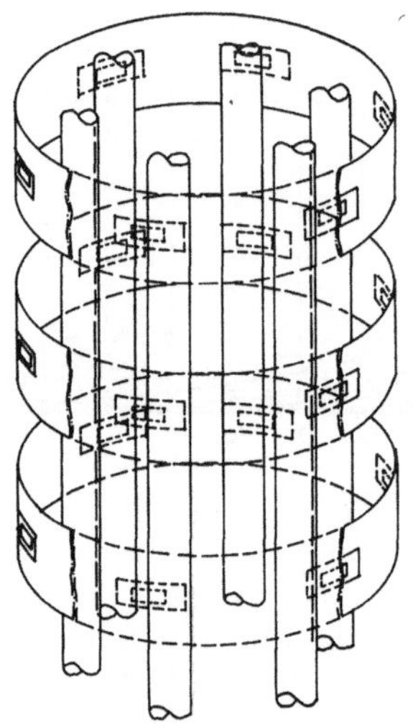

Abb. 17. Meßeinheit zur Messung der radialen Spannungen
Radial stress measuring system
Apparail de mesure des contraintes radiales

mörtel wird jedoch wegen der ihn umgebenden Bohrlochwandung an der Querdehnung gehindert. Als Folge davon ergibt sich ein Zuwachs des radialen Druckes.

Außer den geschilderten Ursachen kann auch die Konstruktion des Ankers zur Entstehung der radialen Spannungen beitragen.

Die Messung dieser Druckspannungen sollte in der Verpreßmörtelschicht, also zwischen dem Anker und der Bohrlochwand stattfinden.

Der enge Raum zwischen Anker und Bohrlochwandung machte die Anwendung von schmalen, dünnen Dehnungsmeßstreifen notwendig. Damit sie verdrahtet und genau fixiert werden konnten, mußten sie auf einen Träger geklebt werden. Dieser stellte einen auf 0,35 mm Wandstärke gedrehten Ring dar. Bei dieser geringen Wandstärke ist das Material sehr deformations-

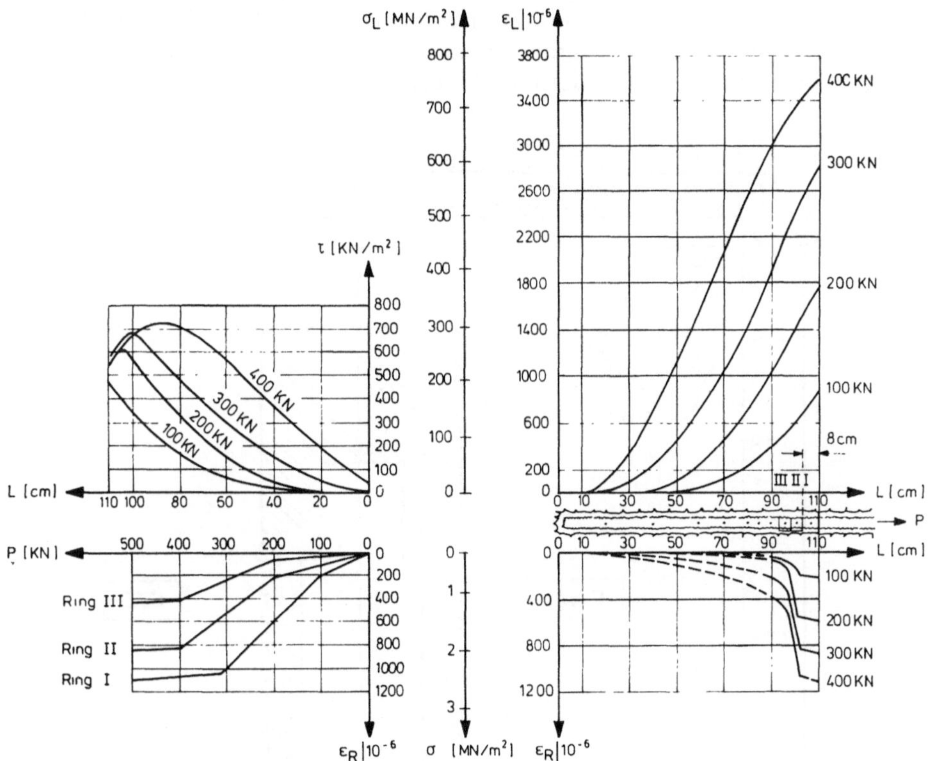

Abb. 18. Zusammenstellung der Meßergebnisse an einem gerippten Spannstab
Summary of measurement results for a ribbed bar
Vue d'ensemble des résultats d'essai avec un tiran en acier tor

fähig, aber noch so stabil, daß sechs Dehnungsmeßstreifen außen aufgeklebt werden konnten. Um den Verbund minimal zu beeinflussen (Abb. 17), wurden nur drei solcher Ringe zu einer Meßeinheit zusammengebaut. Sie wurde mittels Querträgern am Anker fixiert und zusammen mit ihm in das Bohrloch eingeführt.

Zur Eichung war zuvor die Meßeinheit ohne Ankerstab in ein Bohrloch eingebaut und von der Mitte aus allseitig wirkendem Druck unterworfen worden. So konnten direkte Beziehungen zwischen den Dehnungen im Ring und dem dadurch verursachten Radialdruck hergestellt werden.

Die Ergebnisse der Messung auf den Spannstab-Anker stellt die Abb. 18 dar. Rechts unten sind die Radialspannungen entlang der Haftstrecke bei gleicher Belastung, links unten die Radialspannungen auf die drei Meßebenen für unterschiedliche Laststufen dargestellt. Die Spannungen nehmen 8 cm vom Packer entfernt auf der gemessenen Länge von 10 cm rasch ab. Während des Spannungsabfalles im untersuchten Bereich nimmt der Schlupf von Null bis zum Abscheren zu. Der Radius der zylindrischen Scherzone ist durch die Rippen des Stabes, nicht durch den Schaft gegeben.

Die Abb. 19 zeigt den Verlauf der Radialspannung beim Druckrohranker des Types C. Das Spannungsmaximum wurde nicht unmittelbar an der Krafteinleitungsstelle, sondern an Meßring Nr. I gemessen. Der mögliche

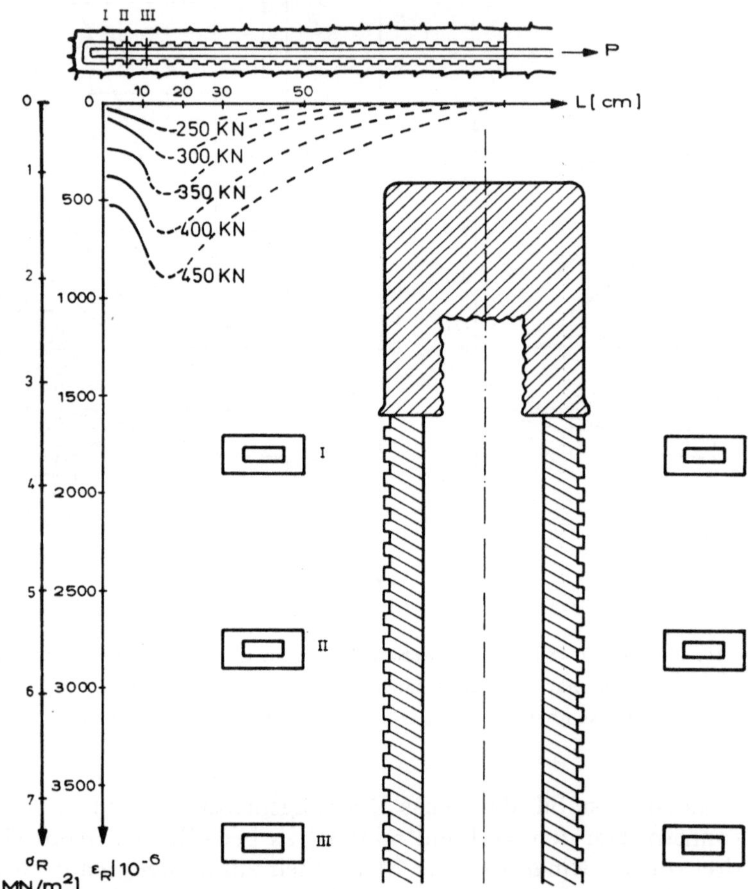

Abb. 19. Verlauf der radialen Spannung an einem Druckrohranker
Radial stress distribution on a pressure pipe anchor
Répartition des contraintes radiales dans un ancrage en forme de tube

Verlauf der radialen Spannung außerhalb des Meßbereiches ist gestrichelt dargestellt. Bis zur Einleitung der Kraft von 300 kN nimmt der Spannungsverlauf im Meßbereich langsamer zu als bei höheren Laststufen. Dies ist auf die Deformation des Druckrohres in Längsrichtung zurückzuführen.

Literatur

Abraham, K. H., Porzig, R.: Die Felsanker des Pumpspeicherwerkes Waldeck II. Baumaschinen und Bautechnik Jg. 20, H. 6 u. 7 (1973).

Egger, P.: Erster Zwischenbericht für das Teilprojekt 6.12. Jahresbericht 1971 des SFB 77, Karlsruhe 1971.

Jirovec, P.: Untersuchungen zur Wirkungsweise von Verankerungen. Jahresbericht 1974 des SFB 77, Karlsruhe 1975.

Jirovec, P.: Methode zur Felsankeruntersuchung. Erd- und Felsankerseminar, Wuppertal 1976.

Müller, H. R.: Erfahrung mit Verankerungen System BBRV im Fels und Lockergestein. Schweizerische Bauzeitung Jg. 84, H. 4 (1966).

Ostermayer, H.: Construction, Carrying Behaviour and Creep Characteristics of Ground Anchors. Conference on Diaphragm Walls and Anchorages, London, Sept. 1974, Institution of Civil Engineers, London, S. 141—151, 1975.

Stump: Der Stump-Duplex-Permanent-Anker, Felsanker. Allgemeine bauaufsichtliche Zulassung 1976.

Anschrift des Verfassers: Dipl.-Ing. Peter Jirovec, Lehrstuhl für Felsmechanik am Institut für Bodenmechanik und Felsmechanik der Universität Karlsruhe, Richard-Willstätter-Allee, D-7500 Karlsruhe 1, Bundesrepublik Deutschland.

Rock Mechanics, Suppl. 7, 157—177 (1978)

Rock Mechanics
Felsmechanik
Mécanique des Roches

Die Vortriebssicherung des Arlberg-Straßentunnels — Anpassung an die Gebirgsverhältnisse mit Hilfe von geotechnischen Messungen

Von

G. Judtmann

Mit 26 Abbildungen

Zusammenfassung — Summary — Résumé

Die Vortriebssicherung des Arlberg-Straßentunnels — Anpassung an die Gebirgsverhältnisse mit Hilfe von geotechnischen Messungen. Der Arlbergtunnel wird nach der neuen österreichischen Tunnelbauweise vorgetrieben, die eine empirisch-wissenschaftliche Methode ist. Dieser Vortrag wird sich mit dem empirischen Teil dieser Methode befassen.

Der Tunnel mit seiner Gesamtlänge von fast 14 km liegt durchwegs in der Phyllit-Gneis-Decke am nördlichen Rand der Zentralalpen. Die Schieferung steht sehr steil und fällt mit 60 bis 80^0 nach Süden ein. Das Streichen ist nahezu parallel zur Tunnelachse. Diese Verhältnisse erwiesen sich als äußerst ungünstig für die Vortriebsarbeiten.

Der Tunnel wird von beiden Seiten im Kalotten- und Strossenvortrieb vorgetrieben, wobei die Ostseite gleislos und die Westseite mit Gleisbetrieb arbeitet. Die Ostseite war mehr auf eine große Kalotte von ca. 6,5 m eingerichtet und nur bei schlechten Gebirgsverhältnissen wurde eine kleine Kalotte vorausgeschickt, während die Westseite einheitlich mit einer Kalotte von 4,5 m Höhe und zwei gestaffelt nacheilenden Strossen gearbeitet hat.

Insbesonders im Westen zeigte sich im Laufe der Vortriebsarbeiten sehr bald, daß das Gebirge stark unter Druck steht, und daß größere Gebirgsbewegungen auftreten. Es zeigte sich, daß die horizontalen Konvergenzbewegungen ein sehr gutes Maß für den benötigten Ausbauwiderstand darstellt. Insbesondere die Verformungsgeschwindigkeit in den ersten zwei Tagen konnte als Kriterium für die Sicherungs- und Stützmaßnahmen herangezogen werden. Wenn die Verformungen nach zwei Tagen unter 5 cm lagen, waren die Sicherungsmaßnahmen ausreichend. Erst wenn dieses Maß überschritten worden ist, dann mußten die Sicherungsmaßnahmen verstärkt werden. Ebenso erlaubte ein deutliches Unterschreiten dieses Maßes eine Abminderung der Sicherungsmaßnahmen.

Nach diesem Kriterium wurde bei dem weiteren Vortrieb vorgegangen und es konnten horizontale Konvergenzbewegungen bis zu 60 cm einwandfrei beherrscht werden. Es kamen dabei 4, 6, 9 und 12 m lange Anker zum Einsatz. Es muß aber ausdrücklich darauf hingewiesen werden, daß die Größe der Verformung von den Gebirgsverhältnissen abhängig ist, und daß dieses Maß keinesfalls verallgemeinert werden kann.

Wie stark der Einfluß der Schieferungsflächen in bezug auf die Vortriebsrichtung ist, zeigt, daß die kurzen Verbindungstunnel die senkrecht auf die Tunnelachse in Richtung zur zweiten Röhre vorgetrieben worden sind, nur einen Bruchteil der Verformungen aufwiesen als in der Haupttunnelröhre.

Support Measures Used in Excavating the Arlberg Tunnel — Adaptation to Rock Mass Conditions by Means of Geotechnical Measurements. Excavation of the Arlberg Tunnel is performed according to the New Austrian Tunnelling Method, which is an empiric-scientific method. This talk concerns the empiric part of the method.

The tunnel with its total length of almost 14 km lies completely in the phyllitgneiss nappe at the northern edge of the Central Alps. The schistosity is very steep and strikes to the south at an angle of 60 to 80^0. Striking is almost parallel to the tunnel axis. These conditions are extremely unfavorable for excavation procedures.

The tunnel is excavated from two sides using heading and bench excavation methods, whereby on the east side transportation is carried out using trucks and on the west side by means of a railway. Excavation on the east side is generally performed with a large calotte with a height of approximately 6.5 m. This height is only reduced under poor rock conditions. Excavation on the west side is generally performed with a 4.5 m high calotte followed by two graduated benches.

Particularly on the west side, it was very soon obvious that the rock mass was under genuine rock pressure resulting in large deformations of the rock mass. It was seen that the horizontal convergence gave a good idea of the amount of support resistance required. The deformation rate in the first two days, in particular, was used as a criterium for the support measures. If the deformation was less than 5 cm after two days, the support measures were sufficient. If the 5 cm were exceeded, additional support measures were installed. In the same manner, if the deformation was less than 5 cm, the support measures could be reduced.

Excavation was performed according to this criterium and horizontal convergences of up to 60 cm were able to be brought under control. For this purpose, 4, 6, 9 and 12-m anchors were used. However, it must be noted that the deformation is dependent on the rock mass conditions and that the results given here can not be generalized.

The influence of the schistosity planes with respect to the excavation direction is emphasized by the fact that deformations measured in the short tunnels excavated at a right angle to the main tunnel's axis as a connection to the second tunnel tube were only a fraction of the amount of deformations measured in the main tunnel.

Les mesures de sécurité employées pendant les travaux d'avancement dans le tunnel de l'Arlberg — Adaptation aux conditions de roche à l'aide de mesures géotechniques. Les travaux de creusement dans le Tunnel de l'Arlberg s'effectuent suivant la nouvelle méthode autrichienne, une méthode empirique-scientifique. Ce traité s'occupe de l'aspect empirique de cette méthode.

Le tunnel, qui a une longueur de 14 km, se trouve complètement dans la nappe phyllite gneiss du côté nord des Alpes centrales. Le clivage est très raide et tombe de 60^0 à 80^0 vers le sud. La direction de la couche est presque parallèle à l'axe du tunnel. Ces conditions sont extrêmement désavantageuses pour les travaux d'avancement.

L'avancement avec calotte et banc s'est fait bilatéral. Le marinage s'effectue du côté est au moyen de camions et du côté ouest en utilisant des trains sur rail. Au côté est on a choisi une calotte plutôt grande avec une hauteur de 6,5 m et seulement là où les conditions de roche etaient mauvaises il était nécessaire d'établir une calotte plus petite. Au côté ouest on travaillait toujours avec une calotte de 4,50 m, suivie de deux stross en gradins.

Surtout à l'ouest on devait constater que la roche subit des fortes pressions et on y trouve des déformations assez grandes de roche. Il se montrait que la convergence horizontale était une bonne indication pour la résistance du support nécessaire. Comme critère pour les mesures de support et de sécurité on a pu surtout prendre en considération la vélocité de déformation pendant les deux premiers jours. Si la déformation après deux jours était moins que 5 cm, les mesures de support étaient suffisantes. Si la déformation était plus grande, les mesures de support devaient être renforcées. Pareillement, une déformation beaucoup plus inférieure permissait une réduction des mesures de support.

Les travaux de creusement ont été effectués d'après ce critère et on pouvait contrôler les convergences horizontales de jusqu'à 60 cm. Pour celà on se servait d'ancres de 4, 6, 9 et 12 m de longueur. Il faut absoluement attirer l'attention sur le fait que l'étendue de la déformation dépend des conditions de la roche et que le critère de 5 cm employé ici ne peut pas être généralisé en aucun cas.

Im vorliegenden Referat über die Vortriebssicherung des Arlberg-Straßentunnels wird nach einer allgemeinen Einleitung kurz auf die Vortriebsarbeiten der beiden Arbeitsgemeinschaften eingegangen. Anschließend wird die Möglichkeit näher erörtert, die dem ausführenden Ingenieur zur Beurteilung des Gebirges und zur Festlegung der Sicherungsmaßnahmen zur Verfügung stehen; als letztes wird an einem Beispiel gezeigt, wie beim Arlberg-Straßentunnel aufgrund geotechnischer Messungen die Anpassung der Vortriebssicherung an die Gebirgsverhältnisse durchgeführt wurde.

Auf die geologischen Verhältnisse und den Gesamtkomplex der geotechnischen Messungen wird nur soweit eingegangen, als es zum direkten Verständnis dieser Ausführungen notwendig ist. Es wird dabei auf die Vorträge von Prof. Weiss und von Dr. John bei früheren Kolloquien verwiesen.

Der Arlberg-Straßentunnel liegt mit seiner Gesamtlänge von fast 14 km durchwegs in der Phyllit-Gneisdecke am nördlichen Rand der Zentralalpen. Diese Phyllit-Gneisdecke ist eine Serie von tektonisch stark beanspruchten und mit Mylonitzonen durchsetzten Gneisen, Glimmerschiefern und Phylliten, die durch das Anpressen der Silvrettadecke an die Nördlichen Kalkalpen im großen und ganzen ihre heutige Ausprägung erhalten hat.

Die Schieferung fällt 60—80^0 steil nach Süden ein, das Streichen verläuft spitzwinklig und teilweise nahezu parallel zur Tunnelachse. Im westlichen Bereich des Tunnels kam zu der tektonischen Beanspruchung der Gesteine noch eine Durchfeuchtung, die eine weitere Verschlechterung der Gebirgseigenschaften zur Folge hatte.

Die Vortriebsarbeiten erfolgten nach der konventionellen Bauweise, d. h. mit Bohren und Schießen, während die Sicherungsarbeiten nach der Neuen Österreichischen Tunnelbauweise durchgeführt wurden.

Auf beiden Seiten wurde eine Kalotte vorangetrieben und die Strossen nachgezogen. Im Osten hatte man sich darauf eingestellt, größtenteils mit einer großen Kalotte von 6,5 m Höhe und einer nacheilenden Strosse zu arbeiten und nur in ganz besonders schlechten Bereichen eine kleine Kalotte vorauseilen zu lassen. An der Westseite wurde einheitlich mit einer Kalotte von 4,5 m Höhe und zwei gestaffelten Strossen gearbeitet (Abb. 1 und 2). Auch in der Abschlagslänge bestanden Unterschiede; die Ostseite tat im

Abb. 1. Westseite — Kalottenvortrieb, Aufstellen eines Tunnelbogens
West side — driving the top heading, installation of a tunnel rib
Côté d'ouest — excavation de la calotte, mise en place d'un arc

Abb. 2. Westseite — Abbohren der Kalotte
West side — drilling the top heading
Côté d'ouest — forage de la calotte

Regelfalle Abschläge bis zu 4 m ab und stellte jeweils 2 Tunnelbögen hinein, während an der Westseite die Abschlagslänge mit dem Bogenabstand übereinstimmte und im Maximum 2 m betrug.

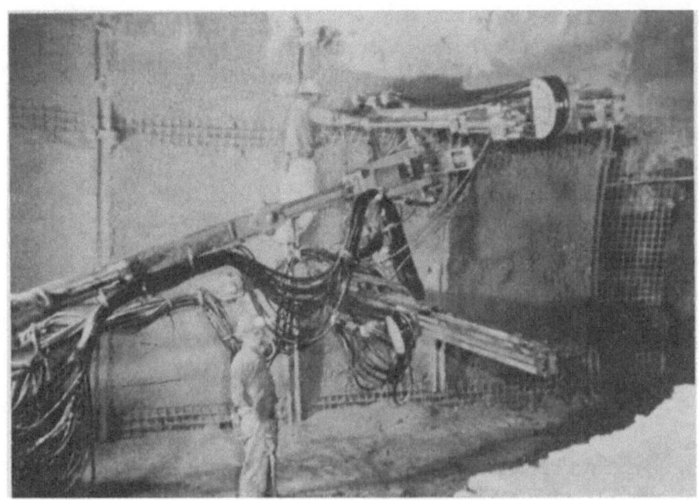

Abb. 3. Westseite — Ankerbohrung
West side — anchor drilling
Côté d'ouest — forage de l'ancre

Abb. 4. Ostseite — Anbringen des Baustahlgitters
East side — application of steel mats
Côté d'est — mise en place du traillis soudé

Das Klassifizierungsschema soll hier noch kurz gestreift werden: Es gibt 6 Gebirgsklassen, wobei die untersten zwei Klassen I und II überhaupt nie und die Klasse IIIa äußerst selten zu Anwendung kam (Abb. 6).

Die Aufteilung der Gebirgsklassen der Ostseite:

Gebirgsklasse III a	1,0%
Gebirgsklasse III b	39,9%
Gebirgsklasse IV	57,1%
Gebirgsklasse V	2,0%

Die Aufteilung der Gebirgsklassen der Westseite:

Gebirgsklasse III b	3,0%
Gebirgsklasse IV	94,3%
Gebirgsklasse V	2,7%

Die sehr bald aufgetretenen großen Verformungen erforderten besondere Maßnahmen. So wurde, um eine Zerstörung der Spritzbetonschale zu

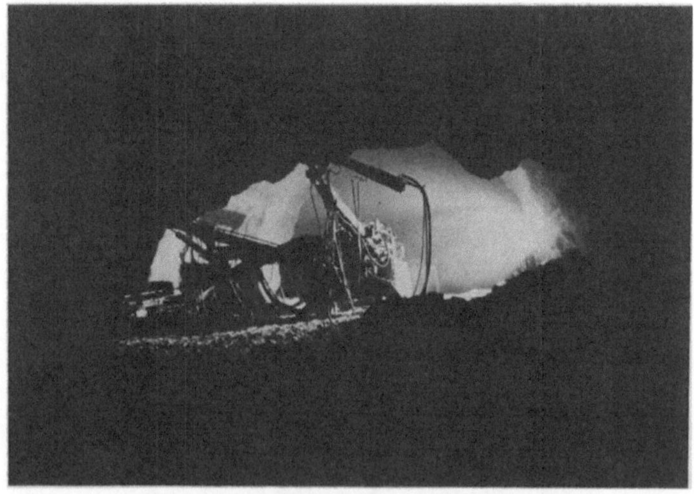

Abb. 5. Ostseite — Aufweitung der Kalotte
East side — enlargement of the top heading
Côté d'est — élargement de la calotte

verhindern, der Spritzbeton durch Schlitze unterteilt (Abb. 7), so daß eine kontrollierte Verformung zugelassen werden konnte.

In den Bereichen mit großen Gebirgsverformungen kam es zum reihenweisen Abreißen von Ankerköpfen. Es wurden daher Überlegungen und Versuche angestellt, wie man den Verformungsweg der Anker vergrößern könnte. Mit relativ weichen Ankerplatten konnte man einen Verformungs-

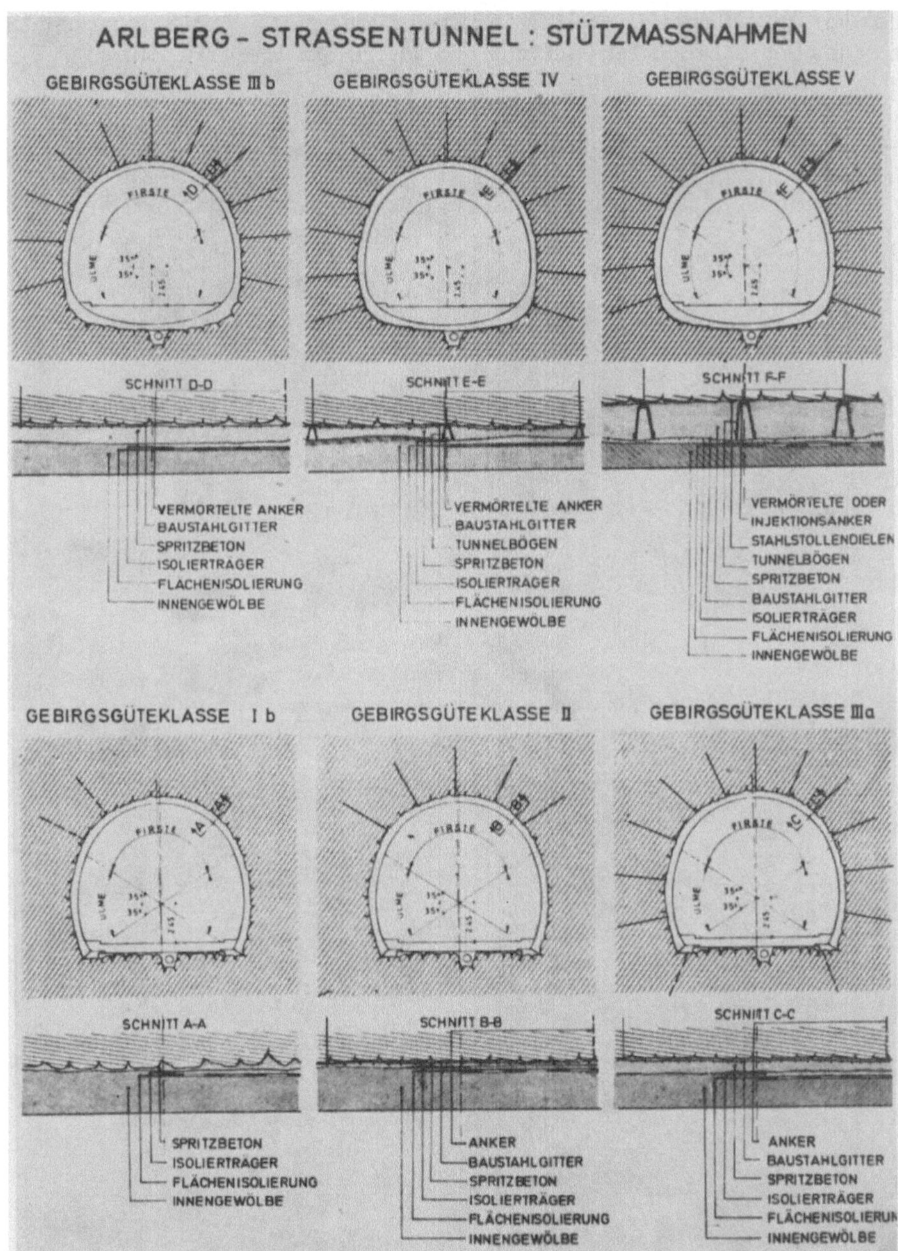

Abb. 6. Gebirgsgüteklassen
Rock quality classes
Classes de qualité de la roche

weg von einigen Zentimetern erreichen (Abb. 8). Einen Verformungsweg von 10—12 cm erlaubte die Zwischenschaltung eines Rohrstückes, die außerdem noch den Vorteil hatte, daß die Kraft im Zuge der Verformung nahezu konstant blieb (Abb. 9, Entwicklung von Herrn Dr. Heierli, Zürich).

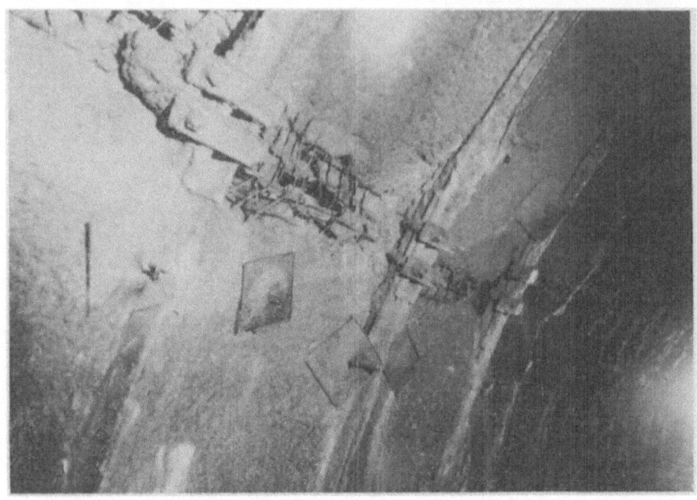

Abb. 7. Westseite — Bewegungsschlitz
West side — convergence slit
Côté d'ouest — fente de convergence

Abb. 8. Westseite — deformierte Ankerplatte
West side — deformed anchor plate
Côté d'ouest — plaque d'ancrage déformée

Von der Arbeitsgemeinschaft West selbst wurde ein einfacheres System entwickelt, und zwar in der Form, daß ein Halbrohr auf die Ankerplatte aufgesetzt wurde (Abb. 10). Dieses System wurde in größerem Umfang und mit gutem Erfolg eingesetzt.

Die im allgemeinen sehr schlechten, aber dennoch stark wechselnden Gebirgsverhältnisse des Arlberg-Straßentunnels erforderten eine ständige Anpassung der Sicherungsmaßnahmen. Die Wirtschaftlichkeit der Neuen Österreichischen Tunnelbauweise liegt im Großteil darin begründet, daß nur soviel Sicherungsmaßnahmen eingebaut werden müssen, wie in jedem Tunnelab-

Abb. 9

Abb. 10

Abb. 9. Verformungsrohr — System Heierli
Pipe for deformation purposes — Heierli system
Tuyau à déformer — système Heierli

Abb. 10. Verformungshalbschale — System ATW
Half pipe for deformation purposes — ATW system
Demi tuyau à déformer, système ATW

schnitt erforderlich sind. Man muß den Ausbau nicht auf die schlechtesten Verhältnisse bemessen. Allerdings erfordert diese Bauweise eine ständige Anpassung durch fortwährende Beobachtung und Abschätzung der Lage. Es ist daher entsprechend geschultes und erfahrenes Personal notwendig, das mit einer gewissen Bandbreite dafür sorgt, daß weder zuviel Sicherungsmaßnahmen eingebaut werden — dies würde zur Unwirtschaftlichkeit führen —, noch, daß zuwenig Sicherungsmaßnahmen zum Einsatz kommen, wodurch der Tunnel gefährdet werden könnte.

Im Fall des Arlberg-Tunnels waren die beiden ausführenden Arbeitsgemeinschaften, die bereits bei der Erstellung des Tauerntunnels und des

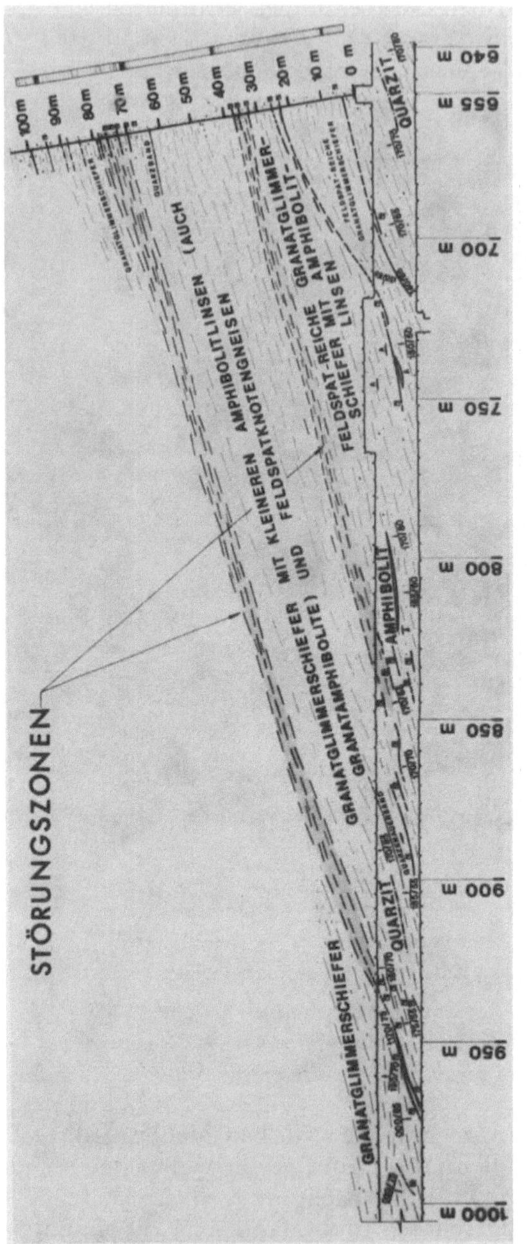

Abb. 11. Westseite — Geologische Erkundung mittels Sondierbohrung
West side — geological exploratory drilling
Côté d'ouest — exploration géologique au moyen du forage de sondage

Katschbergtunnels mitgewirkt hatten, in der Lage, erfahrenes Personal für die Leitung der Tunnelbauarbeiten zur Verfügung zu stellen. In laufenden

Gesprächen zwischen den Herren der Arbeitsgemeinschaft, der örtlichen Bauleitung und der vom Bauherrn abgestellten Geologen, sowie unter fallweiser Zuziehung von Experten wurde versucht, die Möglichkeit dieser Anpassung voll auszunützen. Für die Beurteilung der Gebirgsverhältnisse standen neben der optischen Beobachtung der Ankerplatten und der Verformungsschlitze folgende Hilfsmittel zur Verfügung:

Die geologischen Aufnahmen, die täglich von Geologen des Bauherrn durchgeführt wurden.

Die geotechnischen Messungen, die entsprechend den felsmechanischen Erfordernissen erfolgten.

Um einen echten Aussagewert für eine kurzfristige Prognose der geologischen Verhältnisse zu erhalten, wurden neben den aufgenommenen Brustbildern Grundrisse gezeichnet, die bei dem steilen Einfall der Schieferung anschaulich die Verhältnisse darstellen konnten. Zur Erweiterung des Vorhersagebereiches wurden unter Ausnützung der in weiten Bereichen des Tunnels vorhandenen geometrischen Verhältnisse (Winkel zwischen Tunnelachse und Schieferung ca. 15°) am südlichen Ulm Sondierbohrungen bis zu 100 m Tiefe hergestellt (Abb. 11). Die in den Bohrungen erzielten Aufschlüsse konnten etwa im Verhältnis 1 : 4 vorprojiziert werden. Die Übereinstimmung mit den beim Vortrieb angetroffenen Verhältnissen war recht gut; sie nahm natürlich mit zunehmender Entfernung vom Bohrloch ab.

Diese guten geologischen Prognosen konnten zwar nicht unmittelbar für die Bemessung des Außengewölbes herangezogen werden, erlaubten es aber doch, die Verhältnisse besser zu beurteilen. So waren größere Verformungen zu erwarten, wenn sich der Vortrieb einen der mächtigen, mit Myloniten durchsetzten Störungszonen von Süden her näherte. Wo sich die Störung im Profil befand, war die Wirkung nicht mehr so groß; sie nahm erst wieder zu, wenn sie an der Nordseite das Profil wieder verließ und damit den nördlichen Ulm schwächte. Besonders kritisch waren die Bereiche, bei denen eine Störung nach Norden auskeilte und die nächste sich bereits von Süden her dem Profil näherte.

Welch großen Einfluß die Lage der Schieferungsflächen zur Vortriebsrichtung auf die Verformung hatte, zeigten Konvergenzmessungen, die in Querschlägen und Kavernenröhren parallel zur Tunnelachse durchgeführt wurden. Diese Konvergenzmessungen zeigten etwa nur 5 bis 10% der Verformungen senkrecht zur Tunnelachse.

Als eigentliches Kriterium für die Bemessung der Sicherungen wurden die geotechnischen Messungen herangezogen. Insbesondere zeigte sich, daß die horizontalen Konvergenzbewegungen, die sehr rasch hinter dem Ausbruch gemessen werden konnten, ein gutes Maß für das Gebirgsverhalten waren. Die Konvergenzmessungen wurden ca. 1 m über der Kalottensohle im Abstand von 10 bis 50 m je nach Gebirgsverhältnissen und in den ersten zwei Wochen täglich vorgenommen.

Da es auf die Dauer unbefriedigend war, nur aufgrund der optischen Beobachtungen und ohne besondere Kriterien nach den Verformungen die Stütz- und Sicherungsmaßnahmen zu variieren, wurde versucht, das Gebirgsverhalten systematisch zu erforschen. Durch statistische Auswertung der Verformungskurven erhielt man eine mittlere Verformungskurve, bei deren Verlauf man aus Erfahrung wußte, daß die dabei auftretenden Verformungen nicht zum Bruch führen (Abb. 12). Die anfängliche Deformationsgeschwindigkeit erwies sich als wesentlich für den weiteren Verlauf der

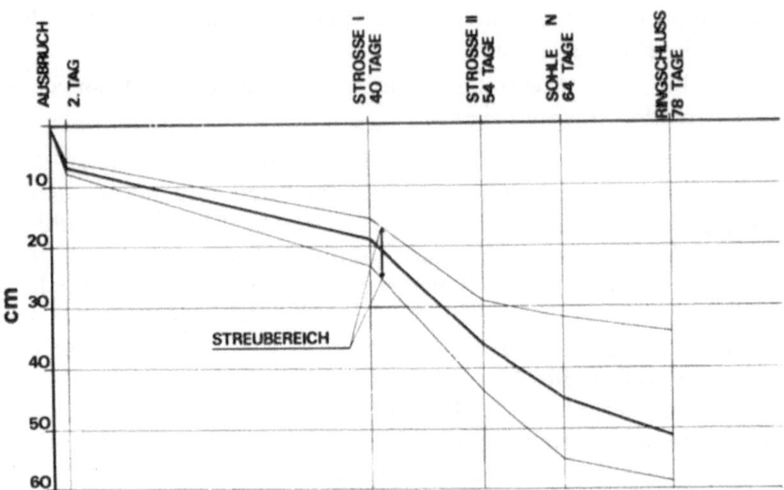

Abb. 12. Westseite — Streubereich der Konvergenzmessungen
West side — range of scatter for convergence measurements
Côté d'ouest — zone de dispersion des mesures de convergence

Kurve. In weiten Bereichen betrugen die Verformungen nach zwei Tagen etwa 1/10 der Gesamtverformung. Daher wurde die Verformung nach zwei Tagen als maßgebend für die Bemessung des Außengewölbes herangezogen.

Da horizontale Konvergenzverformungen bis zu 50 cm vom Ausbau ohne Schäden mitgemacht werden konnten, wurde die Verformung nach zwei Tagen mit 5 cm als Kriterium bestimmt. Das heißt, wenn die Verformung nach zwei Tagen unter 5 cm lag, war die Bemessung des Außengewölbes ausreichend bzw. konnte man den Ausbau so lange vermindern, bis durch größere Verformungen zu ersehen war, daß man sich der Grenze nähert. Bei Verformungen über 5 cm nach zwei Tagen verstärkte man die Sicherungsmaßnahmen, bis sich die Verformungsgeschwindigkeit wieder auf das gewünschte Maß eingependelt hatte.

An Hand eines Beispieles soll nun die Anwendung dieses Verfahrens erläutert werden, und zwar im Baulos West von Station 1920 bis Station 2000.

Es sei hier ausdrücklich darauf hingewiesen, daß die angegebenen Werte nur für das beim Arlberg-Tunnel angetroffene Gebirge gelten. Eine direkte Übertragung auf andere Verhältnisse ist nicht möglich.

In Abb. 13 ist der Stand der Verformungskurven am 24. April 1976 im Bereich von Station 1920 m bis 1960 m angegeben. Bei Station 1940 m ist eine Verflachung zu sehen, die auf eine Arbeitseinstellung während der

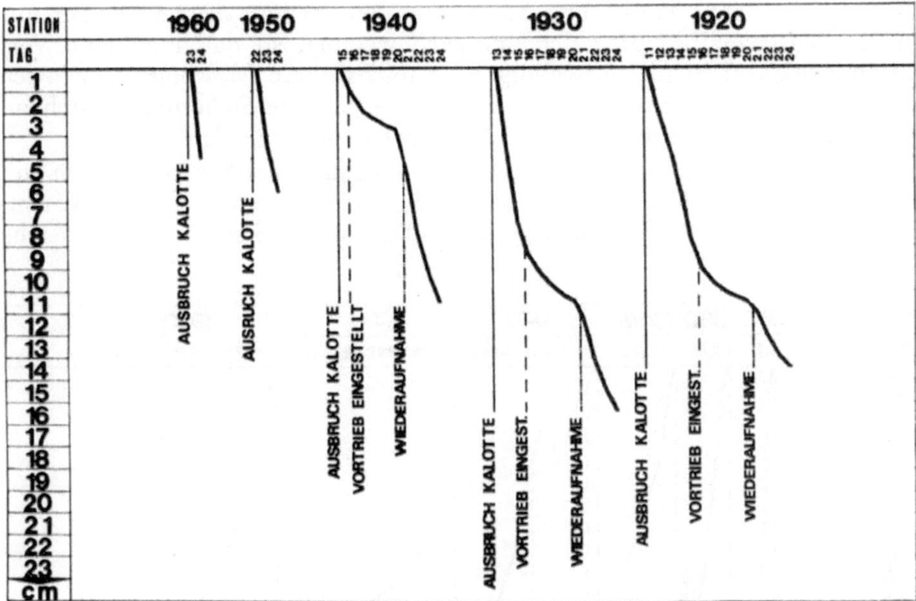

Abb. 13. Konvergenzmessungen am 24. April 1976
Convergence measurements from April 24, 1976
Mesures de convergence du 24 avril 1976

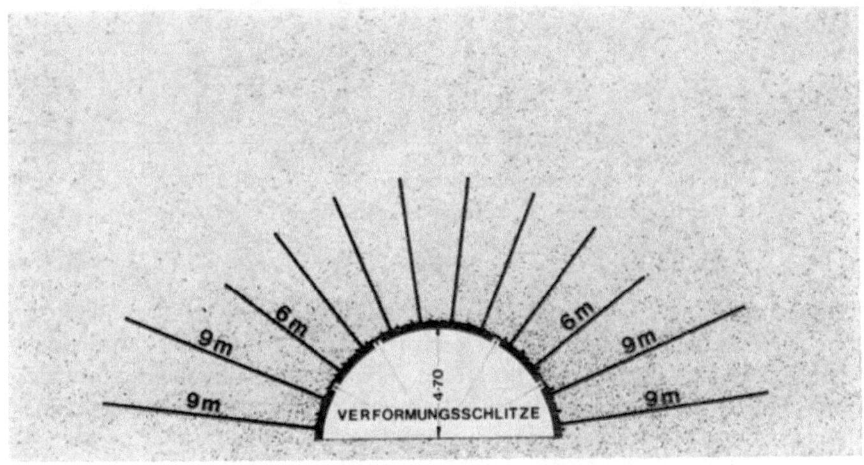

Abb. 14. Sicherung bis Station 1978 m
Supports installed up to station 1978 m
Moyens de support jusqu'à la station 1978 m

Osterfeiertage zurückzuführen ist. Bei den Stationen 1930 m und 1920 m ist die Verflachung auch noch zu erkennen, ist aber nicht mehr so ausgeprägt. Bei Station 1950 m betrug die Verformung nach 2 Tagen 5,5 cm, also knapp über dem gesetzten Limit. Die Messungen bei Station 1960 m zeigten nur eine geringfügige Zunahme der Ersttagsbewegung, so daß man eine Verstärkung noch nicht für notwendig hielt und weitere Messungen abwarten wollte. Bei dieser Station war die in Abb. 14 gezeigte Sicherung, bestehend aus Ankern mit Längen von 9 und 6 m, angewendet worden.

Am 26. April 1976 wurde bereits die Messung bei Station 1970 m mit in die Betrachtungen einbezogen (Abb. 15). Dabei zeigte sich eine wesentliche Zunahme der Verformungsgeschwindigkeit. Bei Station 1960 m betrug

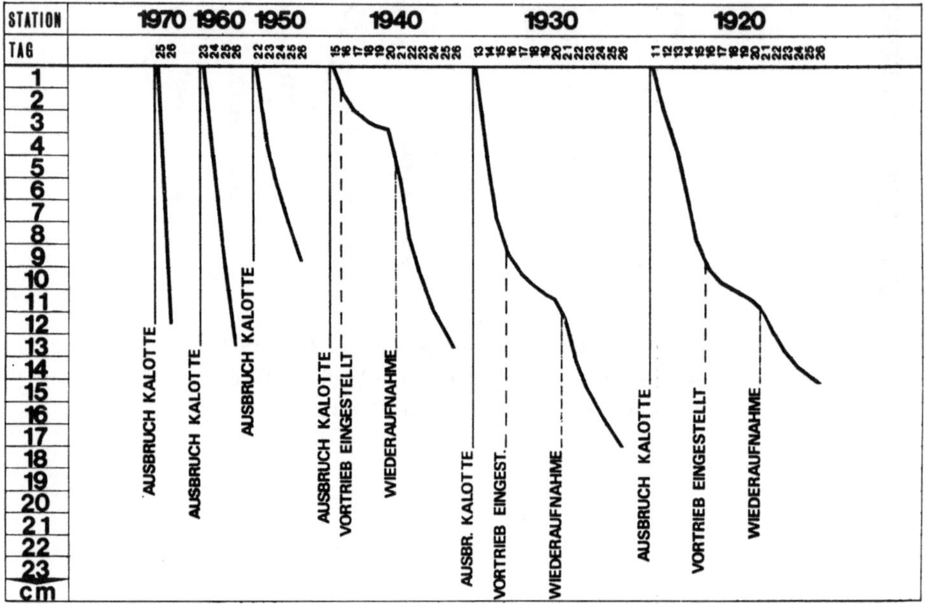

Abb. 15. Konvergenzmessungen am 26. April 1976
Convergence measurements from April 26, 1976
Mesures de convergence du 26 avril 1976

die Verformung nach 2 Tagen 8,4 cm und bei Station 1970 m nach 1 Tag 11,5 cm. Die rasche Zunahme der Verformung ließ erkennen, daß der Ausbauwiderstand zu gering war und eine Verstärkung notwendig wurde. Ab Station 1978 m wurde daher der Ausbau entsprechend der in Abb. 16 angegebenen Sicherung verstärkt. Der Bogenabstand wurde auf 1 m bis 1,20 m verringert und die Ankerlängen auf 12 und 9 m erhöht. Außerdem wurden nicht nur Anker bei den Bögen, sondern auch in den Feldern dazwischen gesetzt.

Darüber hinaus wurde auf eine Länge von 28 m die bereits bestehende Sicherung durch 10 Stück 9 m lange Anker, welche zwischen den Bögen versetzt wurden, verstärkt. Für die Einbringung dieser Nachankerung im unmittelbaren Vortriebsbereich mußten die Vortriebsarbeiten für 2 Tage eingestellt werden (Abb. 17).

Am 27. 4. ergab sich folgendes Bild (Abb. 18): Die Verformungen bei Station 1970 m hatten nur mehr um ca. 4 cm zugenommen, die Ersttagsverformung bei Station 1980 m betrug knapp 5 cm. Die Verminderung der

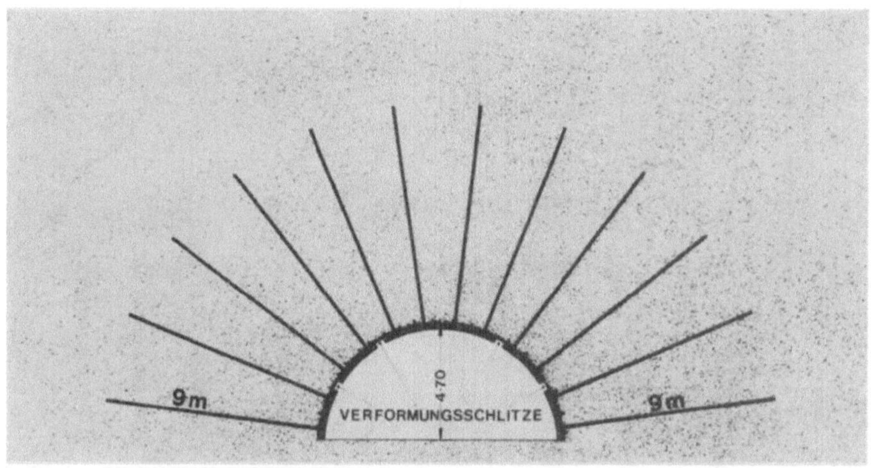

Abb. 16. Nachankerung von Station 1950 m bis 1978 m
Supplementary anchors from station 1950 m to station 1978 m
Ancres supplémentaires de station 1950 m à station 1978 m

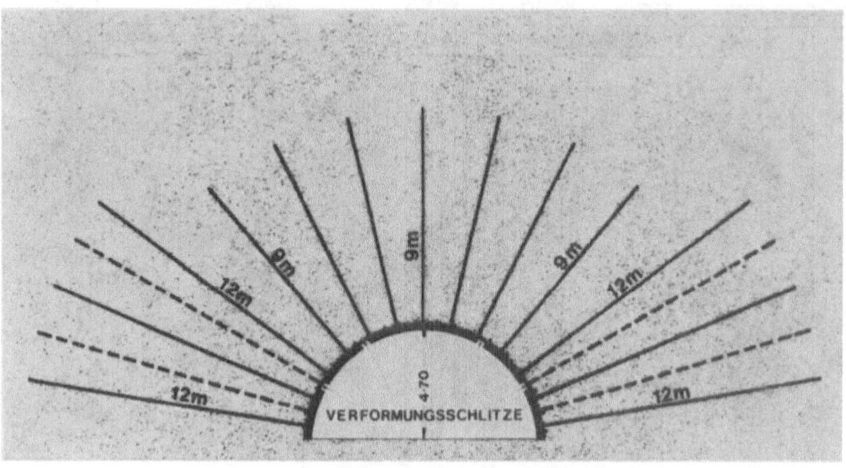

Abb. 17. Sicherung ab Station 1978 m
Supports from station 1978 m on
Moyens de support à partir de la station 1978 m

Verformung war selbstverständlich neben der verstärkten Ankerung auch auf die Arbeitseinstellung zurückzuführen. Nach Wiederaufnahme der Arbeiten am 28. 4. 1976 mit dieser wesentlich verstärkten Sicherung zeigte

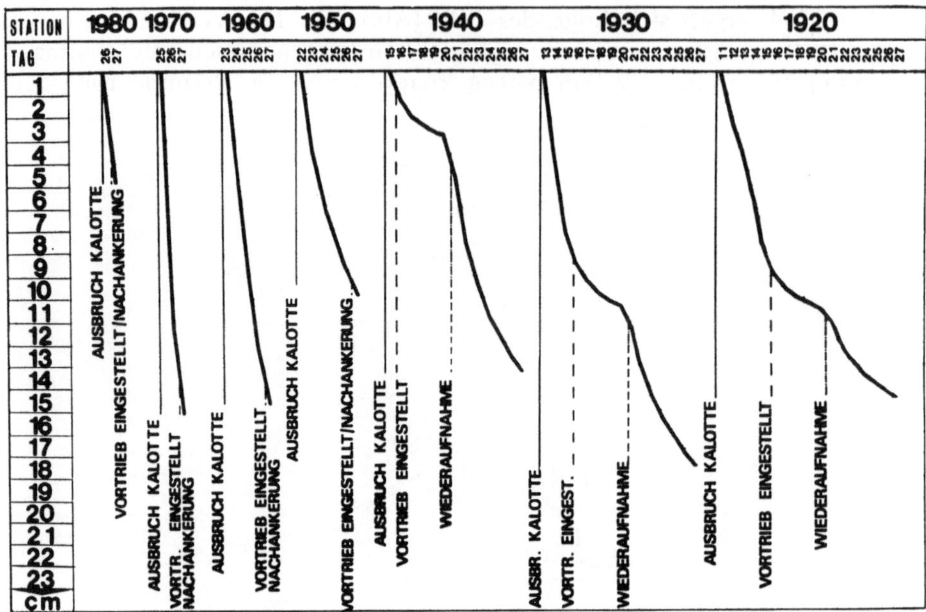

Abb. 18. Konvergenzmessungen am 27. April 1976
Convergence measurements from April 27, 1976
Mesures de convergence du 27 avril 1976

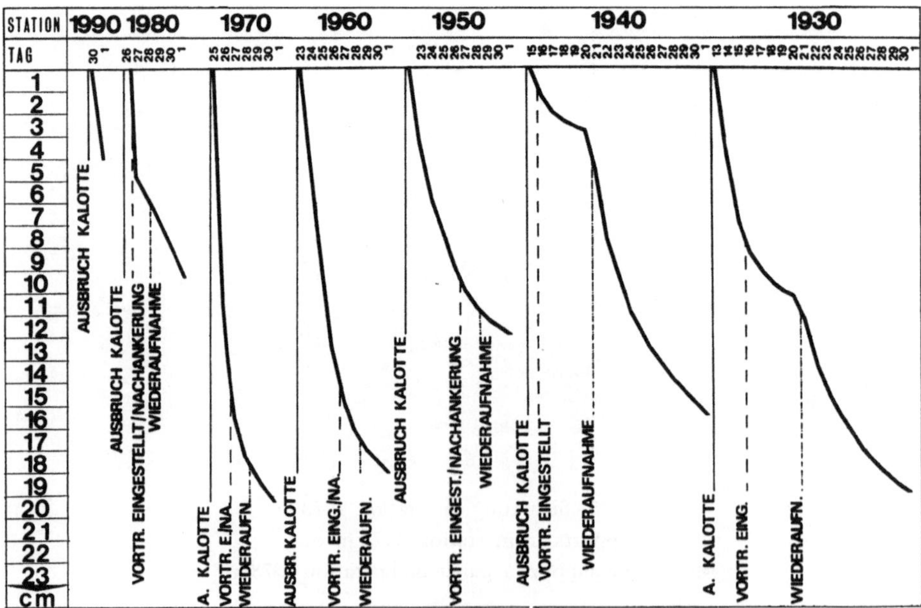

Abb. 19. Konvergenzmessungen am 1. Mai 1976
Convergence measurements from May 1, 1976
Mesures de convergence du 1 mai 1976

sich am 1. 5. 1976 (Abb. 19) eine Verminderung der Verformung sowohl bei Station 1980 m als auch bei Station 1970 m. Die Ersttagsverformung bei Station 1990 m betrug aber immer noch 4 cm, weshalb mit den Sicherungsmaßnahmen noch nicht zurückgegangen werden konnte.

Erst die Verformungen am 4. 5. 1976 (Abb. 20) zeigten, daß die schlechtesten Gebirgsverhältnisse überwunden waren. Die Ersttagsverformungen

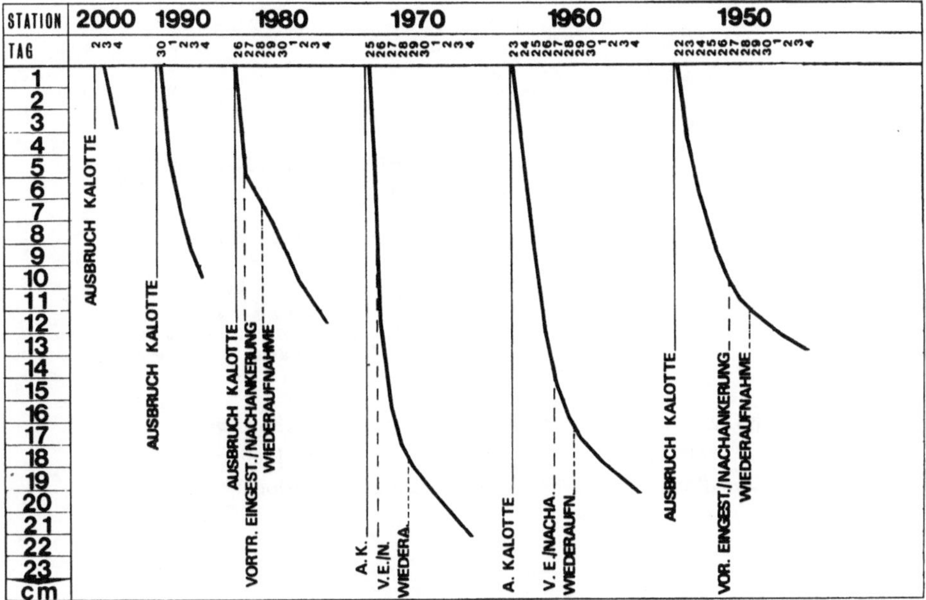

Abb. 20. Konvergenzmessungen am 4. Mai 1976
Convergence measurements from May 4, 1976
Mesures de convergence du 4 mai 1976

betrugen nur mehr 2,7 cm, so daß wieder eine schrittweise Verminderung der Sicherungsmaßnahmen erfolgen konnte.

Wie weitgehend die Anpassung der Sicherungsmaßnahmen an die Gebirgsverhältnisse war, soll Abb. 22 zeigen. Die Gebirgsverformungen, Verformungsgeschwindigkeit und Stützmaßnahmen als lfm Anker je lfm Tunnel sind hier gegenübergestellt.

Die geschilderte Methode hat sich im großen und ganzen recht gut bewährt, mußte aber fallweise etwas angepaßt werden. Da man nicht nur Erfolge bekanntgeben soll und Erkenntnisse aus Fehlschlägen wesentlich zu weiteren Entwicklungen beitragen, wird hier noch kurz auf den zwischen Station 440 m und 470 m an der Westseite aufgetretenen Scherbruch eingegangen.

In Abb. 23 ist die Verformungskurve bei Station 461 m, also im mittleren Bereich des Scherbruches, angegeben. Die Anfangsverformung war nicht groß und gab keinen Anlaß zu Besorgnis. Die anfängliche Beruhigung hielt

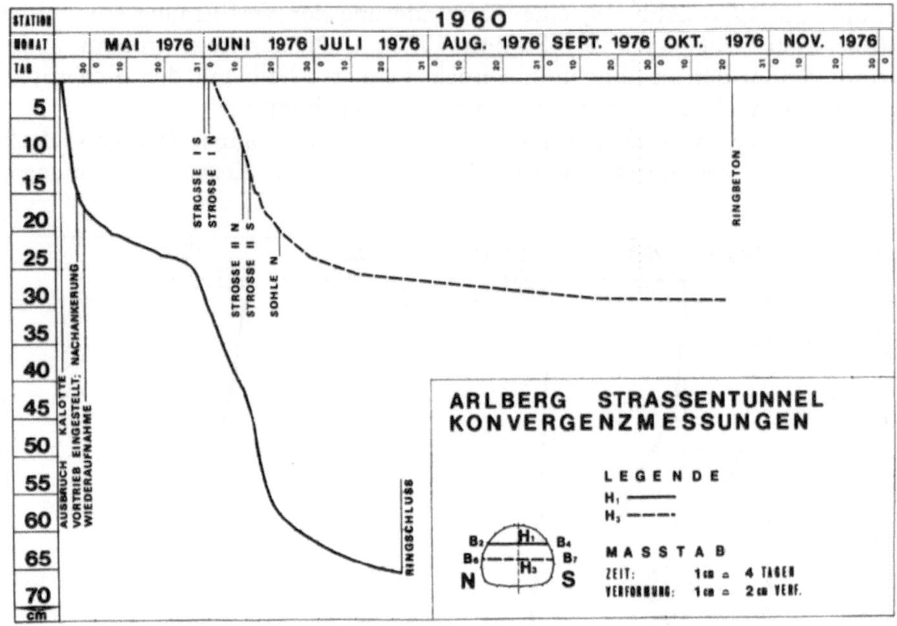

Abb. 21. Westseite — Konvergenzkurven bei Station 1960 m
West side — convergence curves at station 1960 m
Côté d'ouest — courbes de convergence à la station 1960 m

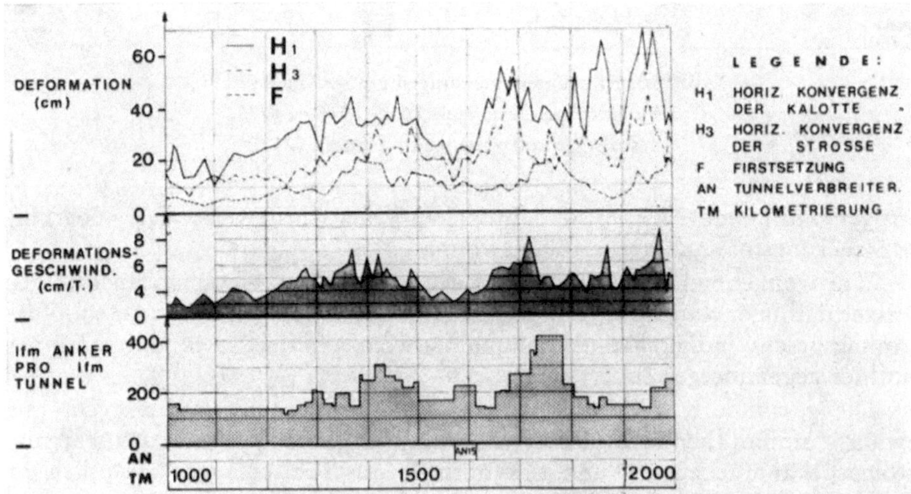

Abb. 22. Westseite — Gegenüberstellung der Verformung, der Verformungsgeschwindigkeit und der angewendeten Ankerung von Station 1000 m bis 2000 m
West side — comparison of deformation, deformation velocity and anchors used from station 1000 m to station 2000 m
Côté d'ouest — Comparaison de la déformation, la vélocité de déformation et des ancres utilisées de la station 1000 m à la station 2000 m

aber nicht an, sondern die Konvergenzgeschwindigkeit nahm zu. Die Sicherung erfolgte in diesem Bereich mit 6 m langen Ankern; eine Nachankerung

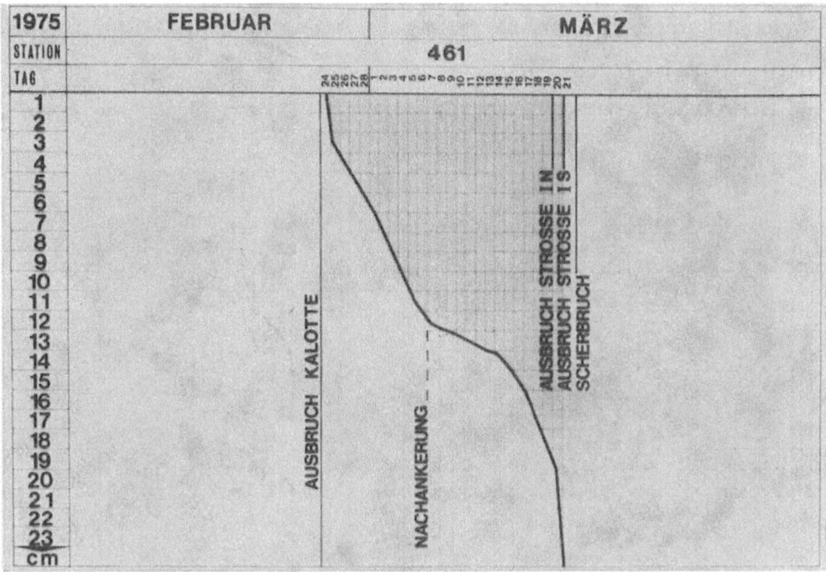

Abb. 23. Westseite — Konvergenzkurve bei Station 461 m
West side — convergence curve at station 461 m
Côté d'ouest — courbe de convergence à la station 461 m

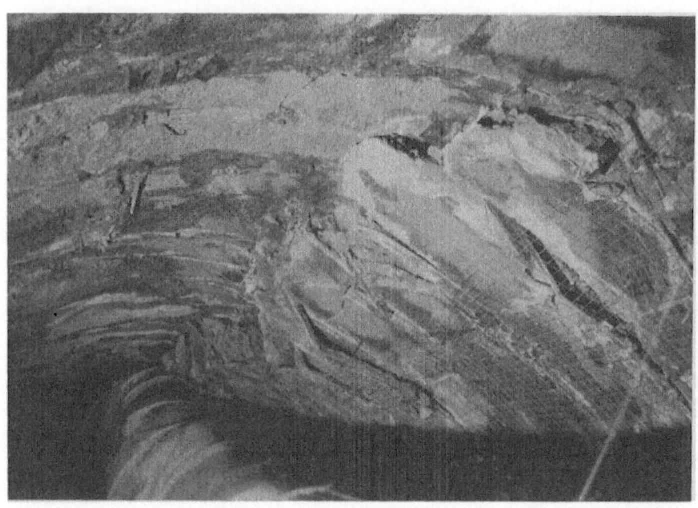

Abb. 24. Westseite — beginnender Scherbruch zwischen Station 440 m und 470 m
West side — start of shear failure between stations 440 m and 470 m
Côté d'ouest — début d'une rupture par cisaillement entre les stations 440 m et 470 m

mit 9 m langen Ankern wurde eingeleitet. Die darauf zurückzuführende Verflachung der Verformungskurve ist deutlich zu sehen.

Das Nachnehmen der Strosse I, die jeweils halbseitig vorgetrieben wurde und auf die in relativ kurzem Abstand der Ausbruch der Strosse II erfolgte, führte zum Auftreten eines Scherrisses in der Firste. In einigen Stunden

Abb. 25. Westseite — Scherbruch zwischen Station 440 m und 470 m
West side — shear failure between stations 440 m and 470 m
Côté d'ouest — rupture par cisaillement entre les stations 440 m et 470 m

Abb. 26. Aufarbeiten des Scherbruches zwischen Station 440 m und 470 m
Repair of the shear failure between stations 440 m and 470 m
Réparation d'une rupture par cisaillement entre les stations 440 m et 470 m

öffnete sich der Riß (Abb. 24) und in kurzer Zeit wurde die komplette rechte Hälfte des Tunnelprofils ca. 2 m weit hereingedrückt (Abb. 25). Der erwartete Einsturz des Tunnels trat nicht ein, vielmehr stellte sich ein neuer stabiler Gleichgewichtszustand ein. Als erste Maßnahme wurden Holzstempel unter den gefährdeten Bereich eingebracht. Die Aufarbeitung der eingedrückten Strecke erfolgte zuerst von der Portal- und dann auch von der Vortriebsseite her. Bei der neuen Sicherung der Verbruchstrecke selbst und der angrenzenden Bereiche kamen erstmalig 12 m lange Anker zum Einsatz (Abb. 26). Nach etwa 3 Wochen konnte die Überfirstung soweit abgeschlossen werden, daß der Vortrieb wieder aufgenommen werden konnte.

Anschrift des Verfassers: Dipl.-Ing. Günther Judtmann, Ingenieurgemeinschaft Lässer-Feizlmayr, Framsweg 16, A-6020 Innsbruck, Österreich.

Rock Mechanics, Suppl. 7, 179—187 (1978)

Rock Mechanics
Felsmechanik
Mécanique des Roches

The Submarine Discharge Tunnel at Forsmark

By

J. Martna

With 5 Figures

Summary — Zusammenfassung — Résumé

The Submarine Discharge Tunnel at Forsmark. In the years 1974—1976, a submarine discharge tunnel was excavated for the cooling-water from units 1 and 2 of the Forsmark nuclear-power plant, situated on the Swedish east coast north-east of Stockholm. The cross-sectional area of the tunnel is about 80 m² and the rock cover 55—60 m. The studies of the rocks made before and during the tunnelling are briefly discussed. Some characteristic properties of the rock mass are given.

Der unterseeische Auslauftunnel in Forsmark. In den Jahren 1974—1976 wurde für das Kühlwasser von Block 1 und Block 2 im Kernkraftwerk Forsmark an der schwedischen Ostküste nordöstlich von Stockholm ein Auslauftunnel unter dem Meer gesprengt. Die Querschnittsfläche des Tunnels beträgt etwa 80 m² und die Mächtigkeit der Felsüberlagerung 55—60 m. Felsuntersuchungen vor und bei dem Tunnelvortrieb werden in Kürze beschrieben. Einige der charakteristischen Eigenschaften der Felsmasse werden angegeben.

Le tunnel d'évacuation sous mer à Forsmark. Au cours des années 1974—1976, un tunnel d'évacuation des eaux réfrigération a été percé à partir des blocs 1 et 2 de la centrale nucléaire de Forsmark située sur la côte orientale de la Suède au Nord-Est de Stockholm. La section du tunnel est d'environ 80 m² et l'épaisseur de la roche au-dessus du tunnel est de 55 à 60 mètres. Les prospections du terrain avant et pendant le percement sont brièvement décrites. Quelques unes des caractéristiques de la masse rocheuse sont indiquées.

Submarine discharge tunnels are rather uncommon in the power industry. However, in connection with the construction of the Forsmark nuclear-power plant, the Swedish State Power Board has gained some experience of submarine tunnelling in pre-Cambrian rocks. In the following pages, a brief presentation of the submarine part of the Forsmark tunnel, from the surge basin to Loven Island, will be given.

The Forsmark nuclear-power plant is under construction in northern Uppland on the Swedish east coast, about 150 km north-east of Stockholm. It is being built as a joint venture between the Swedish State and private interests. The Swedish State Power Board has been commissioned to build and operate the power plant.

The cooling-water from units 1 and 2 will be led out through a tunnel to an artificial "biotest lake", separated by dams from the sea, and from there to the sea in the Gulf of Bothnia (Figs. 1 and 2). The purpose of the "biotest lake" is to study whether and how the fauna and the flora of the sea will be influenced by the cooling-water, which will have a temperature about 10^0 C above that of the sea-water.

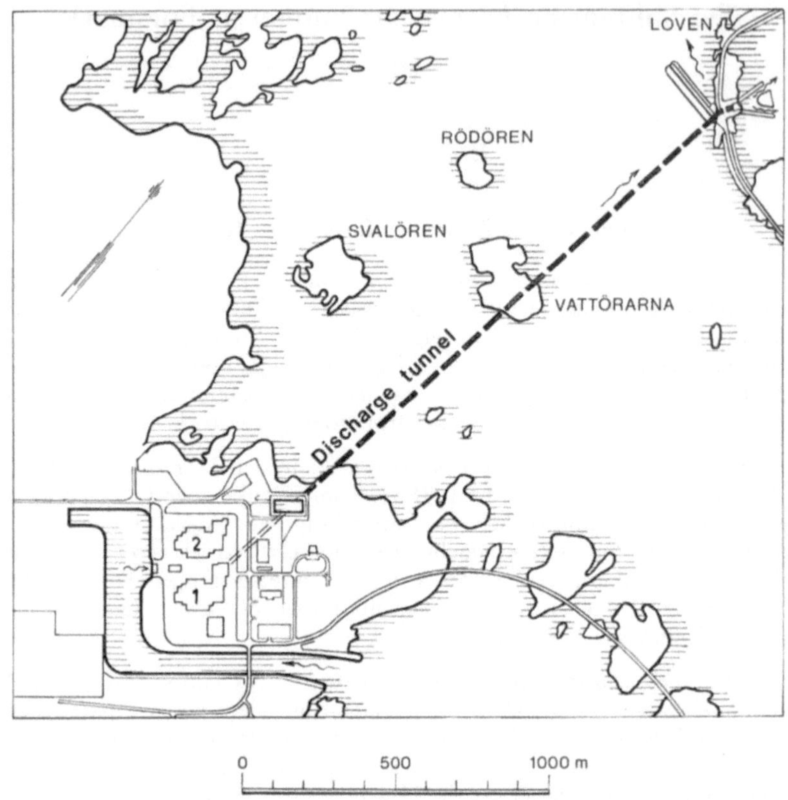

Fig. 1. Map of the Forsmark area with the tunnel line (after Carlsson & Olsson 1977 a)
Karte des Forsmarkbereiches mit Tunnelstrecke (nach Carlsson & Olsson 1977 a)
Carte de la région de Forsmark avec tracé du tunnel (selon Carlsson et Olsson 1977 a)

The total length of the tunnel is about 2300 m and, as Figs. 1 and 3 show, it runs for most of its length, or about 1900 m, underneath the sea. The cross-sectional area is about 80 m^2. The rock cover is 55—60 m and the depth below sea-level approximately 75 m (Fig. 3). In 1977, after two years of tunnelling, the tunnel was completed and filled with water.

The geological and geophysical exploration for the tunnel construction (Larsson 1973, Moberg 1974) included a geological field survey and seismic surveying along the planned tunnel line. Also, nine core drillings were made in sections with low seismic velocities.

The final tunnel line and the depth of the tunnel below the rock surface were determined on the basis of these investigations. The seismic operations and most of the drillings were carried out in winter from the sea ice.

Fig. 2. View from unit 1 along the tunnel line. Photograph by the writer
Aussicht von Block 1 entlang der Tunnelstrecke. Photo des Verfassers
Vue du bloc 1 le long du tracé du tunnel. Photo de l'auteur

The main part of the Forsmark tunnel rocks are made up of old Svecofennian schistose intrusive rocks of tonalitic composition, so-called gneiss granites, as well as mica gneisses and mica schists. The rock traditionally called gneiss granite is from an engineering-geological point of view an orthogneiss, and is so termed here. The mica gneisses and mica schists are collectively termed paragneiss. Dikes of amphibolitic composition in the orthogneiss, as well as dikes and basic layers in the paragneiss are quite frequent, and so are also dikes and small massifs of pegmatite.

Through late tectonic movements, breccias and mylonitic rocks have been developed across a 200-m-wide section of the tunnel (zone G, Fig. 3). In this brecciated orthogneiss there are several clay-filled joints with a maximum width of about 50 cm. The material has been identified by X-ray diffraction as being mainly illite and quartz. Numerous veins of quartz and calcite and large druse cavities with crystallizations of calcite, aragonite and pyrite, are to be found. In these druse cavities, mineral pitch is common.

The tunnel work was started from the two end-points, from the surge basin on the land side and from Loven Island which was made accessible by a causeway. However, the tunnelling operations from the island ran into difficulties in the zones of weakness met with, were slowed down and were eventually stopped. Thus, the excavation work was for the most part done from the access on the land side.

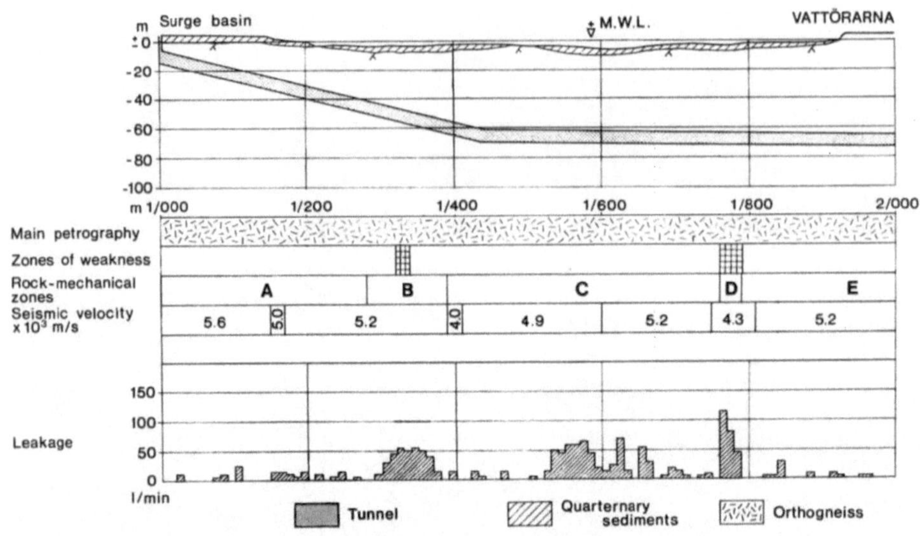

Fig. 3 a

Fig. 3. Longitudinal section of the tunnel.
Längensektion des Tunnels. Geändert
Section longitudinale du tunnel — modifieé

An engineering-geological survey of the tunnel was made during the excavation (Carlsson & Olsson 1976 a). This survey covered a number of characteristic properties of the rock mass and a summary of it is given in Table 1.

The rock-mechanical zones were classified on the basis of these rock properties. As is evident from Fig. 3, some of the rock-mechanical zones include zones of weakness, the existence of which was judged from actual or possible rock falls or an excessive comminution of the rock. The zones of weakness normally require some preliminary support during the advance of the tunnel. In the design of the final supports, however, the possible deterioration of the rock outside these zones must also be considered. Thus, the necessary supports were considered to be a consequence and not a part of the rock conditions and were not included in the classification of the rock-mechanical zones. It is interesting to note, from a fundamental point of view, that the rock conditions change continuously throughout the tunnel and that the same combination of rock properties does not occur twice.

A hydrogeological study was also made in the tunnel during the excavation work (Carlsson & Olsson 1977 a). In this study, the dependence of the water leakage on the predominant geological and tectonic conditions, the different water storages in the rock mass, their mutual relationships and their communications with the sea-water is discussed. With the aid of water-pressure tests made in the bore holes during the geological site ex-

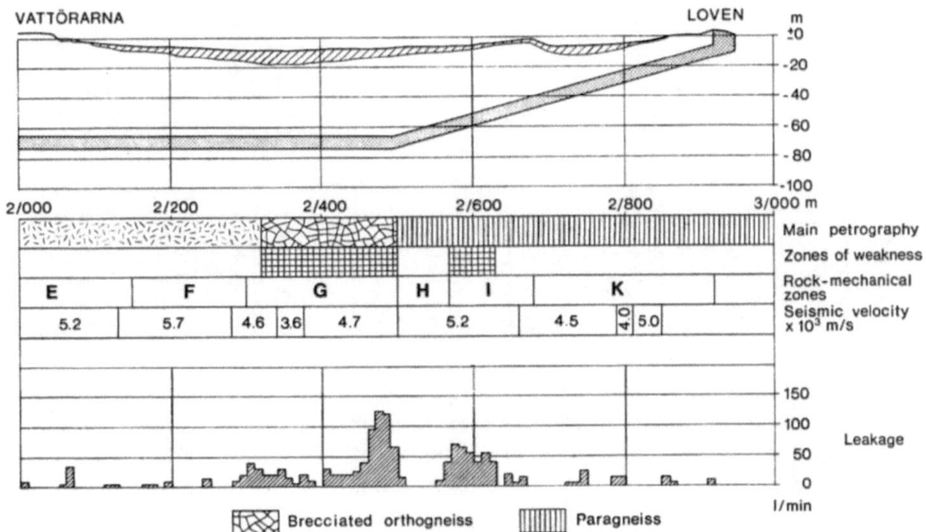

Fig. 3 b

Modified after Carlsson & Olsson 1977 a
nach Carlsson & Olsson 1977 a
d'après Carlsson et Olsson 1977 a

ploration (Moberg 1974), the hydraulic conductivity of the rock mass was calculated by a method described by Carlsson & Olsson (1976b, 1977b and 1977c).

Other investigations during the tunnelling included rock-stress measurements (Hiltscher 1976) and an examination of the possibilities of a photo-documentation of the geology of the tunnel (Carlsson, Martna & Olsson, in press). The result of the rock-stress measurements is shown in Fig. 4.

The tunnel intersects four distinct zones of weak rock (zones B, D, G and I) with widths varying between 30 m and 200 m (Fig. 3, Table 1). Partly on the basis of exploratory percussion drillings from the tunnel face, pre-grouting was carried out in three of them (zones B, G and I), in order to avoid excessive water leakage and to diminish the risk of possible cave-ins.

Two of the four zones (D and G) were approximately vertical faults, with crushed rock extending up to the sea bottom. The other two zones (B and I) were inclined and the tunnelling problems were chiefly caused by a gradual overbreak, mainly due to an unfavourable orientation of open joints.

The supports necessary for the secure advance of the tunnelling operations consisted of pre-grouting, rockbolting and shotcrete. In addition, about

Table 1. Some Characteristic Properties of the Tunnel Rock and the Rock-Supporting Measures Taken During the Advance of the Tunnel

Einige der charakteristischen Eigenschaften des Tunnelfelsens und Felsverstärkungsmaßnahmen während des Tunnelvortriebes

Quelques caractéristiques de la masse rocheuse et les mesures de renforcement prises pendant le percement du tunnel

Rock-mechanical zone	Main petrography and rock structure	Dominant joints	Joint fillings	Estimated RQD	Largest rock stress in horizontal plane	Preliminary supports	Pre-grouting	Estimated water inflow, l/min per m tunnel
A	Orthogneiss, massive	A) N 80^0 W, 80^0 S B) Horizontal		75		Occasional rock bolts		<1
B	Orthogneiss, platy, plate thickness <20 cm	A) E-W, 30^0 S B) Horizontal	Mapping incomplete	45		Rock bolts, shotcrete, reinforced-shotcrete arches	44 holes, 481 m, 53 t cement	3
C	Orthogneiss, massive	A) N 80^0 W, 80^0 S B) E-W, 10^0 S	Sandy material, amphibolite	100	15.1 MPa N 24^0 W	Selective rock bolts		2
D	Orthogneiss, blocky, block size <50 cm	A) N 80^0 W, 70^0 S B) N. 80^0 E, 50^0 W C) Horizontal	Sandy material, amphibolite, non-swelling clay	25		Rock bolts, shotcrete		8
E	Orthogneiss, massive	A) N 80^0 W, 70^0 S B) Horizontal	Sandy material, chlorite, amphibolite	80	12.9 MPa N 60^0 W	Occasional rock bolts		<1
F	Orthogneiss, massive	A) N 70^0 W, 80^{0}S B) N 80^0 E, 40^0 S C) Brecciation, mostly healed	Sandy material, chlorite, calcite, quartz	80	13.7 MPa N 45^0 W	Occasional rock bolts		<1
G	Orthogneiss, brecciated	A) Brecciation B) N 70^0 W, 80^0 S C) N 80^0 E, 40^0 S	Sandy material, chlorite, calcite, quartz, non-swelling clay	30		Rock bolts, shotcrete, reinforced-shotcrete arches	205 holes, 2171 m, 23 t cement	4
H	Paragneiss, platy, plate thickness >20 cm	A) E-W, 70^0 S	Chlorite	95	9.0 MPa N 24^0 W	Occasional rock bolts		1
I	Paragneiss, blocky, block size <50 cm	A) E-W, 60^0 S B) N 20^0 E, 80^0 N	Sandy material, chlorite, non-swelling clay	30		Rock bolts, shotcrete, reinforced-shotcrete arches	140 holes, 1412 m, 46 t cement	4
K	Paragneiss, platy, plate thickness >20 cm	A) N 45^0 W, 70^0 S	Chlorite	100		Selective rock bolts		<1

1.5 m wide and 20 or 30 cm thick reinforced-shotcrete arches were used in the zones of weak rock. In all, 33 such reinforced-shotcrete arches were erected during the excavation of the tunnel.

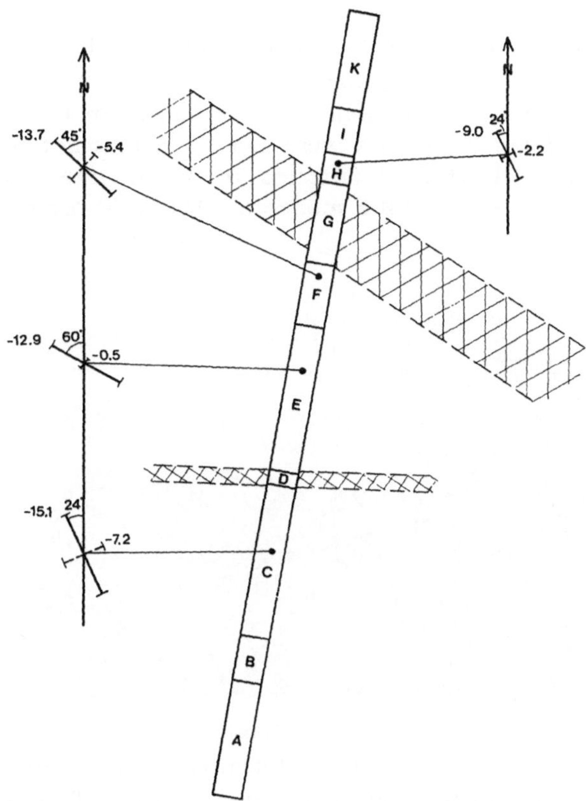

Fig. 4. Horizontal initial stresses of the rock in MPa, according to Hiltscher (1976), and their relationships to the rock-mechanical zones and the two vertical faults (after Carlsson & Olsson 1977 a)

Bezugsspannungen des Felsens (in MPa) in der Horizontalebene, nach Hiltscher (1976), im Verhältnis zu den felsmechanischen Zonen und den beiden vertikalen Verwerfungen (nach Carlsson & Olsson 1977 a)

Contraintes initiales de la roche (en MPa) dans le plan horizontal, selon Hiltscher (1976), par rapport aux zones "mécanique des roches" et des deux failles verticales (selon Carlsson et Olsson 1977 a)

In accordance with the practice of the Power Board, the final decisions on permanent supports were made after the excavation, on the basis of the studies made during the tunnelling and the inspections of the tunnel. Because of the desired high degree of security against cave-ins and rock falls, the design of the supports was very conservative in comparison with those in the headrace and tailrace tunnels of water-power plants.

The permanent supports consisted of grouted rock bolts, which were used both systematically and selectively, and of shotcrete, mainly reinforced and with thicknesses between 5 and 30 cm (Fig. 5). Except for the portals and a gate, no cast-in-place concrete was used in the submarine part of the tunnel discussed here.

Totally, shotcrete was applied in 43 per cent of the tunnel length including the parts where only the roof was shotcreted. No shotcrete was used in the zones C, E and F. The final number of the 20 or 30 cm thick shotcrete arches was 48, and they covered about 4 per cent of the tunnel length.

Although the zones of weakness caused delays in the advance of the tunnel, no exceptional difficulties were met with. It is noteworthy that the water leakage into the tunnel (affected to an unknown extent by pregrouting) turned out to be quite moderate. About 60 per cent of the total

Fig. 5. Part of the completed tunnel, with shotcrete lining. Photograph by G. Hansson

Eine Partie des fertigen Tunnels, mit Spritzbeton verkleidet. Foto von G. Hansson

Une partie du tunnel terminé avec son revêtement de gunite. Photo G. Hansson

water leakage (3000 l/min) into the tunnel came from the above-mentioned four zones of weakness. A comparison between the result of the geological and geophysical exploration and the result of the geological survey in the tunnel showed a fairly good correspondence.

The submarine tunnel was designed and the project was administered by the Power Board. Hagconsult AB was the consultant on permanent supports. The contractor for the tunnelling was AB Forsmarkstunnlar, a joint venture between AB Armerad Betong and AB Vägförbättringar.

References

Carlsson, A., Olsson, T.: Forsmark kraftstation. Geologisk kartering, kylvattentunnel. Internal report, Statens Vattenfallsverk. Stockholm 1976a.

Carlsson, A., Olsson, T.: Bestämning av berggrundens permeabilitet genom vattenförlustmätning. Vannet i Norden, Oslo 9/3, 29—35 (1976b).

Carlsson, A., Olsson, T.: Water leakage in the Forsmark Tunnel, Uppland, Sweden. Sveriges Geol. Unders. C 734, 45 pp. Stockholm 1977a.

Carlsson, A., Olsson, T.: Hydraulic properties of Swedish crystalline rocks. Hydraulic conductivity and its relation to depth. Bull. Geol. Inst. Univ. Uppsala, N. S., *7*, 71—84 (1977b).

Carlsson, A., Olsson, T.: Variations of hydraulic conductivity in some Swedish rock types. — Proceedings of Rockstore — 77. The first International Symposium on storage in excavated rock caverns, Stockholm, *2*, 85—91 (1977c).

Carlsson, A., Martna, J., Olsson, T.: Photodocumentation in the Forsmark Tunnel, Uppland, Sweden. Geol. Fören. Stockholm Förh., Vol. *100* (in press).

Hiltscher, R.: Forsmark kraftstation, block 1 och 2. Mätning av bergets initialspänningar i avloppstunneln. Internal report, Statens Vattenfallsverk. Stockholm 1976.

Larsson, W.: Forsmark kraftstation, aggr. 1 och 2. Avloppstunneln: Berggeologiska förhållanden efter tunnellinjen. Internal report, Statens Vattenfallsverk. Stockholm 1973.

Moberg, M.: Forsmark kraftstation, aggr. 1 och 2. Avloppstunneln: Grundundersökningar 1971—1973. Internal report, Statens Vattenfallsverk. Stockholm 1974.

Address of author: Dr. Jüri Martna, Chief Engineering Geologist. Swedish State Power Board, Civil Engineering, Development, S-162 87 Vällingby, Stockholm, Sweden.

Rock Mechanics, Suppl. 7, 189—201 (1978)

Rock Mechanics
Felsmechanik
Mécanique des Roches

Mechanische Auffahrung des Triebwasserstollens beim Kraftwerk Langenegg der Vorarlberger Kraftwerke AG

Von

E. Schneider

Mit 12 Abbildungen

Zusammenfassung — Summary

Mechanische Auffahrung des Triebwasserstollens beim Kraftwerk Langenegg der Vorarlberger Kraftwerke AG. Der 5,5 km lange Rotenbergstollen im Bregenzer Wald (Vorarlberg) wurde von Dezember 1975 bis März 1977 mit einer Robbins-Tunnelbohrmaschine, Durchmesser 3,90 m, aufgefahren. Bauherr der Wasserkraftanlage Langenegg ist die Vorarlberger Kraftwerke AG in Bregenz. Projektierung und Bauleitung liegen in den Händen der Vorarlberger Illwerke AG.

Die Arbeiten an der Triebwasserführung, die neben dem Rotenbergstollen auch einen 270 m langen Schrägschacht umfaßt, wurden einer Arbeitsgemeinschaft die aus den Firmen Jäger — Schruns, Rothpletz — Aarau, und Zschokke — Zürich, besteht, übertragen.

Bereits im frühesten Planungsstadium wurde mit Hinblick auf die günstigen geologischen Verhältnisse die Auffahrung des Stollens mit einer Vollschnittmaschine ins Auge gefaßt. Die Ausschreibung enthielt aber auch eine Variante mit konventionellem Vortrieb. Da der gefräste Kreisquerschnitt bei einem hochbeanspruchten Druckstollen erhebliche Vorteile bietet, wurde schließlich diese Variante in Auftrag gegeben.

Mechanical Heading of the Penstock Pipe Tunnel at the Langenegg Power Plant of the Vorarlberger Kraftwerke AG. Between December 1975 and March 1977 the 5,5 km long Rotenberg tunnel in the Bregenzer Wald (Vorarlberg) was driven with a Robbins tunneling machine with a diameter of 3,90 m. This tunnel constitutes the core of the penstock of the Langenegg power plant.

The tunnel lies in the subalpine Molasse zone. Approx. 1150 m of the route run through the so-called building stone zone, which consists of highly abrasive and impact-resistant sandstone beds, the remainder of the tunnel runs through alternating sequences of marly sandstone, calcareous sandstone and conglomerate, which — with the exception of a big conglomerate bed — did not constitute any great obstacle to boring. Other influences, such as cracks and fissure water, were of minor importance.

The Robbins machine, Model 124-134, is equipped for this diameter with 26 disccutters and has a maximum driving force of 320 t. The ancillary installations consist of a working platform for shotcrete lining located between the tunneling machine and the conveyor belt, the placing equipment for the bottom segments, the conveyor belt, and a California switch.

Both because of the geological conditions and the future stresses and strains acting on the ground as a result of the high internal pressure in the tunnel, the safety measures had to be selected with the utmost care and consideration. By using well-proven methods (steel arches, roof bolts, shotcrete, and precast reinforced concrete elements) in a new combination it was possible to meet the task and to achieve an optimal support and conservation of the ground. The facilities were designed in such a way that all safety measures could be carried out without any great impairment of the boring operation.

Despite a number of difficulties and problems the use of the tunneling machine for the Rotenberg tunnel was altogether successful. The heading operation could be completed two months ahead of schedule.

Geologische Verhältnisse

Der Stollen liegt zur Gänze im Bereich der subalpinen Molasse. Die erst im Zuge der Auffahrung endgültig festgelegte Trasse führt auf ca. 1150 m durch die sogenannte Bausteinzone, die aus äußerst abrasiven und

Abb. 1. Schutterzug am Stollenportal
Mucking train

zähen Sandsteinbänken mit dünnen Mergelzwischenlagen besteht. Die einachsige Druckfestigkeit dieses Gesteins liegt zwischen 1000 und 1600 kp/cm^2. Der Maximalwert wurde mit 2178 kp/cm^2 ermittelt, das arithmetische Mittel aus allen Proben beträgt rd. 1400 kp/cm^2.

Die übrige Stollenstrecke führt durch eine Wechsellagerung von Mergel und mergeligem Sandstein sowie einzelnen Sandstein- und Konglomerat-bänken geringer Mächtigkeit. Diese Gesteine gehören den sogenannten Weißachschichten an und boten mit Ausnahme einer mächtigen Konglomeratbank dem Fräsvortrieb keinen großen Widerstand.

Abb. 2. Vortriebsmaschine vor dem Portal
Robbins tunneling-machine

In Tabelle 1, 2a und 2b sind die wichtigsten Gesteinskennwerte des Rotenbergstollens im Vergleich mit den Werten anderer bisher in Österreich gefräster Stollen aufgeführt. Die auffallend hohe Spaltzugfestigkeit des Sandsteins ist ein wichtiger Hinweis für die schlechte Bohrbarkeit dieses Gesteins.

Tabelle 1. Gesteinskennwerte
Rock Characteristics

Projekt	Gestein	mittlere einachsige Druckfestigkeit kp/cm2	Spaltzug-festigkeit kg/cm2	Quarz-gehalt %	Nettobohrge-schwindigkeit cm/min.
ROTENBERG-STOLLEN	Sandstein Konglomerat Mergel	1.4oo max. 2.178 1.25o 5oo	18o 8o 9o	3o,5 - -	1,7 2,o 5,9
RICHTSTOLLEN PFÄNDER	Sandstein Mergel	68o 53o	38 54	nicht gemessen	3,4 - 4,o
GLETSCHERSCHIENEN-BAHN KAPRUN	Kalkglimmerschiefer Prasinit mit Epidat 7o %	8oo - 1.8oo 84o - 1.ooo	85 3oo	nicht gemessen	2,8 o,4
HIRZBACHSTOLLEN	Kalkglimmerschiefer	595	7o - 12o	1o,o	1,3 - 3,5
ROTLECHSTOLLEN	Hauptdolomit	1.8oo	nicht gemessen	nicht gemessen	2,5
BEILEITUNG SÖLK	Granatglimmer-schiefer ~9o %	4oo - 1.3oo	5o,4	5o,o	o,9 - 2,5
VOMPERBACH STOLLEN	Hauptdolomit Jura (stark mylonitisiertes	1.o9o wasserführendes Trümmergebirge)	6o	9,4	1,2 - 3,5
ABWASSERSTOLLEN FELDKIRCH	Drusbergmergel Schrattenkalk	3oo 1.4oo	nicht gemessen	nicht gemessen	o,6 - 1,2 1,8 - 2,5
ÖSTERREICHER-STOLLEN	Kalk Flysch	6oo - 1.6oo	nicht gemessen	-,- 7o,-	o,5 - 2,5 -

Tabelle 2 a, b. Übersicht über
Survey of Mechanically

Projekt	Bauherr	Unternehmung
Rotenbergstollen	Vorarlberger Kraftwerke AG, Bregenz	Jäger - Rothpletz - Zschokke
Richtstollen Pfänder	Republik Österreich	Baresel - Moosbrugger
Gletscherschienenbahn Kaprun	Tauernkraftwerke AG	Beton- u.Monierbau, Schachtbau Thyssen
Hirzbach-Beileitungsstollen	Tauernkraftwerke AG	G.Hinteregger & Söhne - Murer
Rotlech-Stollen	E-Werke Reutte	Rella
Beileitung Sölk	Steweag	Universale - Stuag
Vomperbach-Stollen	Stadtwerke Schwaz	Montana
Abwasserstollen Feldkirch	Abwasserverband Region Feldkirch	G.Hinteregger, Bregenz
Österreicher-Stollen	Magistrat der Stadt Wien MA 31 - Wasserwerke	Arge Schneealpenstollen-Nord Rella-P.Auteried-Porr-Universale

Sonstige Einflüsse wie Klüftung und Bergwasserandrang waren von untergeordneter Bedeutung. Von erheblichem Einfluß auf die Bohrbarkeit

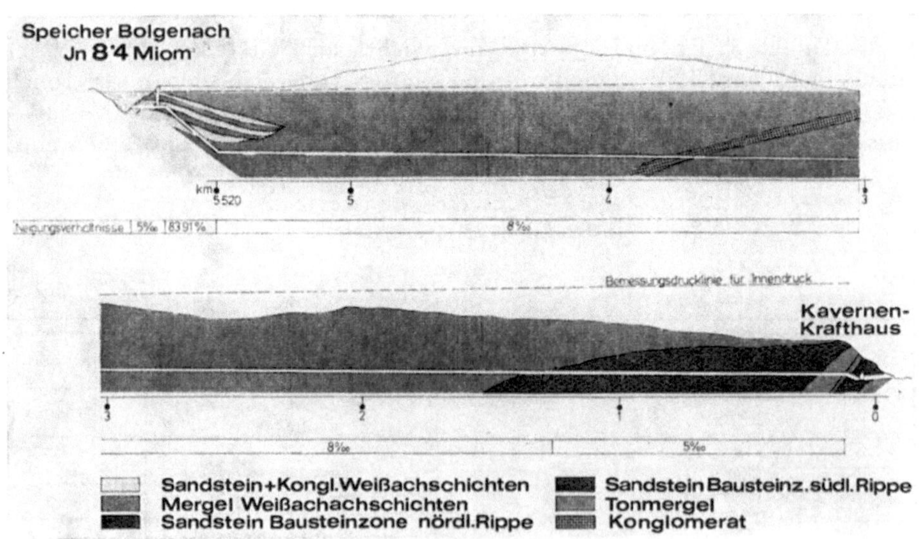

Abb. 3. Baugeologischer Längenschnitt
Geological cross-section

und auch auf die Standfestigkeit des Stollens ist jedoch die Lagerung des Gebirges. Beim vorliegenden Projekt streichen die Schichten annähernd parallel zur Stollenachse bei gleichzeitig steilem Einfallen.

gefräste Stollen in Österreich
Driven Tunnels in Austria

Stollen		Vortriebsmaschine		Bauzeit
Länge	Durchmesser	Type	Maschinenbesitzer	Kalenderjahr
5.5oo m	3,9o m	Robbins, Mod. 124-134	Jäger - Rothpletz - Zschokke	1975 - 1977
6.74o m	3,55 m	Robbins, Mod. 123-133 Wirth TB II	Theiler u. Kalbermatter Murer	1974 - 1975
3.25o m	3,6o m	Wirth TB II - 36o H	B & M - Thyssen	1972 - 1973
4.97o m	2,4o m	Wirth TB I - 24o E	Murer	1971 - 1973
4.656 m	3,4o m	Wirth TB II - 346 H	Kopp	1975 - 1976
3.4oo m	3,98 m	Robbins, Mod. 134-153	Schachtbau Thyssen	1976 - 1977
2.386 m	3,8o m	HRT Ingersoll-Rand (Lawrence) oo8 R	Pitsch	1975 - 1977
7oo m	2,56 m	Robbins, Mod. 81-	Prader	1976
65o m	2,9o m	Robbins, Mod. 81-113	Jäger - Hinteregger	197o

Hohe einachsige Druckfestigkeit, Zähigkeit und ungünstige Lagerung des Gebirges bewirkten, daß die eingesetzte Vortriebsmaschine im Sandstein

Abb. 4. Vortriebsmaschine im Stollen
Tunnel-borer in action

und Konglomerat an den Grenzen ihrer Leistungsfähigkeit anlangte und die Kosten für Meißel- und Maschinenersatzteile überproportional anstiegen.

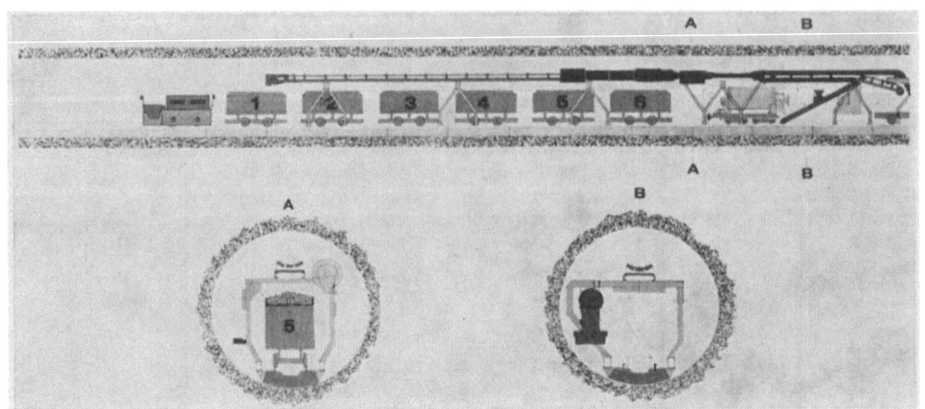

Abb. 5 a, b. Vortriebsmaschine
Tunnel-borer and

Beschreibung des Vortriebsystems

Vortriebsmaschine und Nachgeschaltete Betriebe

Die Robbins-Vortriebsmaschine, Modell 124-134, ist für den Bohrdurchmesser von 3,90 m mit 26 Rollmeißeln bestückt und verfügt über einen maximalen Andruck von 320 to, somit rd. 12 to pro Meißel.

Die Nachgeschalteten Betriebe bestehen aus einer Arbeitsbühne für Spritzbetoneinbau, die unmittelbar anschließend an die Vortriebsmaschine angeordnet ist, der Verlegeeinrichtung für die Sohltübbinge, dem Beladeband und einer nachgeführten California-Weiche für den Zugwechsel. Auf der Arbeitsbühne befinden sich außerdem die Hochspannungskabeltrommel und ein Ankerbohrgerät. Die Entstaubungsanlage (Hölter-Rotovent) ist auf dem Beladeband montiert. Die ganze Einrichtung mußte kurvengängig konstruiert werden, da mehrere Kurven mit einem Radius von 150—200 m geplant waren.

Das Gesamtgewicht der Installationen vor Ort beträgt ca. 130 to.

Der Abtransportes des Ausbruchsmaterial erfolgt in Mühlhäuser-Kippern mit 4 m³ Inhalt. Wegen der beschränkten Platzverhältnisse auf der California-Weiche konnten nur 70 PS-Akkuloks verwendet werden. Trotz der Steigung von 8 ‰ war es möglich, mit einem Zug von sieben Einheiten das Ausbruchmaterial für einen ganzen Hub abzufördern. Der Nachschub an Ausbaumaterial, Spritzbetonmischgut, Sohltübbingen, Injektionsmörtel und Gleismaterial erfolgte nach Bedarf mit den Schutterzügen.

Sicherungs- und Stützmaßnahmen

Die Sicherungsmaßnahmen mußten sowohl wegen der geologischen Verhältnisse als auch wegen der späteren Beanspruchung des Gebirges durch den hohen Innendruck von über 30 atü mit großer Überlegung und Sorgfalt gewählt werden.

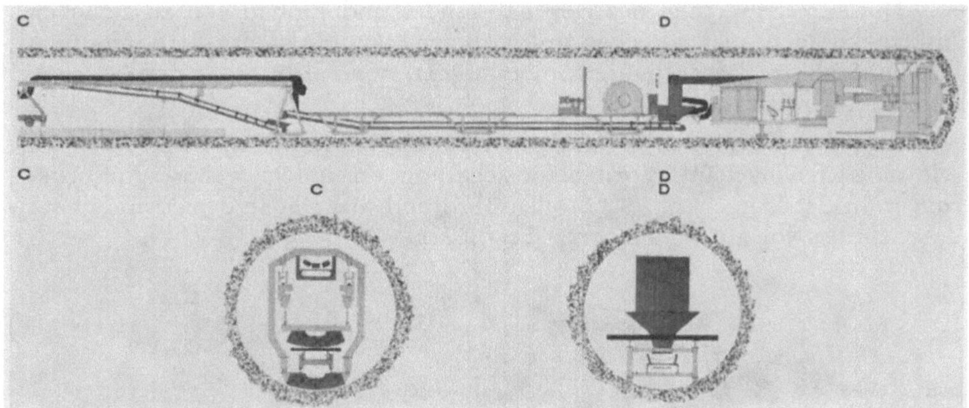

und nachgeschaltete Betriebe
ancillary installations

Nach einigem Experimentieren stellte sich schließlich die folgende Kombination von Stützmaßnahmen als besonders zweckmäßig heraus:

Einbau von geankerten Bogenteilen aus leichten U-Profilen unmittelbar hinter dem Bohrkopf. Diese Maßnahme verhindert Nachbrüche aus der

Abb. 6. California-Weiche
California switch

Auflockerungszone im Firstbereich und dient hauptsächlich dem Schutz von Mannschaft und Maschine.

Durch Verringerung des Bogenabstandes und Einbau von zusätzlichen Ankern konnten auch Strecken mit geringer Standfestigkeit, die u. U. einen leichten Stahlstreckenausbau erfordert hätten, ohne allzu große Behinderung durchfahren werden.

Aufbringen von 5 cm Torkret von der Arbeitsbühne aus. Diese so früh wie möglich eingebaute Spritzbetonsicherung verhindert weitere Auflockerungen in der Umgebung der Stollenröhre und dient außerdem dem Schutz des Gebirges vor atmosphärischen Einflüssen.

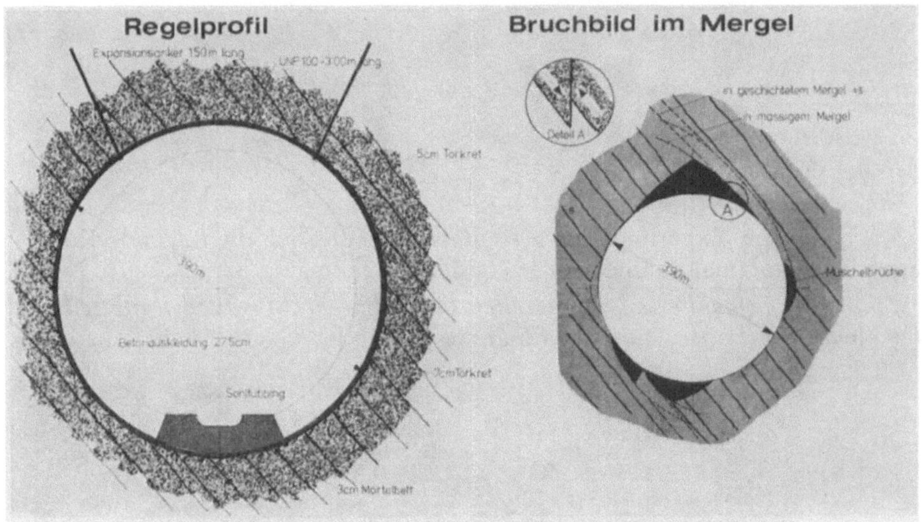

Abb. 7. Regelprofil mit Stützmaßnahmen

Regular cross-section with rock support measurements

Um anfänglich aufgetretene Schäden im unteren Ulmenbereich zuverlässig zu verhindern, wurde der an den Sohltübbing anschließende Spritzbeton später auf 7 cm verstärkt und an kritischen Stellen nachträglich geankert.

Verlegen von Sohltübbingen. Diese Stahlbetonfertigteile wurden exakt nach Richtung und Höhe verlegt und anschließend mit einem Injektionsgerät untermörtelt. Sie dienen dem Schutz des Gebirges gegen mechanische Beanspruchungen durch den Zugsverkehr und sollen die kontrollierte Ableitung der anfallenden Berg- und Betriebswässer gewährleisten. Gleichzeitig sind sie Bestandteil der endgültigen Auskleidung.

Im Großen und Ganzen ist es mit der beschriebenen Methode gelungen, der gestellten Aufgabe gerecht zu werden und mit wirtschaftlichen Mitteln eine optimale Stützung und Konservierung des Gebirges zu erreichen.

Obwohl die Installation des Betriebes so ausgelegt wurde, daß sämtliche Sicherungsarbeiten parallel mit dem Fräsvortrieb durchgeführt werden konn-

Abb. 8. Sicherung mit U-Bogenteilen und Spritzbeton
Supporting steel segments and shot-crete

Abb. 9. Einbau der Sohltübbinge
Placing of the bottom segments

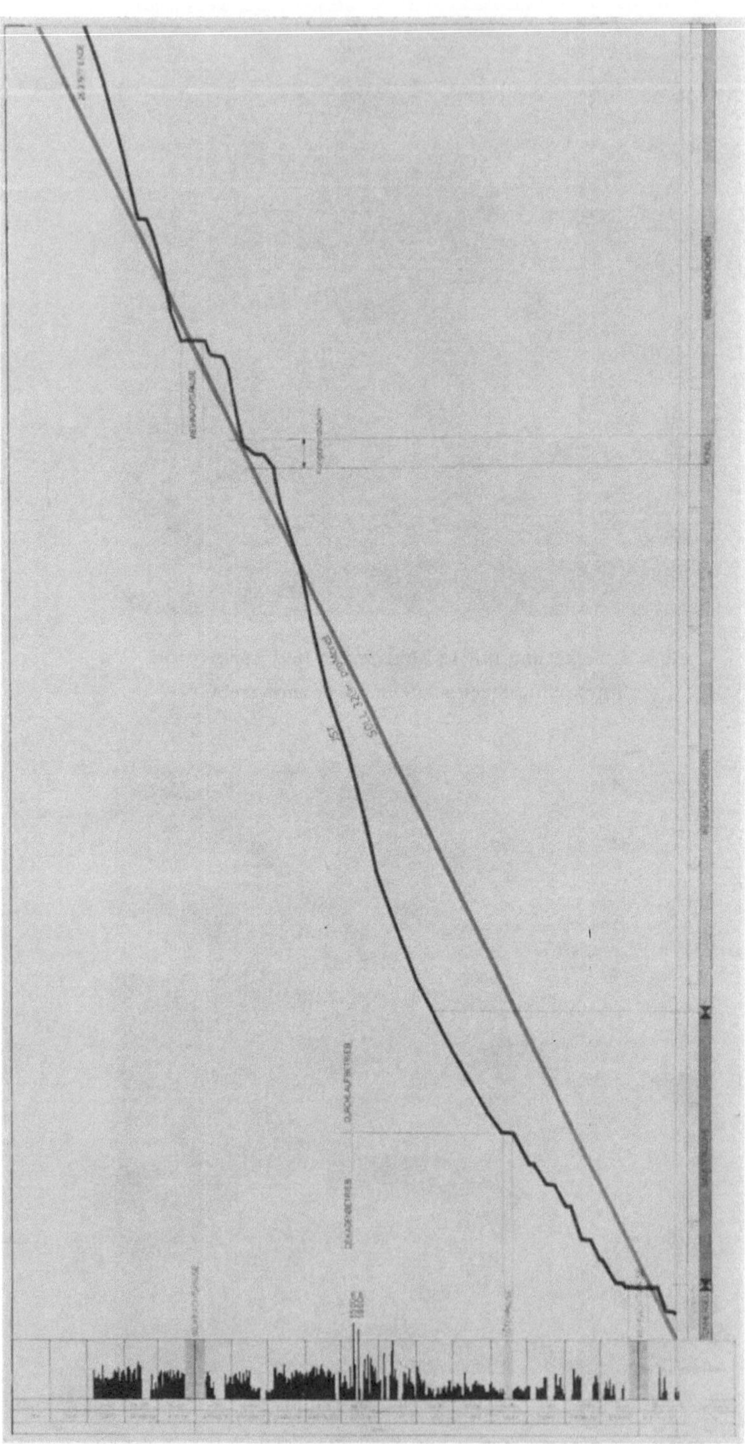

Abb. 10. V-Z-Diagramm
V-Z comparison

ten, entstanden durch die Gleichzeitigkeit Behinderungen, die die Gesamtleistung beeinträchtigten.

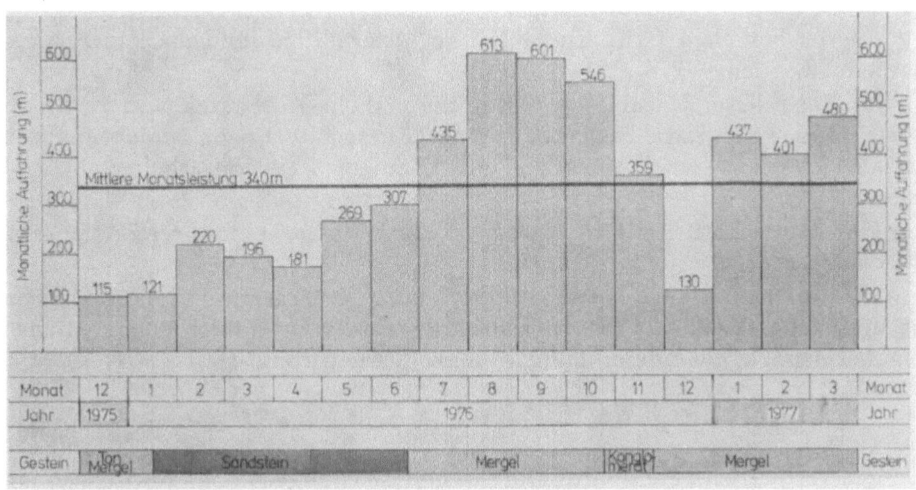

Abb. 11. Monats-Leistungsdiagramm
Monthly performance

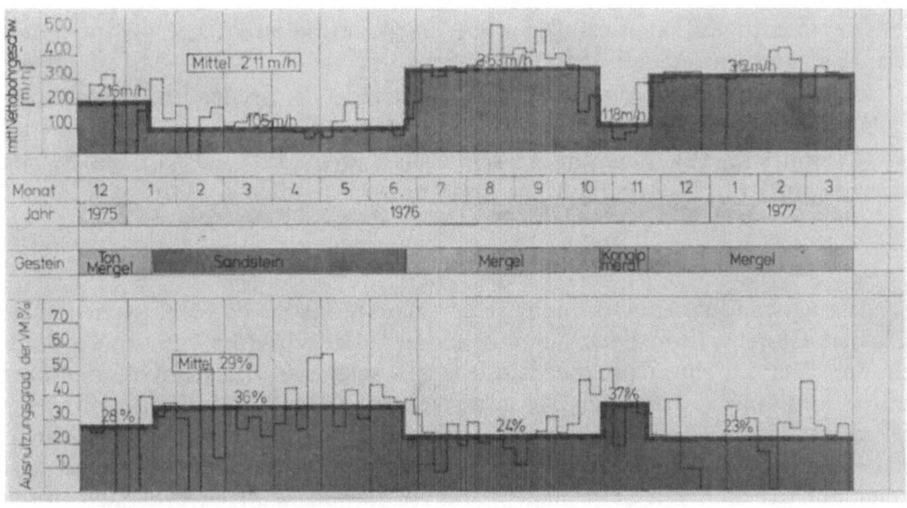

Abb. 12. Spezifisches Leistungsdiagramm
Specific performance

Wie aus dem spezifischen Leistungsdiagramm (Abb. 12) zu entnehmen ist, sank der Ausnutzungsgrad der Vortriebsmaschine in weicherem Gestein stark ab, was nicht zuletzt auf die erwähnte Behinderung durch die Sicherungsarbeiten zurückzuführen ist.

Mechanischer Vortrieb

Organisation

Aus terminlichen Gründen mußte ab Ostern 1976 im Durchlaufbetrieb gearbeitet werden.

Im Zuge dieser Umstellung wurde die tägliche Arbeitszeit von 3 × 8 auf 2 × 10 Stunden reduziert. In den Pausen zu Mittag sowie beim Schichtwechsel wurden jeweils Meißelkontrollen und kleinere Revisionsarbeiten durchgeführt.

Die Belegschaft vor Ort betrug 10 Mann (ohne Lokfahrer, Platz und Werkstätte).

Das anfänglich angestrebte Taktverfahren für Fräsern und Torkretieren hat sich nicht bewährt. Erst die Einführung eines kontinuierlichen Arbeitsablaufes ermöglichte eine erhebliche Leistungssteigerung.

Leistungen

Die vertraglich vereinbarte Durchschnittsleistung von 320 m pro Monat konnte trotz des härteren Gesteins in der Bausteinzone schlußendlich um 20 m überboten werden.

Die Maximalleistung pro Monat betrug im Sandstein 269 m, im Mergel 613 m.

Die maximale Tagesleisutng ohne Spritzbetoneinbau betrug im Sandstein 22 m und im Mergel 65 m.

Die entsprechenden Werte für den Vortrieb mit gleichzeitig aufgebrachter Spritzbetonsicherung betragen im Sandstein 16 m und im Mergel 32 m.

Die mittlere Tagesleisutng einschließlich aller Stützmaßnahmen betrug im Sandstein 9 m, im Mergel 18 m und im Schnitt über alles 16 m.

Spezifische Kosten

Aus verständlichen Gründen ist es nicht möglich, diese Kosten für die einzelnen Gesteinsarten detailliert zu nennen. Die Grenzen, innerhalb derer sich die Werte beim Rotenbergstollen bewegten, sollen jedoch angeführt werden, um eine Vorstellung davon zu vermitteln, wie schwierig es für den Kalkulanten ist, diese Zahlen abzuschätzen. Die Angaben der Maschinenhersteller sind für diesen Zweck nur bedingt zu gebrauchen, da sie fast immer auf der sicheren Seite liegen und im harten Wettbewerb nach unten korrigiert werden müssen.

Meißelkosten: Schneidringe, Gehäuse, Wellen, Lager etc.
(Ersatzteilkosten ohne Arbeit)
S 15,— bis S 220,— pro m^3.

Laufende Instandhaltung der Vortriebsmaschine: Motoren, Getriebe, Hydraulikteile etc. (Ersatzkosten ohne Arbeit)
S 4,— bis S 38,— pro m^3.

Zusammenfassung und Schluß

Trotz mancher Schwierigkeiten und Probleme war der Einsatz der Robbins-Vortriebsmaschine beim Rotenbergstollen erfolgreich. Die Vortriebsarbeiten konnten zwei Monate vor dem geplanten Termin abgeschlossen werden.

An die Kollegen in den Unternehmungen möchte ich zum Abschluß die Bitte richten, mit Ihren Erfahrungen auf diesem Gebiet gleichfalls nicht hinter dem Berg zu halten und vor diesem oder einem anderen Forum darüber zu berichten.

Anschrift des Verfassers: Dipl.-Ing. Eckart Schneider, Fa. Ing. Karl Jäger, A-6780 Schruns, Österreich.

Rock Mechanics, Suppl. 7, 203—224 (1978)

Rock Mechanics
Felsmechanik
Mécanique des Roches

Gegenüberstellung der Spritzbeton- und Ortbetonvorauskleidung beim Abteufen der Schächte Maienwasen und Albona

Von

M. John

Mit 16 Abbildungen

Zusammenfassung — Summary — Résumé

Gegenüberstellung der Spritzbeton- und Ortbetonvorauskleidung beim Abteufen der Schächte Maienwasen und Albona. Für die Belüftung des Arlberg-Straßentunnels werden 2 Lüftungsschächte errichtet, und zwar der Schacht Maienwasen mit einer Tiefe von 206 m und einem Ausbruchdurchmesser von 9,3 m und der Schacht Albona mit einer Tiefe von 724 m und einem Ausbruchdurchmesser von 8,98 m.

Für die Vorauskleidung beider Schächte wurden 2 Varianten geplant und ausgeschrieben: Variante 1 mit Spritzbeton, Ankerung und Tunnelbogen, Variante 2 mit Ortbeton, welchem eine Oberflächenversiegelung durch Spritzbeton vorauseilt. Der Festlegung der beiden Auskleidungsarten wurde eine Vergleichsrechnung des Ausbauwiderstandes nach Rabcewicz-Golser unter Berücksichtigung unterschiedlicher Gebirgskennwerte zugrundegelegt. Auf Grund des Angebotsergebnisses kam beim Schacht Maienwasen eine Vorauskleidung entsprechend Variante 1 und beim Schacht Albona jene entsprechend Variante 2 zur Ausführung.

Beim Schacht Maienwasen wurde vorerst ein Pilotschacht mittels Alimakgerät aufgebrochen, bevor der Schacht auf sein endgültiges Profil abgeteuft wurde. Wegen der Situierung des Schachtes Albona und der knappen Bautermine mußte dieser über die ganze Länge ohne Pilotschacht abgeteuft werden.

Die Abteufarbeiten verliefen bei beiden Schächten aufgrund des Einsatzes von geotechnischen Messungen ohne besondere Schwierigkeiten. Der Zeitaufwand für die Vorauskleidung war bei Variante 1 niedriger als bei der Variante 2.

Comparison of the Shotcrete Lining Used While Sinking Maienwasen Shaft With the Cast-Concrete Lining Used as Primary Support at Albona Shaft. Two ventilation shafts are being constructed for the ventilation of the Arlberg expressway tunnel, namely Maienwasen shaft, with a depth of 206 m and an excavation diameter of 9.3 m, and Albona shaft with a depth of 724 m and an excavation diameter of 8.98 m.

Two alternatives were planned and included in the tender documents for the primary support of the two shafts, i. e. using shotcrete, anchors and steel ribs or using cast concrete including a thin shotcrete lining preliminary to the concrete lining. The comparison of these two types of support was based on calculations of the support resistance according to Rabcewicz-Golser under consideration of different rock mass characteristics. As a result of the tenders submitted, shotcrete, anchors and steel ribs were chosen for the Maienwasen shaft and cast concrete for the Albona shaft.

For the Maienwasen shaft a pilot shaft was first driven upward using an Alimak machine before the shaft was sunk in its final profile. Due to its situation and the short construction period, the Albona shaft had to be sunk along its entire length without a pilot shaft.

The excavation work for both shafts was successfully completed thanks to the geotechnical measurements used to control rock mass behaviour. The following excavation rates were achieved:

Maienwasen shaft:	class S II	1.75 m per day
	class S III	1.10 m per day
Albona shaft:	class S II	1.45 m per day
	class S III	1.00 m per day

The time needed for installation of the primary support was as follows:

For rock class S II:	Maienwasen shaft	5.9 h/m
	Albona shaft	6.8 h/m
For rock class S III:	Maienwasen shaft	11.6 h/m
	Albona shaft	11.3 h/m

Regarding the different outer diameters of the two shafts, time consumption for the primary support was less for Maienwasen shaft in both classes, namely 20% less in S II and 5% less in S III.

Comparaison entre les méthodes de revêtir les puits de Maienwasen et d'Albona avec béton projeté ou avec béton coulé en place. Pour aérer le Tunnel de l'Arlberg on a construit deux puits de ventilation — celui de Maienwasen avec une profondeur de 206 m et un diamètre d'excavation de 9,3 m et celui d'Albona avec une profondeur de 724 m et un diamètre d'excavation de 8,98 m.

On avait prévu deux méthodes afin à faire revêtir les deux puits, pour lesquelles on nous a commis de préparer le projet et l'adjudication. La première méthode prévoit l'application de béton projeté, des ancres et des arcs. La seconde prévoit celle de béton coulé en place avec une couche mince de béton projeté au-dessous. La décision d'appliquer l'une ou l'autre méthode de revêtement a été basée sur les résultats des calculs de la résistance de support selon Rabcewicz-Golser tenant comptes des caractéristiques différentes du rocher encontré sur place. Par suite des offres reçues on a choisi la première méthode pour le puits de Maienwasen et la seconde pour celui d'Albona.

Pour la construction du puits de Maienwasen un puits pilote a été foncé de bas en haut en utilisant an appareil Alimak, avant de le foncer avec son diamètre définitif. À cause de sa situation particulière et le délai très court fixé pour sa construction celui d'Albona a dû être foncé complètement sans puits pilote. Les travaux d'excavation des deux puits ont réussi sans difficultés grâce aux mesures géotechniques utilisées.

Le progrès dans l'excavation a été le suivant:

Puits de Maienwasen:	en classe S II	1,75 m le jour
	en classe S III	1,10 m le jour
Puits d'Albona:	en classe S II	1,45 m le jour
	en classe S III	1,00 m le jour

Le temps nécessaire pour installer les premiers supports a été comme suit:

Pour la classe S II:	Puits de Maienwasen	5,9 h/m
	Puits d'Albona	6,8 h/m
Pour la classe S III:	Puits de Maienwasen	11,6 h/m
	Puits d'Albona	11,3 h/m

Vue la différence dans les diamètres externes des deux puits, moins de temps a été nécessaire pour exécuter les premiers supports dans le puits de Maienwasen en toutes les deux classes, i. e. 20% moins en classe S II et 5% moins en classe S III.

1. Übersicht

Für die Belüftung ist der aus einem Vor- und Haupttunnel bestehende rd. 14 km lange Arlbergtunnel als eine Einheit anzusehen (Abb. 1). Das nur 70 m lange Zwischenstück zwischen Vor- und Haupttunnel wird aus wirtschaftlichen und umweltbedingten Gründen mit einer Röhre verkleidet.

Die Luftversorgung des Tunnels erfolgt von 4 Lüftungsstationen aus, von denen zwei an den Portalen und zwei in Kavernen angeordnet sind. Die Kavernenstationen sind für die Luftansaugung und -ausblasung durch Vertikalschächte mit obertag verbunden (Abb. 2). Die Schächte haben bis zur Kavernenfirste folgende Endteufe:

Schacht Maienwasen	206 m
Schacht Albona	724 m

Das Projekt des Arlberg-Straßentunnels einschließlich aller Nebenanlagen und Einrichtungen wird von der Arlberg-Straßentunnel AG als Bauherr verwirklicht. Die Gesamtplanung und Bauleitung obliegt der Ingenieurgemeinschaft Lässer-Feizlmayr, als deren Projektsleiter der Verfasser tätig ist.

Der Schacht Maienwasen wurde von den Firmen Oberranzmeyer, I-L-Bau, Innerebner & Mayer, Soravia, mit der Spezialfirma Österreichische Schacht- und Tiefbauunternehmung, der Schacht Albona von folgenden Firmen in Arge ausgeführt: Deilmann-Haniel, Gebhardt & König, Jäger, Mayreder, Porr, Hinteregger, Rella, Union, Universale. Die geotechnischen Messungen wurden von der Interfels Ges. m. b. H. als Subunternehmer der beiden Argen vorgenommen.

2. Regelquerschnitte

Die Festlegung der Regelquerschnitte für die beiden Schächte erfolgte aufgrund einer Optimierungsrechnung, bei welcher die Baukosten des jeweiligen Schachtes und die Kosten der Ventilatoren der zugehörigen Kavernenstation den Betriebskosten für die Lüftung gegenübergestellt wurden. Diese Berechnung hat für die beiden Schächte unterschiedliche Abmessungen ergeben (Abb. 3 und 4).

Der Innendurchmesser beträgt beim Schacht Maienwasen 8,3 m, beim Schacht Albona 7,68 m. Beide Schächte sind durch eine exzentrisch ange-

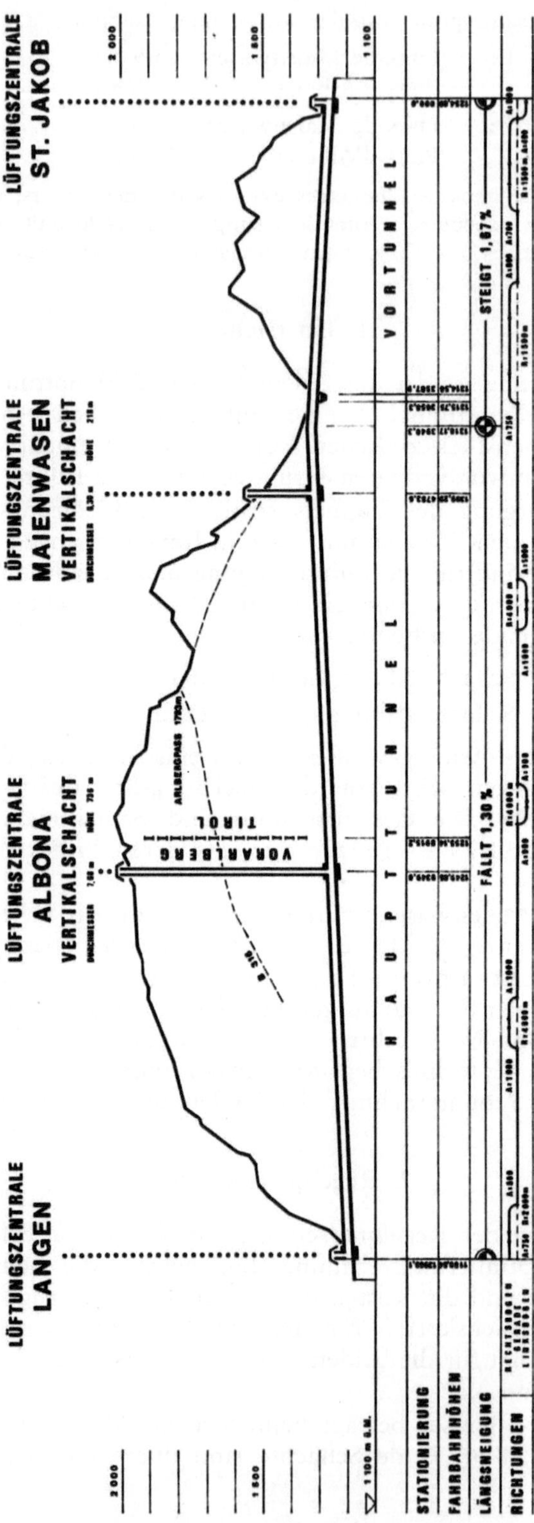

Abb. 1. Längenschnitt des Arlberg-Straßentunnels
Longitudinal section of the Arlberg expressway tunnel
Coupe longitudinale du tunnel routier de l'Arlberg

ordnete Trennwand in einen Zu- und Abluftteil getrennt. Dies ist dadurch bedingt, daß weniger Abluft ausgeblasen als angesaugt wird; die Differenzmenge entweicht bei den beiden Tunnelportalen.

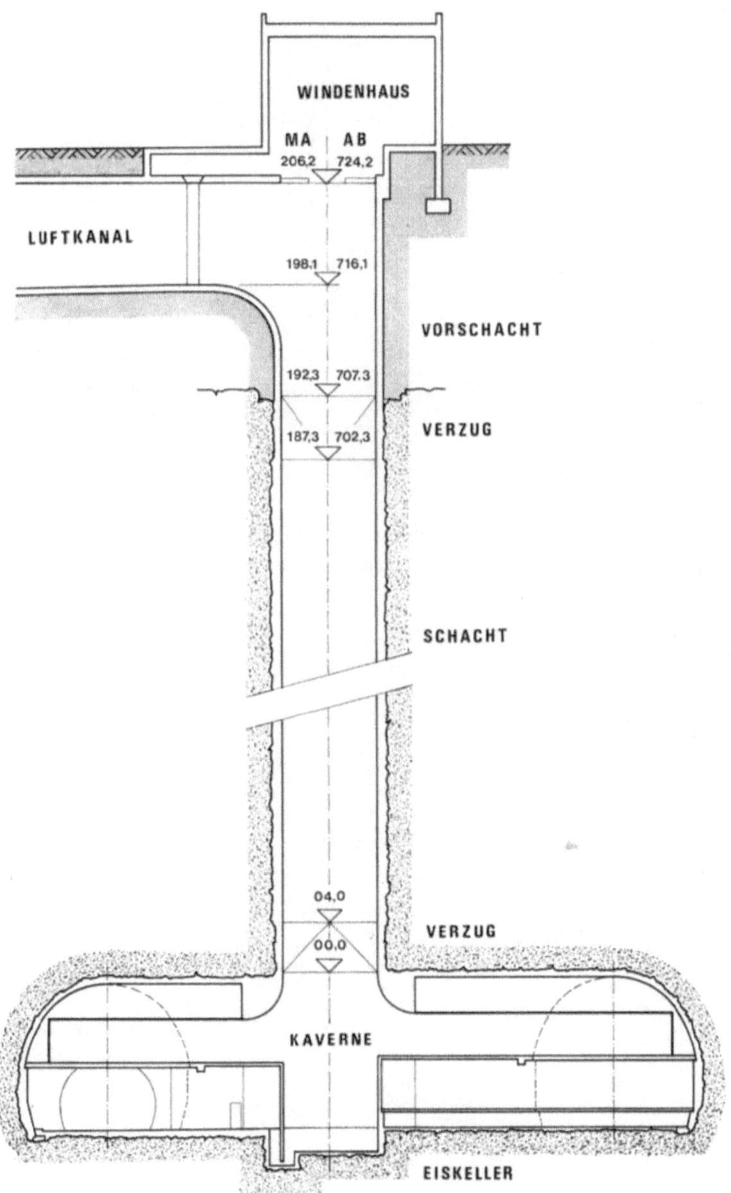

Abb. 2. Schematischer Längenschnitt für die Schächte Maienwasen (MA) und Albona(AB)
Schematic longitudinal section of Maienwasen (MA) and Albona (AB) shafts
Coupe longitudinale schématique des puits de Maienwasen (MA) et d'Albona (AB)

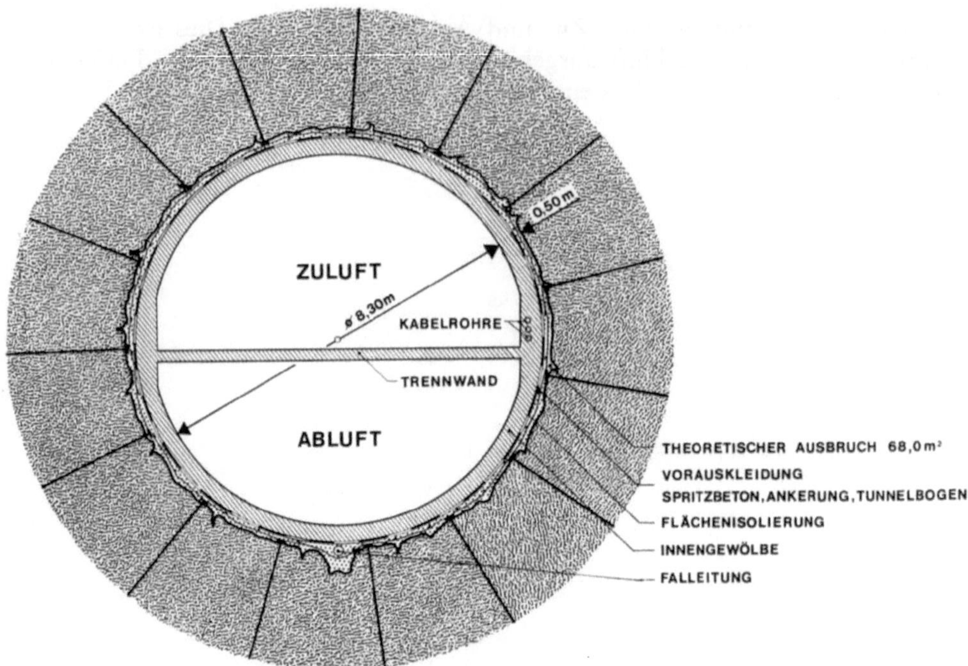

Abb. 3. Regelquerschnitt — Schacht Maienwasen
Cross section of Maienwasen shaft — Section transversale du puits de Maienwasen

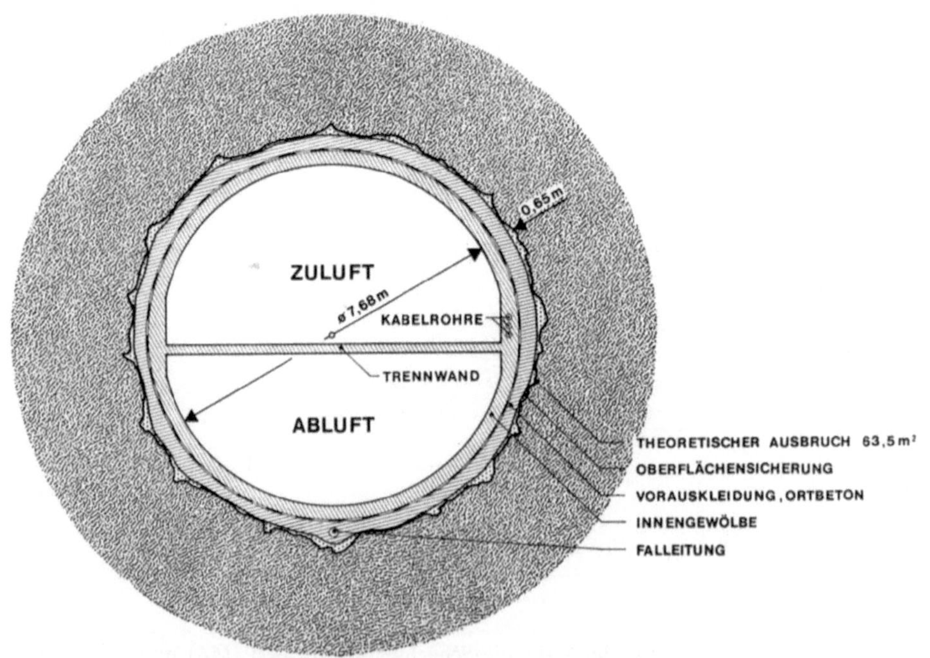

Abb. 4. Regelquerschnitt — Schacht Albona
Cross section of Albona shaft — Section transversale du puits d'Albona

3. Längsschnitte

Bei der Dimensionierung der Auskleidung wurde ein Wasserdruck auf den durch eine Isolierung abgedichteten Innenring nicht berücksichtigt, da das Bergwasser über Entwässerungslöcher, die in einem Horizont alle 10 m angeordnet sind, abgeleitet wird. Um einem Verlegen der Entwässerungsleitungen begegnen zu können, wurde alle 50 m eine Putzmöglichkeit geschaffen. Bei einer Wassersäule von 50 m hat der mind. 30 cm starke Innenring eine Bruchsicherheit von über 2. Die in den Abb. 5 und 6 dargestellten Ausschnitte haben für die gesamte Schachttiefe Gültigkeit.

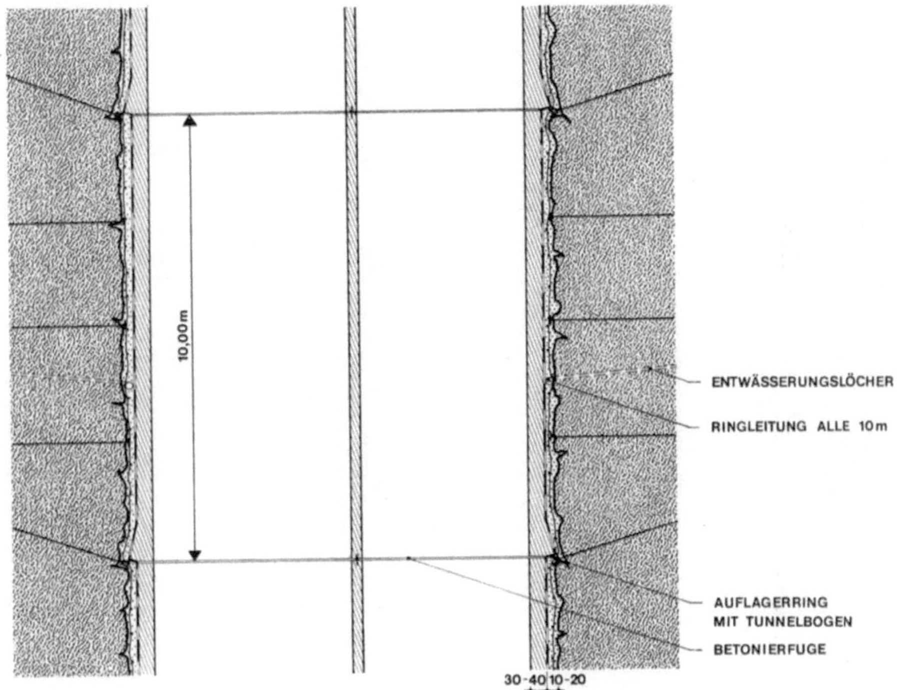

Abb. 5. Längenschnitt — Schacht Maienwasen
Longitudinal section of Maienwasen shaft
Coupe longitudinale du puits de Maienwasen

Der Innenring stützt sich alle 10 m auf einem Auflagerring ab. Dieser erfordert bei der Spritzbetonvorauskleidung eine geringfügige Vergrößerung des Ausbruches. Die Maßhaltigkeit des Auflagerringes wird durch einen eingespritzten Tunnelbogen gewährleistet. Durch den Auflagerring wird verhindert, daß durch die Bewegungsmöglichkeit aufgrund der Isolierung zwischen Vorauskleidung und Innenauskleidung hohe örtliche Ringspannungen entstehen, welche eine Bewehrung des Innenringes erfordern würden.

Der Ortbetonring stützt sich an der unregelmäßigen Felsoberfläche, welche erforderlichenfalls durch Spritzbeton versiegelt ist, ab. Von einer Ausbildung eines Auflagers wurde abgesehen, da Risse im Ortbeton belanglos sind. Bei den Betonierfugen des Ortbetons wird ein 40 cm hoher Abschnitt beim Betonieren freigelassen; in diesen Entwässerungsschlitz münden die Entwässerungsbohrungen. Die Bergwässer werden durch eine Ringlei-

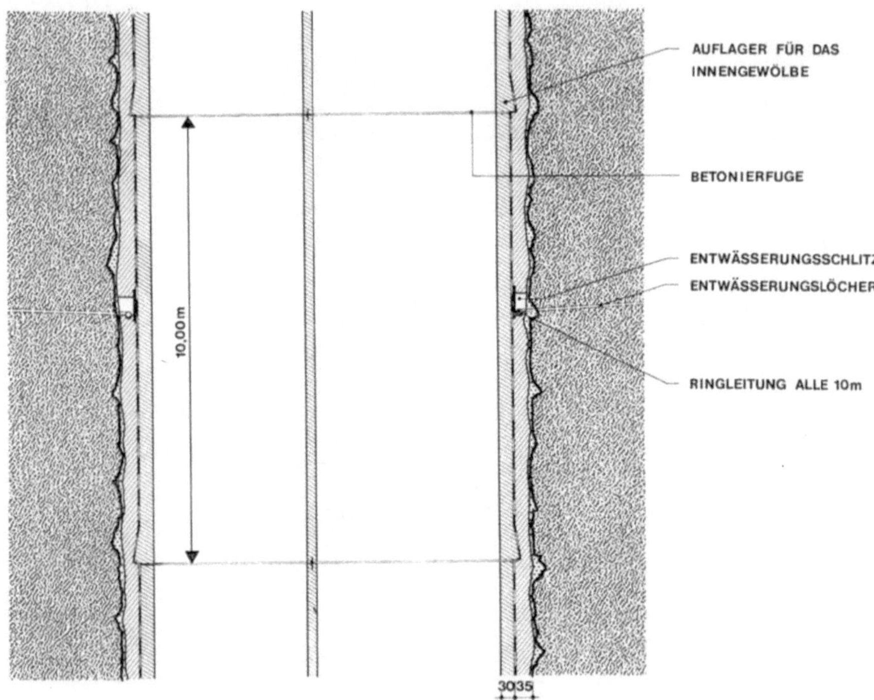

Abb. 6. Längenschnitt — Schacht Albona
Longitudinal section of Albona shaft
Coupe longitudinale du puits d'Albona

tung in die Falleitung eingeleitet, welche in der Kaverne mit der Tunnelhauptentwässerung verbunden wird. Im Ortbeton selbst sind alle 10 m Aussparungen für die Auflagerung des Innengewölbes vorgesehen.

4. Vorauskleidungen

Für beide Schächte waren 2 Varianten der Vorauskleidung geplant und ausgeschrieben worden. Beim Schacht Maienwasen kam die nachfolgend beschriebene Variante 1, beim Schacht Albona jene entsprechend Variante 2 zur Ausführung.

Variante 1 — Spritzbetonvorauskleidung (Abb. 7)

Das Prinzip dieser Vorauskleidung besteht darin, oberflächliche Ablösungen durch eine bewehrte Spritzbetonschale, größere Nachbrüche durch eine Ankerung zu verhindern und echten Gebirgsdruck durch einen Ankertragring verstärkt durch Tunnelbogen zu bewältigen.

Variante 2 — Ortbetonvorauskleidung (Abb. 8)

Zur Stützung des Gebirges wird der Schacht mit einem Ortbetonring ausgekleidet. Da dieser aus arbeitstechnischen Gründen erst in einiger Entfernung von der Schachtsohle eingebaut werden kann, muß der freie Stoß bei nachbrüchigem Gebirge mit Spritzbeton versiegelt werden; bei druckhaftem Gebirge ist eine zusätzliche Verstärkung mit Ankern notwendig.

Um vergleichbare Angebote für die beiden Arten der Vorauskleidung zu erhalten, ist es notwendig, eine Vergleichsbasis für die Dimensionierung aufzustellen. Für standfestes und nachbrüchiges Gebirge ist ein Vergleich nicht angebracht. Für druckhaftes Gebirge wurden unter Heranziehung der Berechnung des Ausbauwiderstandes nach Rabcewicz und Golser (1973) Vergleichsrechnungen durchgeführt. Bei diesen Berechnungen wurde die Gebirgstragmitwirkung durch den Scherwiderstand einer ideellen Bruchfläche berücksichtigt. Aus Versuchsergebnissen ist bekannt, daß beim Bruchvorgang hauptsächlich ein Verlust der Kohäsion auftritt, während der Reibungswinkel nur gering abnimmt (Abb. 9). Durch eine Ankerung des Gebirges, welche vor Beginn der Bruchvorgänge eingebaut wird, wird die Abnahme der Kohäsion verringert (Abb. 10). Ein Ankertragring bewirkt somit eine höhere Restscherfestigkeit. Diese Überlegungen zum Bruchmechanismus, welche durch Messungen beim Bau des Arlbergtunnels bestätigt wurden (John 1976 und 1977), wurden durch folgende Annahmen in die Ermittlung des Ausbauwiderstandes einbezogen:

Reibungswinkel: $\varphi = 30^0$

Kohäsion, im Bereich des Ankertragringes $c_1 = 10$ kp/cm²

Kohäsion, im Bereich ohne Ankertragring $c_2 = 5$ kp/cm²

Die Berechnung nach Rabcewicz und Golser wurde insoferne abgeändert, als der Scherwiderstand der gesamten Bruchfläche, d. h. auch außerhalb des Ankertragringes berücksichtigt wurde (Abb. 11 und 12). Für den Schacht Maienwasen wurden folgende Werte ermittelt:

	Spritzbetonvorauskleidung		Ortbetonvorauskleidung	
	S II	S III	S II	S III
Innendruck	4,4 kp/cm²	7,3 kp/cm²	9,9 kp/cm²	10,8 kp/cm²
Gebirgstragring	14,2 kp/cm²	16,8 kp/cm²	7,7 kp/cm²	13,4 kp/cm²
Ausbauwiderstand	18,6 kp/cm²	24,1 kp/cm²	17,6 kp/cm²	24,2 kp/cm²

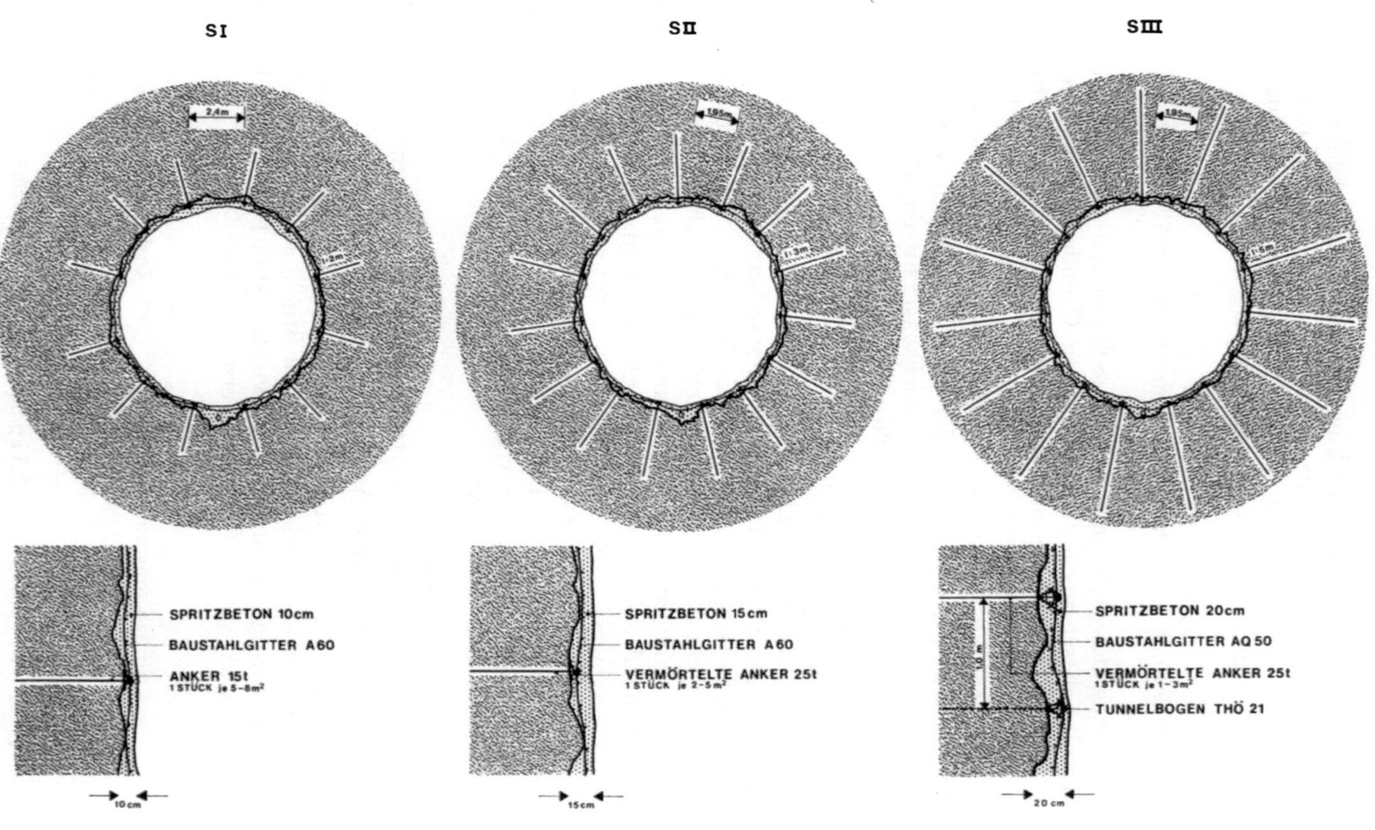

Abb. 7. Spritzbetonvorauskleidung — Schacht Maienwasen
Shotcrete lining as primary support for Maienwasen shaft
Revêtement en béton projeté — puits de Maienwasen

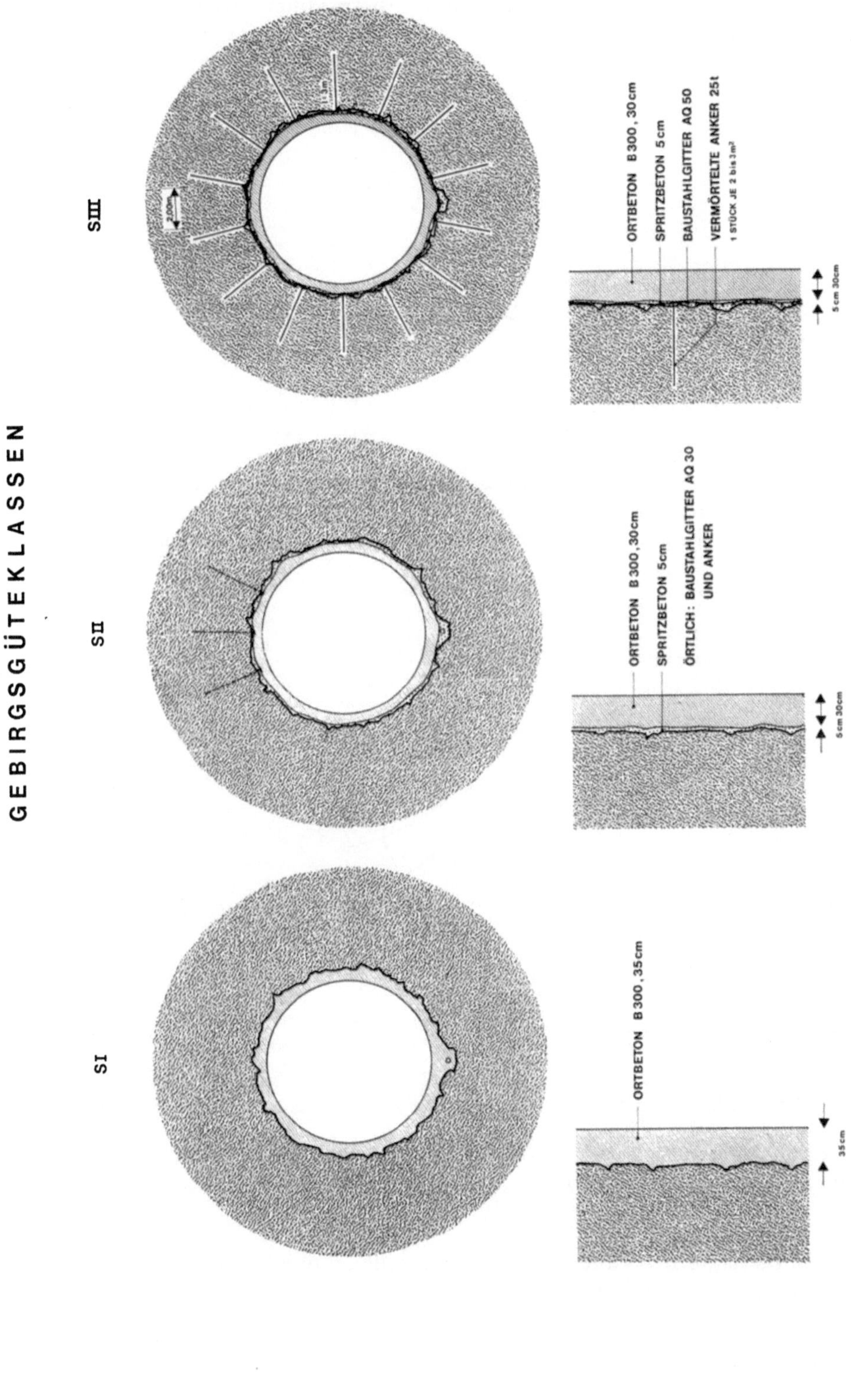

Abb. 8. Ortbetonvorauskleidung — Schacht Albona
Cast-concrete lining as primary support for Albona shaft
Revêtement en béton projeté coulé en place — puits d'Albona

Bei der Spritzbetonvorauskleidung überwiegt die Gebirgstragwirkung, bei der Ortbetonvorauskleidung übernimmt diese einen Großteil des Gesamtausbauwiderstandes.

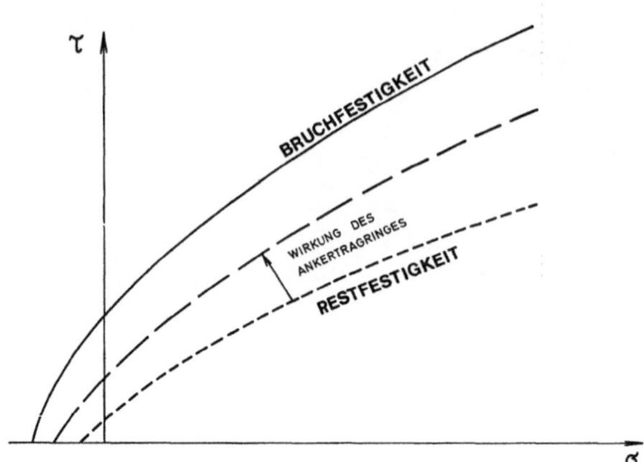

Abb. 9. Schematische Darstellung des Bruchvorganges anhand der Mohrschen Hüllkurve
Schematic illustration of failure process using Mohr's diagram
Illustration schématique de la rupture au moyen de la courbe intrinsèque

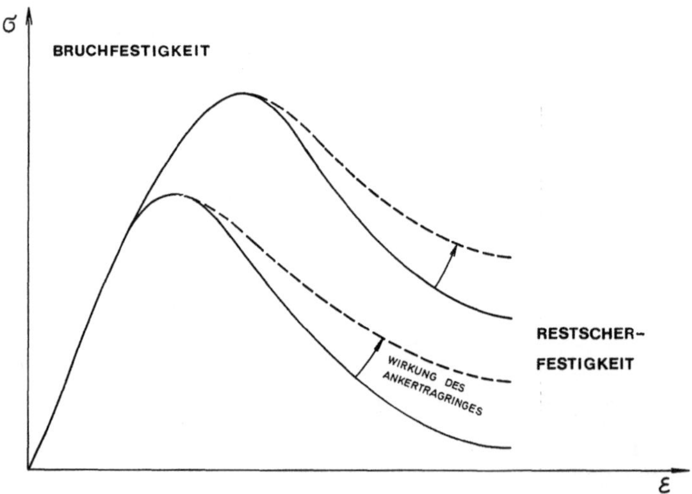

Abb. 10. Schematische Darstellung der Wirkungsweise der Anker anhand des Spannungs-Dehnungsdiagrammes
Schematic illustration of the effect of anchoring using the stress-strain diagram
Illustration schématique de l'efficacité des ancres au moyen de la graphique contrainte-déformation

Die aufgezeigte Berechnung der Ausbauwiderstände berücksichtigt nicht die unterschiedliche Nachgiebigkeit der Vorauskleidungen. Dies wurde je-

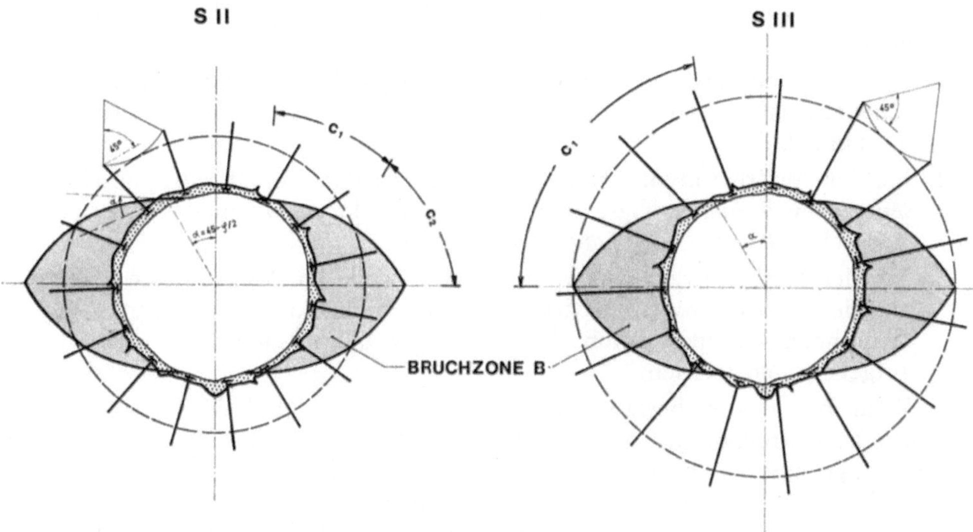

Abb. 11. Bruchzone (B) für die Berechnung des Ausbauwiderstandes der Spritzbetonvorauskleidung beim Schacht Maienwasen

Failure zone (B) for calculation of the support resistance of the shotcrete lining for Maienwasen shaft

Zone de rupture (B) pour la calculation de la résistance de support du revêtement en béton projeté au puits de Maienwasen

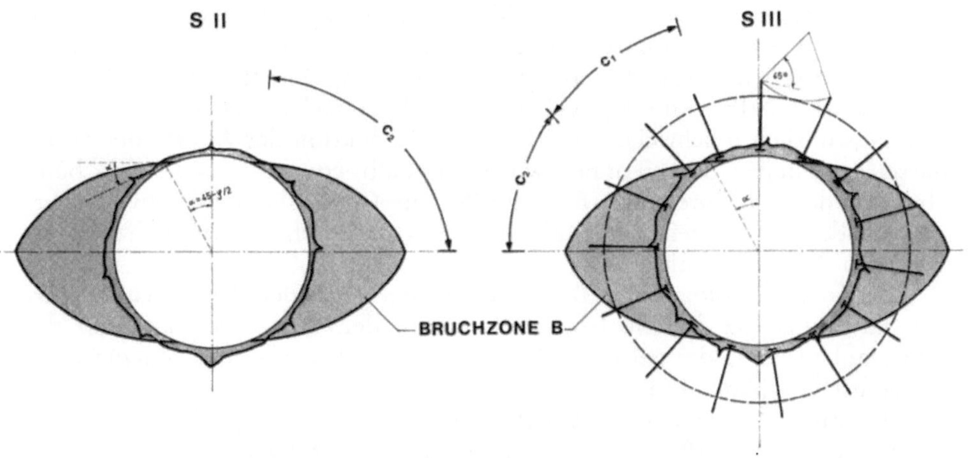

Abb. 12. Bruchzone (B) für die Berechnung des Ausbauwiderstandes der Ortbetonvorauskleidung beim Schacht Maienwasen

Failure zone (B) for calculation of the support resistance of the cast concrete lining for Maienwasen shaft

Zone de rupture (B) pour la calculation de la résistance de support du revêtement en béton coulé en place au puits de Maienwasen

doch aus nachfolgenden Gründen nicht in den Mittelpunkt der Betrachtung gezogen:

1. Beim Schachtabteufen verursachen herabfallende Gesteinsteile große Schäden; daher sind die Bewegungen in engeren Grenzen zu halten als im Tunnel.
2. Bei der Spritzbetonvorauskleidung ist durch den Einbau eines geschlossenen Ringes die Nachgiebigkeit eingeschränkt. Bewegungsschlitze sind aus Sicherheitsgründen nicht zu empfehlen.
3. Bei der Ortbetonvorauskleidung ist ein Teil der Bewegungen bereits vor dem Wirksamwerden des Betonringes abgeklungen. Weitere Verformungen könnten durch den Einbau von nachgiebigen Materialien aufgenommen werden.
4. Beim Bau des Arlbergtunnels wurde festgestellt, daß das Gesamtmaß der Endverformungen nicht als Kriterium der Festlegung der Stützmittel herangezogen werden soll, sondern die Verformungsgeschwindigkeit während der Spannungsumlagerung (John 1977, Judtmann 1977). Eine rechnerische Ermittlung der zu erwartenden Verformungsgeschwindigkeiten liegt bisher noch nicht vor.

5. Gebirgsklassifizierung

Die Einteilung des Gebirges für die Schachtbauarbeiten in 4 Güteklassen wurde vom Tauernschacht übernommen (Rabcewicz und Pacher 1970). Die Güteklasse S IV stellt eine Sonderklasse für Verhältnisse dar, die ohne Sondermaßnahmen, wie z. B. Injektionen, Gefrierverfahren o. ä., nicht bewältigt werden können. Für diese Verhältnisse, die nicht erwartet wurden, wurden keine Preisvereinbarungen getroffen.

Die Kriterien der Unterteilung der einzelnen Güteklassen entsprechen jenen für den Arlbergtunnel, welche an anderer Stelle (Weiss 1976, John 1976) ausführlich beschrieben wurden. Eine Reduktion der Anzahl der Güteklassen gegenüber dem Tunnel wurde deshalb vorgenommen, weil beim Abteufen der Schächte auch bei standfestem Gebirgsverhalten aus Sicherheitsgründen umfangreichere Maßnahmen erforderlich sind und die Unterteilung in Teilquerschnitte beim Ausbruch entfällt.

Neben der geologischen Beschreibung der einzelnen Güteklassen wurden die Stützmittel für die beiden Varianten der Vorauskleidung und der Arbeitsablauf, d. i. die zeitliche Abfolge der Ausbruchs- und Stützarbeiten vorgeschrieben (Abb. 13a und 13b).

Aufgrund der geologischen Verhältnisse — siehe Baugeologischer Längenschnitt (Weiss 1976) — wurde gemeinsam folgende Klassifizierung prognostiziert:

	Schacht Maienwasen	Schacht Albona
S I	70%	36%
S II	20%	46%
S III	10%	18%

Bei der Ausführung hat sich gezeigt, daß die Sicherheit der Mannschaft gegen herabfallende Gesteinsteile schwierig zu beurteilen ist. Aus diesem Grund kam die Klasse S I nicht zur Anwendung. Eine Unterscheidung der Klassen S I und S II aufgrund objektiver Kriterien ist kaum möglich. Kurze Baufristen und entsprechender Konkurrenzkampf und damit knapp kalkulierte Preise sind Voraussetzung, die Klassifizierung bei günstigen Gebirgsverhältnissen zu erleichtern.

Beim Schacht Maienwasen stellte sich heraus, daß die Verwitterungszone tiefer reichte als ursprünglich angenommen. Außerdem wurde wegen der Gefahr von Wassereinbrüchen vorsichtiger abgeteuft. Aus diesen Gründen ergab sich eine Verschiebung zu den ungünstigeren Klassen. Die Stützmaßnahmen in den einzelnen Klassen konnten verringert werden und es entfielen die Wassererschwernisse, so daß trotz der Änderung der Klassifizierung die Kosten für die Abteufarbeiten gegenüber dem Urangebot um weniger als 5% anstiegen.

Das Gebirge wurde tatsächlich wie folgt eingeteilt:

	Schacht Maienwasen	Schacht Albona
S I	0%	0%
S II	58%	74%
S III	42%	26%

6. Meßprogramm

Das Meßprogramm wurde auf die unterschiedlichen Vorauskleidungen abgestimmt. Bei der Spritzbetonvorauskleidung (Abb. 14) ist bei größeren Verformungen mit einem Reißen der Spritzbetonschale zu rechnen. Um diese Gefahr kontrollieren zu können, wurden die Verformungen mit Extensometern gemessen, deren Meßwerte durch Fernübertragung am Schachtkopf abgelesen wurden. Dadurch war es möglich, die Messungen ohne Unterbrechung der Abteufarbeiten auch dann durchführen zu können, wenn die Meßpunkte von der Sohle aus nicht mehr erreichbar waren. Die Wirkungsweise der Anker wurde durch mechanische Meßanker ermittelt. Diese entsprechen den verwendeten SN-Ankern mit dem Unterschied, daß sie bei gleicher Querschnittsfläche innen hohl sind. Im Inneren der Stangen sind in 1,5, 3,0, 4,5 und 6,0 m Tiefe Meßgestänge mit der Ankerstange verbunden. Es werden die Verformungen zwischen diesen Festpunkten und dem Ankerkopf gemessen. Die erste Ableitung der Meßwerte ist ein Maß für die Beanspruchung, die zweite Ableitung ein Maß für die Scherkraftübertragung zwischen Anker und Gebirge.

Bei der Ortbetonvorauskleidung (Abb. 15) ist die Beobachtung der Verformungen bis zum Einbau des Ortbetonringes von Bedeutung. Da die Ablesung der Meßwerte von der Sohle aus durchgeführt werden kann, wurden im allgemeinen Konvergenzmessungen durchgeführt, welche vereinzelt durch Messungen mit Einfach-Extensometern ergänzt wurden. Um die Sicherheit des Ortbetonringes gegen Bruch feststellen zu können, wurden die Tangentialspannungen am Innenrand des Ortbetonringes gemessen.

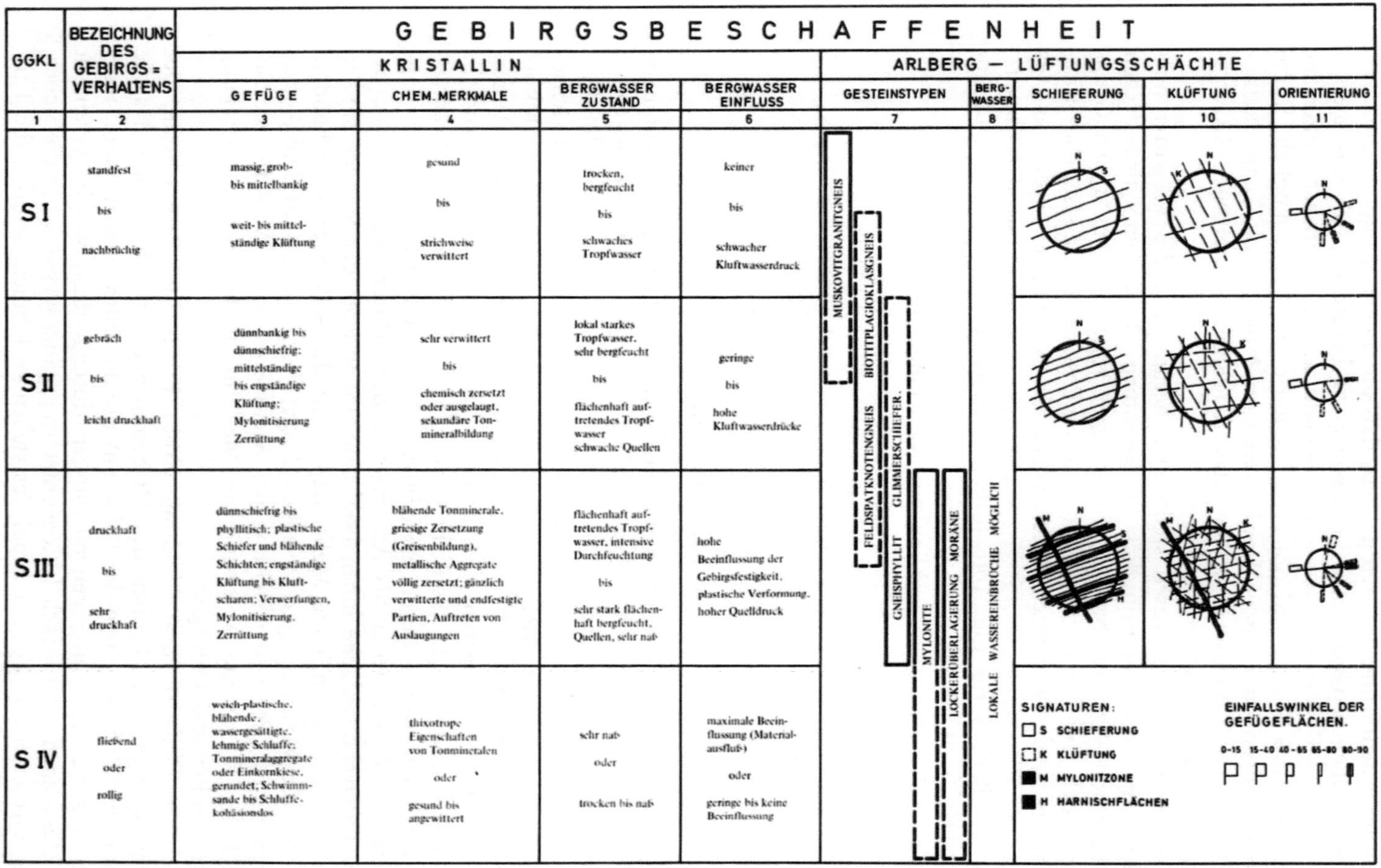

GGKL	BEZEICHNUNG DES GEBIRGS = VERHALTENS	GEBIRGSBESCHAFFENHEIT								
		KRISTALLIN				ARLBERG — LÜFTUNGSSCHÄCHTE				
		GEFÜGE	CHEM. MERKMALE	BERGWASSER ZUSTAND	BERGWASSER EINFLUSS	GESTEINSTYPEN	BERG-WASSER	SCHIEFERUNG	KLÜFTUNG	ORIENTIERUNG
1	2	3	4	5	6	7	8	9	10	11
S I	standfest bis nachbrüchig	massig, grob- bis mittelbankig weit- bis mittelständige Klüftung	gesund bis strichweise verwittert	trocken, bergfeucht bis schwaches Tropfwasser	keiner bis schwacher Kluftwasserdruck	MUSKOVITGRANITGNEIS FELDSPATKNOTENGNEIS BIOTITPLAGIOKLASGNEIS	LOKALE WASSEREINBRÜCHE MÖGLICH			
S II	gebräch bis leicht druckhaft	dünnbankig bis dünnschiefrig; mittelständige bis engständige Klüftung; Mylonitisierung Zerrüttung	sehr verwittert bis chemisch zersetzt oder ausgelaugt, sekundäre Tonmineralbildung	lokal starkes Tropfwasser, sehr bergfeucht bis flächenhaft auftretendes Tropfwasser schwache Quellen	geringe bis hohe Kluftwasserdrücke	GNEISPHYLLIT GLIMMERSCHIEFER.				
S III	druckhaft bis sehr druckhaft	dünnschiefrig bis phyllitisch; plastische Schiefer und blähende Schichten; engständige Klüftung bis Kluftscharen; Verwerfungen, Mylonitisierung, Zerrüttung	blähende Tonminerale, griesige Zersetzung (Greisenbildung), metallische Aggregate völlig zersetzt; gänzlich verwitterte und endfestigte Partien, Auftreten von Auslaugungen	flächenhaft auftretendes Tropfwasser, intensive Durchfeuchtung bis sehr stark flächenhaft bergfeucht, Quellen, sehr naß	hohe Beeinflussung der Gebirgsfestigkeit, plastische Verformung, hoher Quelldruck	MYLONITE LOCKERÜBERLAGERUNG MORÄNE				
S IV	fließend oder rollig	weich-plastische, blähende, wassergesättigte, lehmige Schluffe; Tonmineralaggregate oder Einkornkiese, gerundet; Schwimmsande bis Schluffe-kohäsionslos	thixotrope Eigenschaften von Tonminetalen oder gesund bis angewittert	sehr naß oder trocken bis naß	maximale Beeinflussung (Materialausfluß) oder geringe bis keine Beeinflussung			SIGNATUREN: S SCHIEFERUNG K KLÜFTUNG M MYLONITZONE H HARNISCHFLÄCHEN	EINFALLSWINKEL DER GEFÜGEFLÄCHEN. 0-15 15-40 40-65 65-80 80-90	

Abb. 13 a. Baugeologische Beschreibung der Güteklassen
Engineering geological description of rock mass classification
Description de la qualité des classes de roches

GGKL	BEZEICHNUNG DES GEBIRGS=VERHALTENS	GEBIRGSVERHALTEN GEBIRGSMECHANISCHE BESCHREIBUNG	SCHACHTBAUARBEITEN AUSBRUCH	STÜTZUNG SPRITZBETONBAUWEISE	 ORTBETONBAUWEISE	ARBEITSABLAUF	ANMERKUNGEN
1	2	12	13	14	15	16	17
S I	standfest bis nachbrüchig	Die auftretenden Spannungen überschreiten nur lokal die Gebirgsfestigkeit, entsprechend der Schieferung und Klüftung treten schwache bis starke Nachbrüche auf. Bei hoher Überlagerung besteht Bergschlaggefahr	schonendes Sprengen, Abschlagslänge max. 3,0 m	10 cm Spritzbeton, Baustahlgitter, Anker l = 2,0 m 1 Stk je 5 - 8 m²	35 cm Ortbeton B 300	Bei Bergschlaggefahr hat eine Stützung mit Anker sofort nach dem Ausbruch zu erfolgen, sonst keine zeitliche Beschränkung, soferne die Sicherheit der Mannschaft gewahrt ist	1) Die Darstellung des Gebirgsgefüges in bezug auf den Schacht stellt eine schematische Verallgemeinerung der zu erwartenden Verhältnisse dar. Geologische Besonderheiten (Verwerfer, Harnischflächen etc.) wurden nicht berücksichtigt. 2) Die Stützmaßnahmen sind den örtlichen Verhältnissen anzupassen! 3) Injektionen zwischen Außenring und Gebirge sind bei etwa auftretenden Hohlräumen und schädlichen Auflockerungen erforderlich
S II	gebrach bis leicht druckhaft	Der Scherwiderstand der Schieferungs- und Kluftflächen ist geringer als die auftretenden Scherspannungen. Es kommt zu starken Nachbrüchen und leichten Druckerscheinungen	schonendes Sprengen örtlich Schrammarbeit, Abschlagslänge 2,5 m	15 cm Spritzbeton Baustahlgitter vermörtelte Anker l = 3,0 m 1 Stk je 2 - 5 m²	5 cm Spritzbeton-oberflächensicherung + Baustahlgitter, 30 cm Ortbeton B 300	Die Stützmaßnahmen sollen möglichst rasch nach dem Ausbruch in Angriff genommen werden. Wenn der Stützring 4 m über der Schachtsohle noch nicht eingebaut ist, muß der Stoß entsprechend gesichert sein	
S III	druckhaft bis sehr druckhaft	Die Druckspannungen überschreiten die Gebirgsdruckfestigkeit. Es treten Druckerscheinungen an der Schachtlaibung auf, die durch Wassereinflüsse verstärkt werden können. Es kann auch zu Sohlhebungen kommen	vornehmlich Schrämmen, Abschlagslänge max. 1,5 m	20 cm Spritzbeton, Baustahlgitter, vermörtelte Anker l = 5,0 m 1 Stk je 1 - 3 m² Tunnelbogen THO 21 alle 0,8 bis 1,5 m	5 cm Spritzbeton-oberflächensicherung + Baustahlgitter und vermörtelte Anker l = 3,0 m, 1 Stk je 2 - 3 m²; 30 cm Ortbeton B 300	Die freigelegten Oberflächen müssen sofort gesichert werden. Die Stützmaßnahmen sind zeitlich auf die Meßergebnisse abzustimmen	
S IV	fließend oder rollig	Sofort nach der Hohlraumbildung treten starke Druckerscheinungen sehr rasch auf, das Gebirge oder der Boden kann mit konventionellen Methoden nicht beherrscht werden	Ausbruch entsprechend den Sondermaßnahmen	Sondermaßnahmen wie chemische Verfestigung, Gefrierverfahren, Elektroosmose etc.		Der Arbeitsablauf richtet sich nach den gewählten Sondermaßnahmen	

Abb. 13 b. Gebirgsmechanische Beschreibung der Güteklassen
Rock mechanic description of rock mass classification
Description de la qualité de roches en termes mécanique de roches

7. Baudurchführung — Schacht Maienwasen

Beim Bau des Schachtes Maienwasen wurden im oberen Bereich Glimmerschiefer, welche bis zu rd. 30 m Tiefe verwittert waren, angetroffen. Im unteren Bereich wurde ein geschieferter Granitgneis aufgefahren. Die Schieferung ist im ganzen Schacht stark ausgeprägt, die Schieferflächen fallen mit rd. 70° ziemlich steil nach Süden. Diese Verhältnisse birgen die Gefahr des Ausgleitens von Schieferpaketen insbesonders im Liegenden in sich.

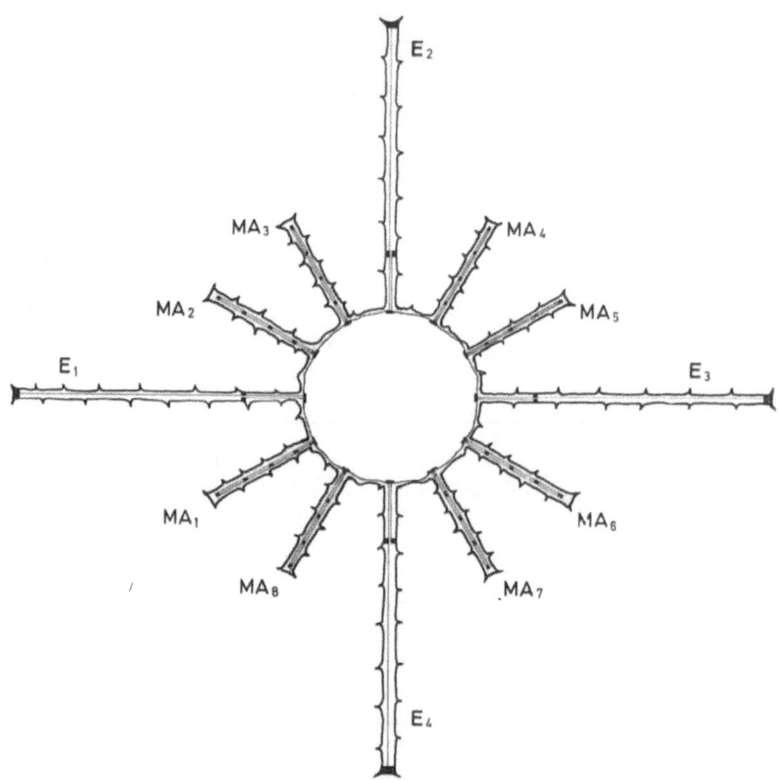

Abb. 14. Meßquerschnitt — Schacht Maienwasen mit Zweifach-Extensometer (E) und Meßanker (MA)

Measuring cross section for Maienwasen shaft including double extensometer (E) and measurement anchors (MA)

Section transversale de mesures — puits de Maienwasen avec double extensomètres (E) et ancres de mesures (MA)

Auf Grund des Baukonzeptes für den Haupttunnel Ost, für welchen sofort ein Richtstollen bis über die Kaverne Maienwasen hinaus vorgetrieben wurde, ergab sich die Möglichkeit, vorerst einen Pilotschacht von der Kaverne aus aufzubrechen. Der Pilotschacht wurde in der Größe von 2 × 2 m mit einem Alimakgerät in 30 Arbeitstagen fertiggestellt. Dies ergibt eine tägliche Leistung von 7,0 m.

Beim Abteufen des Schachtes wurde das Ausbruchsmaterial in die Kaverne abgeworfen. Aus diesem Grund war es möglich, mit einer verhältnismäßig einfachen Förderanlage das Auslangen zu finden. Für das Abteufen

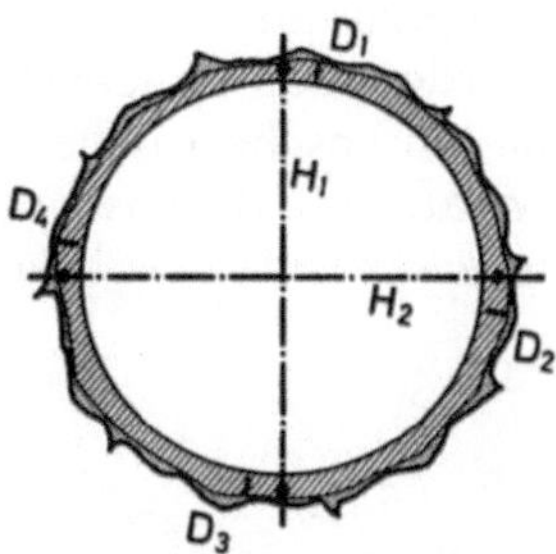

Abb. 15. Meßquerschnitt — Schacht Albona mit Konvergenzmessungen (H) und Spannungsmessungen (D)

Measuring cross section for Albona shaft including convergence measurements (H) and stress measurements (D)

Section transversale de mesures — puits d'Albona avec mesures de convergence (H) et mesures de contrainte (D)

des Schachtes einschließlich Baustelleneinrichtung und Herstellen des Vorschachtes wurden 5 1/2 Monate benötigt. Der Schacht selbst wurde in 132 Arbeitstagen fertiggestellt, das ergibt eine durchschnittliche Abteufleistung von 1,4 m je Arbeitstag; in der Klasse S III betrug die durchschnittliche Leistung 1,1 m, in der Klasse S II 1,75 m je Arbeitstag.

Die Innenauskleidung wurde in 29 Tagen mit Hilfe einer Gleitschalung eingebaut, das entspricht einer Leistung von 6,3 m je Arbeitstag.

8. Baudurchführung — Schacht Albona

Die geologischen Verhältnisse des Schachtes Albona waren insbesondere im oberen Bereich ungünstig. Bis auf eine Tiefe von 200 m sind in die mürben, tektonisch stark beanspruchten Gneise Mylonitzonen mit rd. 5 m Mächtigkeit eingeschaltet. Da diese ebenso wie die Schieferung mit 50 bis 80^0 nach Süden einfallen, waren neben der Gefahr des Auskeilens größerer Gesteinspakete Druckerscheinungen zu bewältigen. Mit zunehmender Tiefe nahm die Häufigkeit und Ausdehnung der Mylonitzonen ab, es wurden teilweise Glimmerschiefer überwiegend jedoch Feldspatknotengneise aufgefahren. Das Einfallen der Schieferung änderte sich unwesentlich.

Der Schacht Albona mußte von einer Seehöhe von rd. 2000 m aus konventionell abgeteuft werden. Das Ausbruchsmaterial mußte obertag abgefördert werden. Dazu wurde eine leistungsstarke zweitürmige Schachtförderanlage (Abb. 16) für die Förderung mit zwei Bergekübeln mit 3 m^3 Inhalt installiert. Aufgrund der Situierung des Schachtes nahe der Baulosgrenze im Haupttunnel mußte auf eine möglichst kurze Bauzeit gedrängt werden. Aus diesem Grund wurde die gesamte Einrichtung wintersicher ausgestattet, so daß mit Ausnahme des Weihnachtsabganges im sogenannten 4/3 Betrieb durchgearbeitet werden konnte.

Für die Baustelleneinrichtung und die Herstellung des Vorschachtes wurden $6^{1}/_{2}$ Monate, für das Abteufen des Schachtes 19 Monate benötigt. Abzüglich der witterungsbedingten und durch den Weihnachtsabgang bedingten Unterbrechungen wurden 538 Arbeitstage für die Abteufarbeiten aufgewendet, das ergibt eine durchschnittliche Leistung von 1,3 m je Arbeitstag. In der Klasse S III wurde eine durchschnittliche Leistung von 1,0 m, in der Klasse S II von 1,45 m je Arbeitstag erbracht.

Abb. 16. Förderanlage — Schacht Albona (Photo: Albrecht, Innsbruck)
Shaft hoist installation — Albona shaft
Installation d'extraction — puits d'Albona

Beim Einbringen der Innenauskleidung wurde eine durchschnittliche Leistung von 6,6 m je Arbeitstag erbracht, dies entspricht bei Berücksichtigung des unterschiedlichen Umfanges der beiden Schächte jener des Schachtes Maienwasen.

9. Vergleich der Vorauskleidungen

Bei der *Spritzbetonvorauskleidung* wurden alle Stützmittel entsprechend der jeweiligen Güteklasse spätestens bis zum nächsten Abschlag eingebaut. Daraus resultiert eine hohe Anfangssicherheit. Die Standsicherheit der Vorauskleidung im weiteren Bauablauf wurde durch Beobachtung der Auskleidung in Hinsicht auf Rißbildung und durch die geotechnischen Messungen kontrolliert. Aufgrund dieser Beobachtungen wurden die Stützmittel in den einzelnen Güteklassen den örtlichen Verhältnissen angepaßt. Es hat sich gezeigt, daß die radialen Verformungen in der Regel ein Gesamtmaß von 1 cm nicht überstiegen und nach 2 Wochen abgeklungen waren. Die Auswertung der Ergebnisse der Ankermessungen ergaben, daß Bewegungen entlang der Schieferflächen auftreten. Die Anker wurden daher teilweise auf Zug, teilweise auf Druck beansprucht. Die Beanspruchung der Anker blieb unter deren Fließgrenze, die Anker haben somit den Bewegungen standgehalten und die geringen Gesamtverformungen bewirkt.

Für die *Ortbetonvorauskleidung* wurden von der ausführenden Firma Betonierabschnitte mit 10 m Länge gewählt. Dadurch konnte der Ortbeton frühestens 12 bis 14 m oberhalb der Sohle eingebracht werden. Der Stoß zwischen Schachtsohle und Ortbetonring mußte durchwegs durch Spritzbeton und Ankerung gesichert werden. In der Klasse S III reichte die in der Ausschreibung vorgesehenen, dem Ortbeton vorauseilende Oberflächensicherung deshalb nicht aus, weil der zugelassene Maximalabstand des Einbaues des Ortbetonringes von der Schachtsohle wesentlich überschritten wurde. Die daraus resultierenden zusätzlichen Stützmittel (Verstärkung des Spritzbetons und der Ankerung) waren in den Kosten des Ortbetonringes einzukalkulieren und wurden daher nicht gesondert vergütet. Aus diesem Grund wurde versucht, den Umfang der vorauseilenden Sicherung niedrig zu halten.

Dies bedingte eine geringe Anfangsicherheit und Radialverformungen bis zu 8 cm. Zum Zeitpunkt des Einbaues des Ortbetonringes betrug die Verformungsgeschwindigkeit der Konvergenz 5 mm je Tag. Die Tangentialspannungen im Ortbetonring stiegen bis zu 60 kp/cm^2 an. 4—6 Wochen nach dem Einbau des Ortbetonringes stiegen die Spannungen nur mehr unwesentlich an. Diese Messungen zeigen, daß die Stärke des Ortbetonringes von 30 cm den Anforderungen entspricht. Für die Verhältnisse entsprechend der Klasse S II bedeutet die 30 cm starke Ortbetonvorauskleidung eine Überdimensionierung, die arbeitstechnisch bedingt ist.

Ein Vergleich des *Zeitaufwandes* der Schachtarbeiten bei den beiden Schächten muß sich auf die Vorauskleidungen beschränken, da beim Schacht Maienwasen ein Pilotschacht die Abteufarbeiten wesentlich erleichtert hat.

Der mittlere Zeitaufwand für den Einbau der Vorauskleidung betrug je Schachtmeter,

in der Klasse S II:	für Maienwasen:	5,9 Stunden
	für Albona:	6,8 Stunden
in der Klasse S III:	für Maienwasen:	11,6 Stunden
	für Albona:	11,3 Stunden

Wenn man den unterschiedlichen Umfang der beiden Schächte berücksichtigt, war der Zeitaufwand für die Spritzbetonvorauskleidung in der Klasse S II um 20% und der Klasse S III um 5% niedriger als bei der Ortbetonvorauskleidung. Zu berücksichtigen ist dabei ferner, daß beim Schacht Maienwasen wegen des Zeitvorsprunges durch den Pilotschacht eine einfache Baustelleneinrichtung gewählt worden war.

Beim Schacht Albona wäre der Zeitaufwand für eine Spritzbetonvorauskleidung noch kürzer ausgefallen. Die Verwendung der Spritzbetonvorauskleidung soll daher bei vergleichbaren Projekten in die Planung einbezogen werden, wie dies von Kleiner (1974), bereits angeregt wurde. Selbstverständlich soll damit die Ortbetonvorauskleidung nicht von vornherein ausgeschlossen werden, da auch der Spritzbetonvorauskleidung wirtschaftliche Grenzen gesetzt sind (siehe Ries 1974).

10. Schlußbemerkung

Bei beiden Schächten wurde die vorgesehene Bauzeit unterschritten. Beim Schacht Albona hat sich gezeigt, daß bei einem eingespielten Arbeitsrhythmus noch wesentlich höhere Abteufleistungen, u. zw. bis zu 35%, als die erwähnten Durchschnittswerte erzielt werden können. Es ist somit für die Zukunft eine Steigerung der Abteufleistungen zu erhoffen, welche dementsprechende wirtschaftliche Folgen hat. Für die richtige Wahl der Vorauskleidung ist das Abschätzen der zu erwartenden Gebirgsverhältnisse von ausschlaggebender Bedeutung. Bei nachbrüchigem und leicht druckhaftem Gebirge sowie bei einem Gebirge mit hoher Mobilität, d. h. wenn hohe Verformungsgeschwindigkeiten zu erwarten sind, ist derzeit die Spritzbetonvorauskleidung aus wirtschaftlichen Gründen der Ortbetonvorauskleidung überlegen.

Literatur

John, M.: Die geotechnischen Messungen im Arlbergtunnel und deren Auswirkungen auf das Baugeschehen. Rock Mechanics, Suppl. 5, pp. 157—177. Wien – New York: Springer 1976.

John, M.: Adjustment of Programs of Measurements Based on the Result of Current Evaluations. International Symposium on Field Measurements in Rock Mechanics, Zürich 1977, pp. 639—656.

Judtmann, G.: Die Vortriebssicherung des Arlbergtunnels — Anpassung an die Gebirgsverhältnisse mit Hilfe von geotechnischen Messungen. Rock Mechanics, Suppl. 7, pp. 157—177. Wien – New York: Springer 1978.

Kleiner, J.: Der Spritzbeton als Endausbau in Schächten. Rock Mechanics 6, 237—246 (1974).

Rabcewicz, L. v., Golser, J.: Principles of Dimensioning the Supporting System for the "New Austrian Tunneling Method", Water Power, March 1973.

Rabcewicz, L. v., Pacher, F.: Ausschreibungsunterlagen für den „Lüftungsschacht — Tauerntunnel", 1970.

Ries, A.: Abteufen des Lüftungsschachtes für den Tauerntunnel. Rock Mechanics 6, 167—183 (1974).

Weiss, E. H.: Die baugeologische Prognose für den Schnellstraßentunnel durch den Arlberg, Tirol — Vorarlberg. Rock Mechanics, Suppl. 5, pp. 133—156. Wien – New York: Springer 1976.

Anschrift des Verfassers: Dipl.-Ing. Dr. techn. Max John, Ingenieurgemeinschaft Lässer-Feizlmayr, Framsweg 16, A-6020 Innsbruck, Österreich.

Rock Mechanics, Suppl. 7, 225—248 (1978)

Rock Mechanics
Felsmechanik
Mécanique des Roches

Anwendung von Stahlfaserspritzbeton im Tunnelbau — derzeitiger Stand der Verfahrenstechnik und konstruktive Bewertung

Von

B. Maidl

Mit 27 Abbildungen

Zusammenfassung — Summary

Anwendung von Stahlfaserspritzbeton im Tunnelbau — derzeitiger Stand der Verfahrenstechnik und konstruktive Bewertung. Über die Materialeigenschaften von Stahlfaserbeton liegen bereits genügend gesicherte Aussagen vor, die sich auf theoretische Modelle und Versuche stützen können. Eine Anwendung von Stahlfaserspritzbeton im Tunnelbau wird von vielen Seiten in Betracht gezogen, doch gibt es kaum ausgeführte Objekte mit größerer Bedeutung. Es fehlen anerkannte Bemessungsregeln, aber auch die Verfahren für eine störungsfreie und wirtschaftliche Anwendung.

Im Bericht wird kurz auf den derzeitigen Stand der Materialtechnologie eingegangen, um dann die Verfahren mit den Problemen für das Herstellen darzustellen. Die Anwendung des Spritzbetons und Ortbetons im Tunnel- und Bergbau wird für einschalige Tunnelbauweisen untersucht. Ein Vorschlag für die Bemessung soll die Möglichkeit eines Dimensionierungsansatzes zeigen.

Abschließend wird über laufende Forschungsprojekte am Lehrstuhl für Bauverfahrenstechnik und Baubetrieb berichtet, bei dem im Vordergrund die Verbesserung des Rückprallverhaltens und die Weiterentwicklung der Verfahren stehen. Also Fragen, die die wirtschaftliche Anwendung wesentlich beeinflussen.

Application of Fiber-Wire Shotcrete in Tunnelling — Present Situation of Research Technique and Constructive Rating. There are sufficiently secured statements about material properties of fiber-wire concrete basing on theoretic models and experiments. Use of fiber-wire shotcrete at tunnelling has been drawn into consideration from many directions, but there scarcely exist carried out objects of more importance. There is lack of approved rules for proportioning but procedures for static-free and economic application, too.

During the report it is entered into the present situation of technology of material to describe then procedures with problems for production.

There are investigations executed about application in tunnelling and mining of shotcrete and in situ concrete for univalve lining. A proposition for rating shall show the possibility of a dimensioning start.

Finally it will be reported about running research projects at the professorship of constructions methods and constructions management, at which the improvement of the rebound behaviour and development of procedure are well to the fore, namely questions which substancially influence the economic application.

1. Allgemeine Übersicht

Eine Verstärkung von Werkstoffen durch Einlagerung hochfester Fasermaterialien wird schon seit langem mit Erfolg praktiziert. Ich erinnere an glasfaserverstärkte Kunststoffe, bewehrtes Gummi und Asbest. Die Verbesserung von Eigenschaften einer Materialmatrix kann nur mit Fasern bestimmter Eigenschaften verwirklicht werden. So sind E-Modul und Bruchdehnung von größter Bedeutung. Abgesehen von der Verträglichkeit der Werkstoffe gelten für eine spröde Matrix (z. B. die Zement- oder Betonmatrix) folgende Kriterien für die Wahl:

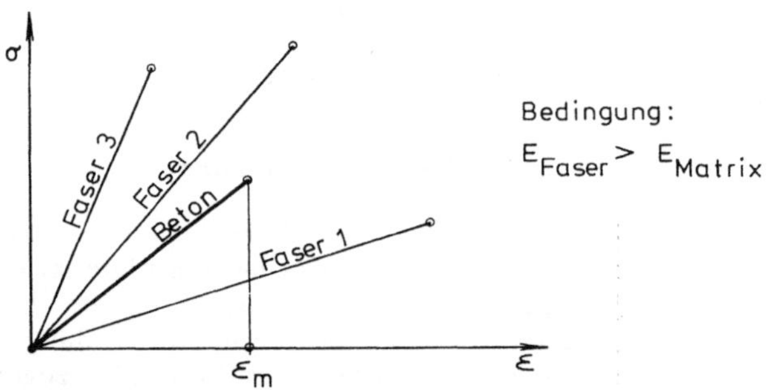

Abb. 1. Spannungs-Dehnungsverhalten von Betonmatrix und Fasern
Behavior of strain and stress of matrix and fibres

Der Beton folgt idealisiert dem Hookschen Gesetz bis zum Bruch ε_m.

Fasertyp 1: Größere Bruchdehnung, jedoch ein wesentlich kleinerer E-Modul; führt in der Matrix zu keiner bedeutenden Materialverbesserung (Beispiel: Kunststoffasern).

Fasertyp 2: Größere Bruchdehnung, größerer E-Modul; stellt eine echte Verstärkung dar und hat bis zum Bruch Reserven (Beispiel: Stahlfasern).

Fasertyp 3: Kleine Bruchdehnung, größerer E-Modul, verstärkt wie Faser 2, bricht aber bevor die Matrix bricht (Beispiel: Glasfasern).

Diese Überlegung ist auch auf Werkstoffe übertragbar, die sich nicht linear verhalten. Auch Beton hat ein viskoelastisches Verhalten, nämlich, daß bei jeder Spannung neben der elastischen Verformung ein zeitabhängiges plastisches Fließen eintritt (Kriechen).

Bezüglich der Orientierung können Fasern gerichtet oder unregelmäßig verteilt werden. Die größte Ausnutzung mit 100% ist natürlich bei paralleler Lage zur Beanspruchungsrichtung. Bei einer willkürlichen räumlichen Verteilung werden bei einachsiger Beanspruchung nur noch ca. 20—30% an Ausnutzung erreicht. Diese Betrachtung läßt erkennen, welche enormen Einsparungen bei Faserbeton durch Ausnutzung anderer Vorteile erreicht werden müssen, um diesen erhöhten Mehraufwand an Materialeinsatz gegenüber Stahlbeton zu kompensieren.

Hinzu kommt auch noch der derzeitige Preisunterschied der Stahlfasern zum Bewehrungsstahl.

Aus folgender Grafik sehen Sie die derzeitigen Kostenrelationen von Fasern. Es gibt jedoch Anzeichen, daß die Stahlfasern in einigen Jahren bei ca. 1000,— DM/t liegen werden.

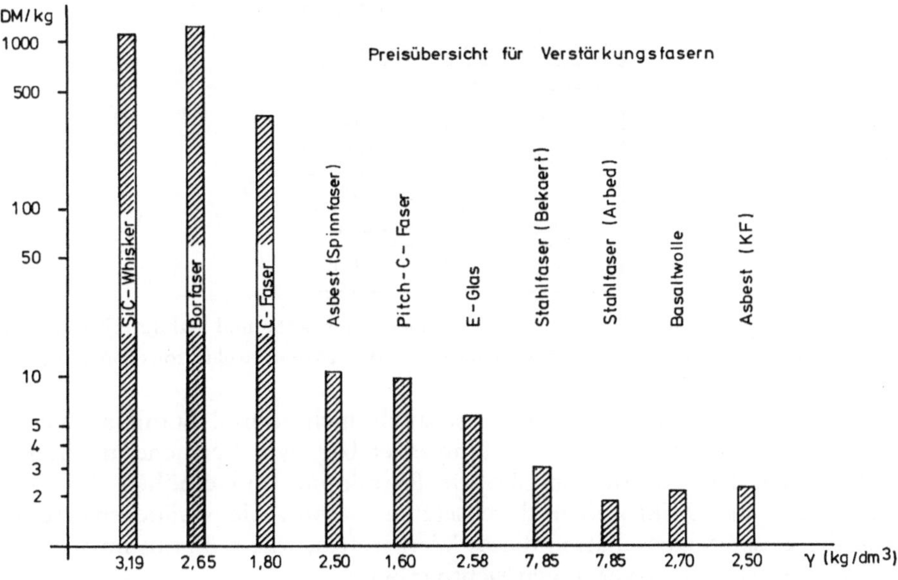

Abb. 2. Kostenrelation von Fasern
Cost relation of fibres

Nach dieser kurzen Einführung will ich darauf verweisen, daß die materialtechnologische und verfahrenstechnische Weiterentwicklung auf dem Gebiet des Stahlfaserbetons, speziell auch des Stahlfaserspritzbetons weltweit sehr intensiv betrieben wird.

2. Eigenschaften von Stahlfaserbeton

Ich will hier die Eigenschaften von unbewehrtem und stahlfaserbewehrtem Beton gegenüberstellen; die Werte wurden experimentell ermittelt. Es gibt bereits eine sehr große Anzahl von experimentellen Untersuchungen, die in der Tendenz übereinstimmen, im Detail voneinander erheblich abweichen.

Umfangreiche Untersuchungen liegen an unserem Institut für Konstruktiven Ingenieurbau der Ruhr-Universität vor, die vom Lehrstuhl von Professor Zerna zum großen Teil mit der Bundeswehr durchgeführt wurden. Ich gebe eine kurze Übersicht der wichtigsten derzeitigen Materialkennwerte im Vergleich zu unbewehrtem Beton.

Zugfestigkeit und Druckfestigkeit werden im Grunde nicht wesentlich erhöht, doch könnte eine garantierte Zugfestigkeit in einer Dimensionierung angesetzt werden. Stahlfasern haben eine andere Wirkungsweise im Beton als die Bewehrung im Stahlbeton. Eine erforderliche statische Bewehrung,

Eigenschaft	Veränderungen
Zugfestigkeit	ca. + 40 %
Druckfestigkeit	ca. + 15%
Arbeitsvermögen	bis+2000%
Elastizitätsmodul	ca. + 5%
Schwinden	ca. - 30%
Kriechen	ca. + 20%
Wärmedehnzahl	ca. ± 0%
Wärmeleitfähigkeit	ca. + 40 %
Dichte	ca. + 7%

Abb. 3. Vergleich der Materialkennwerte von unbewehrtem Beton und Stahlfaserbeton [1, 2]
Comparison of the characteristic values of material of concrete and steel fibrous concrete [1, 2]

z. B. bei einem langen Biegeträger, ist auch nicht durch Stahlfaserzugabe ersetzbar. Allenfalls ist die Verbügelung ersetzbar, wie Versuche an Pfetten gezeigt haben. Die Fasern hemmen die Rißbildung und erhöhen die Zugfestigkeit des mit Mikrorissen durchsetzten Betons. Sie verändern grundlegend die Gesamteigenschaften dieses Werkstoffes, was z. B. gut im wesentlich verbesserten Arbeitsvermögen sichtbar wird.

Weiterhin sind zu nennen:

— Erhöhte dynamische Beanspruchbarkeit, dies ist ein ganz interessanter Faktor, z. B. für den Druckstollenbau oder bei der Auskleidung großer Ölkavernen.
  Hier treten durch Füllen und Entleeren Wechselbeanspruchungen auf.
— Verbesserte Schlagfestigkeit, Verschleißfestigkeit.

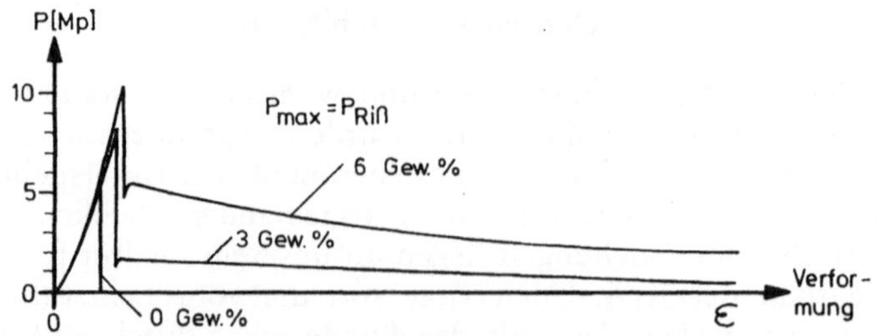

Abb. 4. Typisches Lastverformungsdiagramm aus Zugversuchen [1, 2]
Typical load deformation diagram from tensile tests [1, 2]

— Bessere Frostbeständigkeit.
— Vermutlich verbesserte Wasserdichtigkeit, darüber liegen aber noch keine systematischen Untersuchungen vor.

Auf einige wichtige Eigenschaften will ich besonders verweisen:

1. Arbeitsvermögen

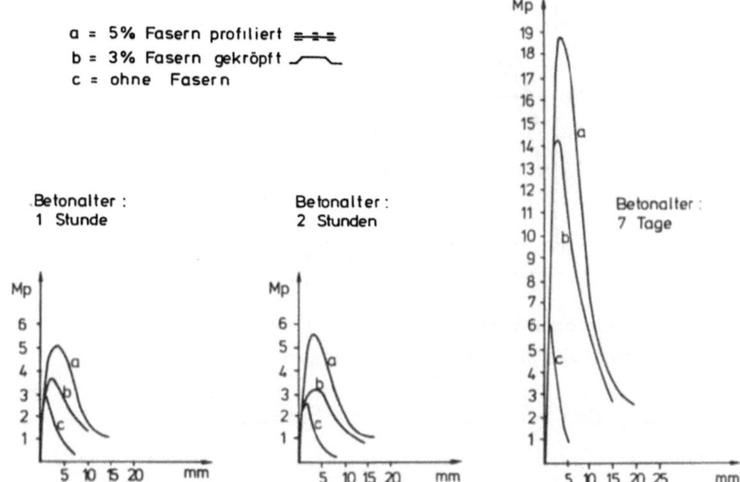

Abb. 5. Lastverformungsverhalten aus Druckversuchen [5]
Stress-strain curve form our tests [5]

2. E-Modul

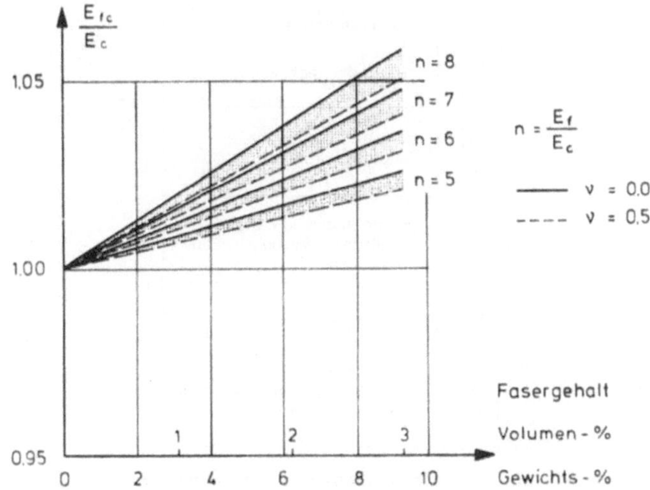

Abb. 6. Einfluß der Fasern auf den Elastizitätsmodul [1, 2] (ohne Berücksichtigung des Längeneinflusses)
Influence of fibres on the Young's modulus [1, 2] (without consideration of the influence of length)

3. Schwinden

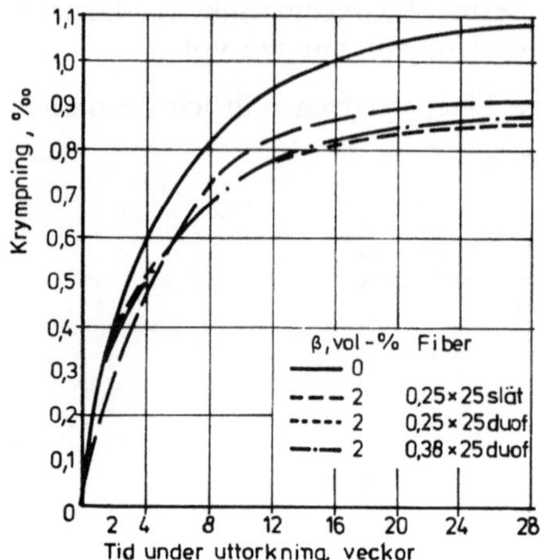

Abb. 7. Zeitlicher Verlauf des Schwindens nach Malmberg [4]
Time slope of shrinkage according to Malmberg [4]

Verringerung des Schwindens durch Stahlfasern.

3. Ausgangsmaterialien

1. Stahlfasern

Stahlfasertypen: Fasergüte (Stahlgüte) ohne Einfluß wegen der erforderlichen Haftung.

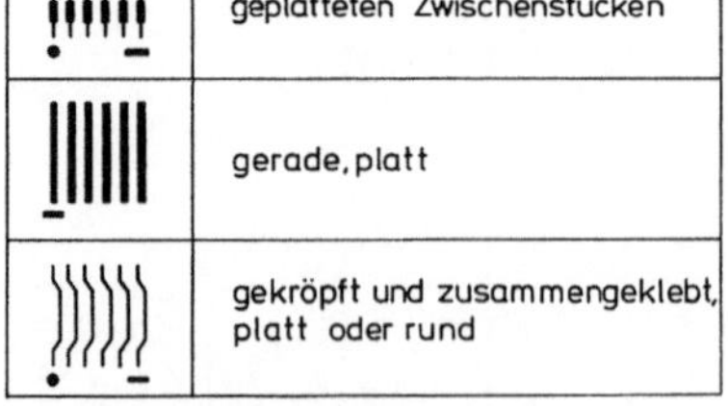

	Fasergeometrie:
	gerade, rund oder rechteckig
	gewellt, rund oder rechteckig
	gerade und rund mit abgeplatteten Zwischenstücken
	gerade, platt
	gekröpft und zusammengeklebt, platt oder rund

Abb. 8. Gebräuchlichste Fasergeometrie
Usual geometry of fibres

Der Einfluß des *l*/*d*-Verhältnisses (aspekt ratio) beeinträchtigt die Verarbeitbarkeit des Stahlfaser-Frischbetons und ist zu beachten. Als geeignet gilt ein Verhältnis $l/d = 60$—100; bei $l/d > 100$ können sich vermehrt Stahlfaserigel bilden.

Umfangreiche Untersuchungen über das Haftverhalten der Fasern liegen in einem Bericht des Zement- und Betoninstitutes Stockholm 1977 vor [4].

Faseranteil: Höherer Fasergehalt führt zu höheren Festigkeiten; unter 4 Gew.% ist bei glatten Fasern nur geringe Festigkeitserhöhung zu erwarten. Fasern mit Endverankerung ergeben die besten Versuchsergebnisse bezüglich des Arbeitsvermögens, jedoch nicht in bezug auf Verbesserung der Festigkeitsentwicklung. Dies ist so zu erklären, daß die Betondichte durch die aufgebogenen Anker gestört wird. Bekanntlich gibt es eine Relation zwischen der Betondichte und der Festigkeit.

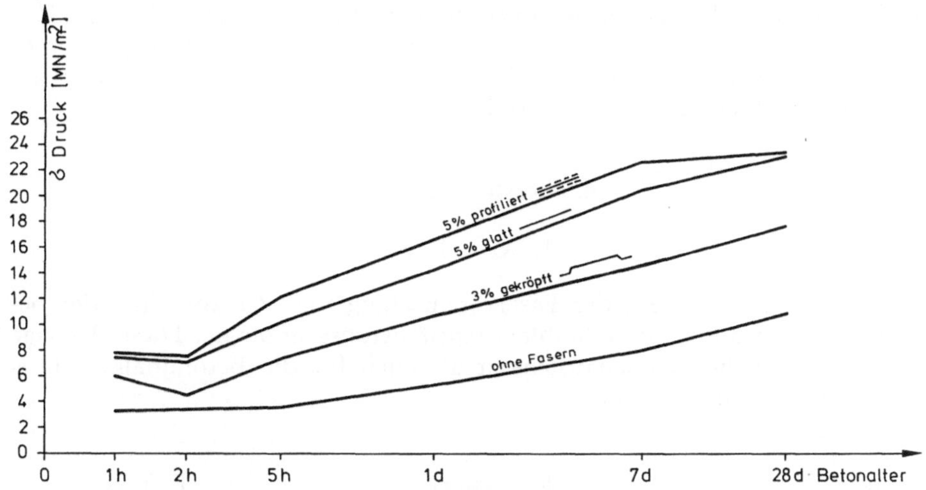

Abb. 9. Abhängigkeit der Druckfestigkeit von Fasertyp und Fasermenge nach eigenen Versuchen [5]

Relationship of compression strength to type and quantity of fibres according to our tests [5]

Sicher hat die Faserlänge einen erheblichen Einfluß, doch sind hier aus der Verfahrenstechnik der Spritzbeton-Herstellung Grenzen gesetzt.

2. Zuschlagstoffe

Bisher überwiegend nach Sieblinie B 8 verwendet, doch zeigen unsere Rückprallversuche, daß einige Abweichungen aus verfahrenstechnischen Gründen vorzunehmen sind. Eine geeignete Zusammensetzung ist beim Faserbeton noch wichtiger als beim normalen Spritzbeton. Grobes Korn > 8 mm ist aus der Matrixgeometrie nicht zu empfehlen, da Hohlräume entstehen.

3. Zement

Hierzulande vorwiegend HOZ 350 F, Menge ca. 400—450 kg/m³ auch wegen der geringeren Korrosionsgefahr.

Im Ausland werden Versuche mit Feinzementen gefahren. Es besteht hier die Meinung, daß dadurch auf den Schnellbinder verzichtet werden kann.

4. Wasser

Üblich wie beim Normalbeton oder Spritzbeton bekannt. WZ-Faktor ca. 0,4.

5. Zusatzmittel

Hier gibt es natürlich Unterschiede zwischen dem Naß- und dem Trockenspritzverfahren. Auf jeden Fall sind nur chloridfreie Zusatzmittel brauchbar. Die Zusatzmittel sind bisher wie beim üblichen Spritzbeton verwendet worden. Hier bietet die Entwicklung mit der Beigabe flüssiger Zusätze zusammen mit dem Wasser sicher einen Beitrag in Richtung homogener Zusammensetzung. Andere Versuche mit Neoprenlatex bewirkten geringere Festigkeiten, doch auch verminderten Rückprall.

4. Herstellvorgänge

1. Mischen

Hier ist das Problem der Faservereinzelung und Zugabe eines der bedeutendsten für die gesamte Stahlfaserspritzbetontechnologie. Diese Aussage gilt sowohl für die Wirtschaftlichkeit als auch für die Betonqualität. Drei verschiedene Arten sind möglich.

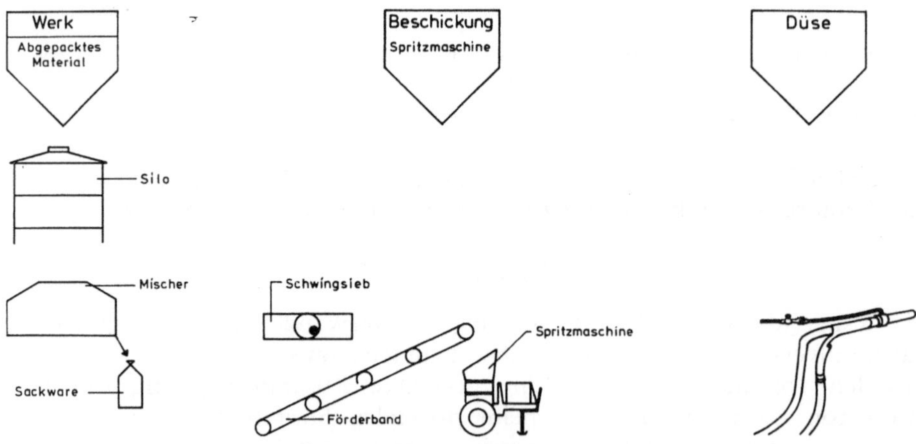

Abb. 10. Möglichkeiten für die Zugabe von Stahlfasern
Different techniques concerning addition of fibres

Es gibt bisher noch keine in großen Maßen erprobte Technologie, die Fasern zu vereinzeln und zuzugeben.

Wir haben uns bei unseren Versuchen mit Sackmaterial geholfen. In einem Betonwerk wurden Zuschlagstoffe, Zement und Schnellbinder zusammen mit den Fasern abgepackt. Die Fasern wurden chargenweise abgewogen und in die fertige Mischung gestreut. Als Mischer wurde ein Turbo-Schnellmischer mit 30 sec. Mischdauer verwendet. Das abgepackte Material, ca. 17 m³, war für den Versuch gut, ist aber für den Großeinsatz zu teuer.

Es gibt eine große Anzahl weiterer Versuche zur Vereinzelung. Von uns wurden vor 2 Jahren Versuche mit einem Schwingsieb zusammen mit der Bergbauforschung Essen betrieben, die jedoch wenig Aussicht auf Erfolg zeigten. Eine neue technologische Entwicklung zeigt sich in folgenden Verfahren, doch sind diese Einrichtungen auf dem Markt noch nicht zu finden.

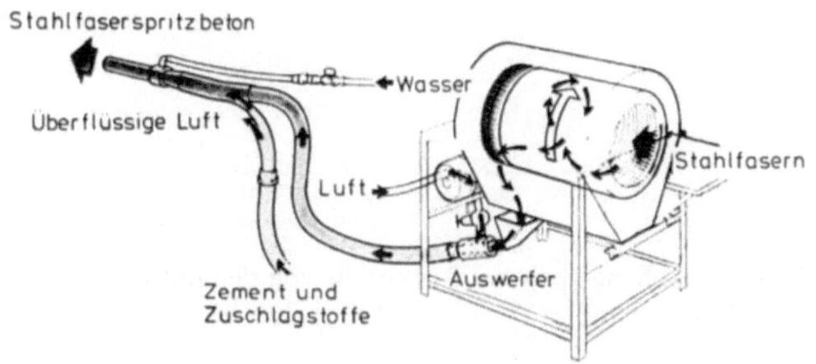

Abb. 11. Gesonderte Faserzugabe an der Düse
Separate addition of fibres at the nozzle

2. Fördern

Das Fördern von faserbewehrtem Pumpbeton ist weitgehend gelöst. Umfangreiche Versuche für einen Einsatz bei einem Sammler in Hamburg bei dem ca. 11000 m³ Faserbeton gepumpt werden sollen, wurden erfolgreich abgeschlossen. Das Größtkorn ist mit 16 mm vorgesehen. W/Z-Faktor = 0,4, 400 kg Zement, ca. 100 kg/m³ Fasern, geschnittener Bindedraht ⌀ 1 mm, $L = 45$ mm. Ein Betonverflüssiger von ca. 1% Gewichtsprozent (bezogen auf das Zementgewicht) wird zugegeben. Verdichten ist wegen Faserkonzentration nicht zu empfehlen. Ich komme auf diesen Einsatz bei den Beispielen nochmals zurück.

Unsere Versuche zusammen mit der Bergbauforschung Essen zeigten, daß beim Spritzen von Faserbeton weit mehr Rücksicht auf einzelne Komponenten zu nehmen ist als beim normalen Spritzbeton. So z. B.:

- Das Spritzgut muß ofentrocken sein.
- Die Druckluft muß trocken und kühl sein. Diese Luft neigt weniger zum Wasserabscheiden.
- Luftmeßinstrumente sind zur Steuerung der Förderluft (konstanten Beigabe) erforderlich. Die Lufthähne sollten besser regulierbar sein.

Folgendes Diagramm zeigt unsere Ergebnisse der Abhängigkeit zwischen Luftmenge und Rückprall. Es zeigt sich, daß der Rückprall bei 7—8 m³/min am geringsten ist und ein Optimum aufweist, an der Firste bei 9 m³/min. Vergleichbar ist das mit dem optimalen Andruck beim Bohren.

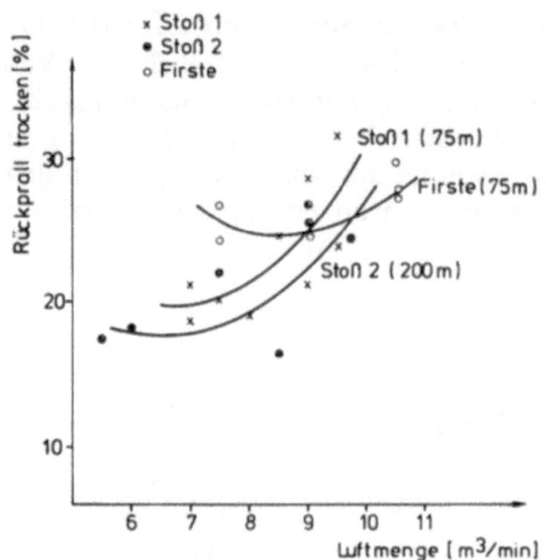

Abb. 12. Abhängigkeit des Rückpralls von der Luftmenge [5]
Relationship of the rebound to the air volume [5]

Unsere Erfahrungen mit Schlauchlängen von 200 m und 70 m zeigten, daß die derzeitigen Einrichtungen für das Faserspritzen Schwachstellen an

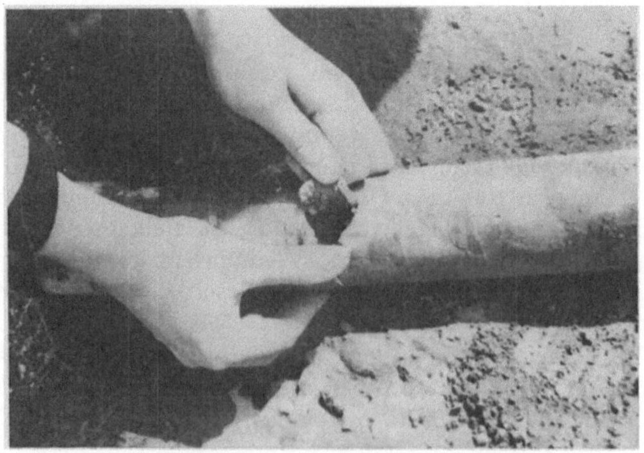

Abb. 13. Durch Verschleiß beschädigte Förderleitung
Damaged feed pipe caused by wear

den Krümmern und am Wasserzuleitungsring haben. Die 200 m Leitung zeigte natürlich wegen der größeren Anzahl solcher Schwachstellen größere

Schwierigkeiten. Zu empfehlen ist hier das Nachziehen einer Stahlleitung mit verhältnismäßig kurzer flexibler Leitung an der Düse. Hinter der Maschine sollte die Förderleitung gerade sein, das führte zu weniger Stopfern. Auch Leitungsschäden durch größeren Verschleiß sind hier gegenüber normalem Spritzbeton aufgezeichnet worden.

Insgesamt zeigte sich bei der von uns verarbeiteten Mischung, daß der zu hohe Feinanteilgehalt, der vielleicht für die niedrige Rückprallentwicklung ganz wünschenswert ist, neben Staub und Anfeuchtungsschwierigkeiten an der Düse, zusammen mit einem niedrigen Fasergehalt, zu vielen Stopfern führte (wegen Ablagerungen).

Das 4—8 mm Korn sollte mehr als 20% Anteil sein, wegen der besseren Förderfähigkeit. Ein größerer Faseranteil führte dagegen nicht zu mehr Stopfern, da die Fasern mit dem Größtkorn zusammen als Fließmittel dienen.

3. Spritzdüse

Meist werden gewöhnliche Spritzdüsen verwendet. Auch hier treten hohe Verschleißerscheinungen und Schwierigkeiten auf. Staubentwicklung, Rückprall und auch Betonqualität können durch Änderung der Düsenkonstruktion und der Wasserbenetzung verbessert werden.

4. Spritzverfahren und Spritzgeräte

Unsere Versuche werden nach dem Trockenspritzverfahren durchgeführt.

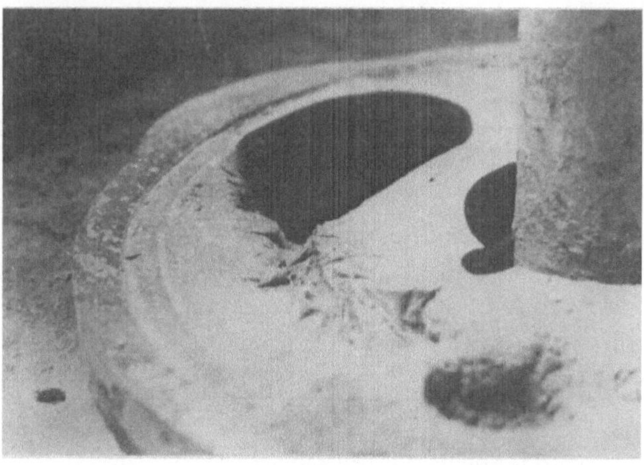

Abb. 14. Schäden an der Gummidichtung
Damaged rubber gasket

Es wird eine marktübliche Spritzmaschine verwendet, die einige Umbauten erfuhr und in Zukunft sicher noch weiter auf den speziellen Einsatz des Faserspritzens verbessert wird. So wurden bereits Veränderungen an Trichtern vorgenommen.

Es zeigte sich, daß eine höhere Materialsäule den Luftausstoß vermindert und dadurch die Beschickung der Förderleitung kontinuierlich wird. Unsere Verschleißbeobachtungen zeigten, daß am stärksten die oberen und unteren Gummiabdichtungen betroffen sind. Besonders bei geköpften Fasern treten erhebliche Schäden an den Rändern auf, die sogar die Fragen aufwerfen, ob ein wirtschaftliches Spritzen überhaupt noch möglich ist.

Eine Verstärkung mit Stahlkreissegmenten wurde schon ausprobiert. Die aufgebrachte Pressung der Gummiabdichtung auf die Reibscheibe sollte nur so groß sein, daß keine Fasern dazwischen gelangen. Dies ist sicher durch Umkonstruktionen zu erreichen. Wichtig ist auch, daß die Maschine an diesen Stellen täglich gereinigt werden muß, denn jedes Absetzen von Material stört den weiteren Betrieb.

5. Arbeitssicherheit

Das Spritzen mit Stahlfasern ist wegen des Rückpralls der Fasern für den Mann an der Düse und die Mannschaft in der Nähe — insbesondere für das Gesicht — sehr gefährlich. Selbst im Versuch konnten diese Gefahrenstellen erkannt werden.

Seit einiger Zeit wurden sogenannte Roboter eingesetzt, bei denen der Düsenführer in gesicherter Position arbeiten kann. Mal ganz abgesehen, daß schon bei normalem Spritzbeton die Tätigkeit des Düsenführers für den Stand unserer Technik nicht sehr erfreulich ist, meine ich, daß sich ein derartiger Einsatz, wenn er z. Z. auch noch nicht ganz befriedigend sein sollte, gar nicht zu umgehen ist.

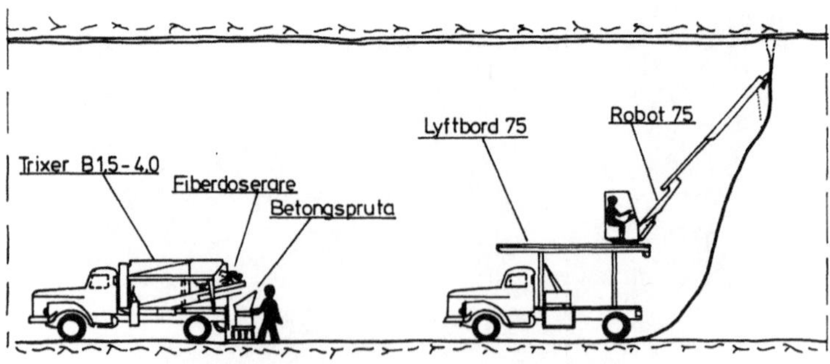

Abb. 15. Roboter für den Einsatz beim Spritzen [4]
Use of roboters at shotcreting [4]

Die Entwicklung wird auch hier weitergehen in Richtung Mechanisierung, z. B. Einrichtungen zur Regulierung des optimalen Düsenabstandes, Verbesserung in der Wasserzugabe.

6. Rückprall

Ich möchte hier auch noch einiges zum Rückprall sagen. Unsere Versuche haben folgendes Ergebnis gezeigt:

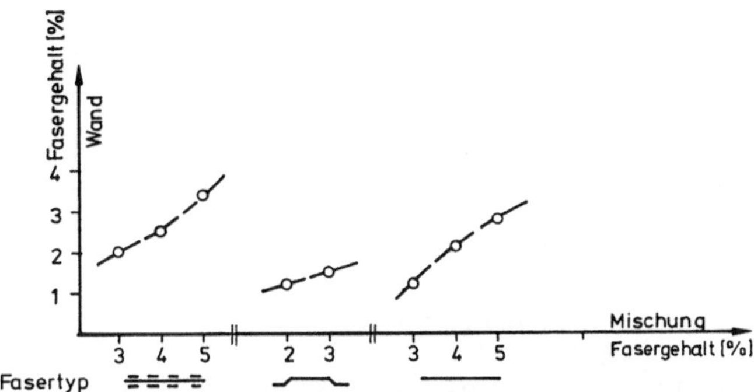

Abb. 16. Veränderung des Fasergehaltes durch den Spritzvorgang [5]
Variations in fibre content causes by shotcreting [5]

d. h., daß der Anteil der Fasern im Rückprall weit höher ist; dies wirkt sich natürlich maßgeblich auf die Wirtschaftlichkeit aus.

Das Hauptziel von weiteren Versuchen wird die Klärung der Einflußfaktoren auf den Rückprall sein, um die Verfahren wirtschaftlicher zu gestalten.

Unsere Erfahrungen zeigten schon, daß der größere Rückprall beim Anfangsspritzen gegen eine harte Fläche durch Vorspritzen einer unbewehrten Schicht erheblich abgemindert werden kann. Die Ausrüstungen müssen deshalb noch so flexibel sein, daß dies möglich ist. Abb. 20 zeigt einen Versuch auf einer Mülheimer Baustelle.

7. Übersicht des derzeitigen Standes

a) Betontechnologisch sind genügend Ergebnisse vorhanden, um zu sagen, daß die Materialeigenschaften des Stahlfaserbetons für den Tunnelbau interessant sind. Das Arbeitsvermögen wird erheblich verbessert, eine Zugfestigkeit kann garantiert werden. Die Risseverteilung wird günstig beeinflußt.

b) Die Technologie für eine wirtschaftliche Herstellung ist noch nicht in größeren Maßen erprobt. Hohe Rückprallwerte, hoher Verschleiß und häufige Störungen an der Maschine führen zu Unsicherheiten für die Anwendung.

c) Die Kosten sind noch zu hoch. Nicht nur die Faserkosten, sondern auch die Gesamtverarbeitungskosten und die Faservereinzelung und Zugabe spielt dabei eine bedeutende Rolle.

— Für unsere Versuche mit abgepacktem Sackmaterial à 50 kg kostet die Tonne DM 600,— bei ca. 4 Gew.% Faserzugabe.

— Ein silomäßiger Umschlag würde sich auf 200,— DM/t ermäßigen. Bei diesen Kosten ist zu berücksichtigen, daß hier das Spritzgut ofentrocken sein muß gegenüber ca. 2% Eigenfeuchte bei normalem Spritzbetongut.

d) Es müssen also Einsatzbereiche gesucht werden, die aufgrund der anderen Vorteile den höheren Preis rechtfertigen.

e) Es gibt noch keine allgemeine Zulassung und keine bekannten Berechnungssätze.

5. Berechnungsansatz [9]

Das im Hochbau übliche Bemessungsverfahren sieht vor, dem Bruchzustand die für den Gebrauchslastfall errechneten Schnittkräfte zugrundezulegen, indem man diese mit dem Sicherheitsfaktor ν multipliziert.

$$F_e = \frac{\nu}{\sigma_{eU}} \left(\frac{M_{eb}}{z} + \frac{\Delta M_e}{h - h'} + N \right)$$ BK 1. 1976 Bemessung der Stahlbetonbauteile

Die vereinfachende Annahme, daß die beiden Schnittkräfte M_e und N bis zum Bruch zueinander proportional bleiben, ist bei einem Hochbautragwerk insofern berechtigt, als normalerweise die Übertragung beider Schnittkräfte durch die einzelnen Elemente des Tragwerks für dessen Standfestigkeit erforderlich ist.

Die statische Wirkungsweise einer Tunnelsicherung unterscheidet sich hiervon in vieler Hinsicht. Für die Standsicherheit ist es meist lediglich erforderlich, daß die vorübergehende Sicherung Normal- bzw. Querkräfte überträgt. Ein Versagen des Tragwerks auf Biegung führt zum Beispiel bei Böden zu Bettungsreaktionen durch das umgebende Gebirge, so daß sich eine resultierende Belastungsform ergibt, für die die Tunnelschale ungefähr die Stützlinie bildet. Im Gegensatz zum Hochbau ist hier die Belastung mittelbar. Es erscheint naheliegend, bei der Bemessung von Tunnelschalen auf Bruchlast das Verhältnis des Biegemomentes zur Normalkraft offen zu lassen und stattdessen Proportionalität der Verformungen, also von Stauchungs- und Krümmungsänderung anzunehmen, zumal der Verformungszustand weitgehend durch die Bewegungen des umgebenden Gebirges bestimmt wird.

Ein Näherungswert für das Verhältnis Stauchung – Krümmungsänderung kann z. B. bei einer Berechnung am elastisch gebetteten Ring erhalten werden. Dabei ist für die Biegesteifigkeit ein geeigneter Wert anzusetzen. Nach Möglichkeit so, daß das aus der Ringberechnung erhaltene Biegemoment ungefähr dem entspricht, das sich bei der Deformation im Bruchzustand unter Berücksichtigung des Betonverhaltens ergibt.

Die Werte x, ε-Dehnung und ε-Stauchung ergeben sich aus dem Verhältnis Stauchung – Krümmung.

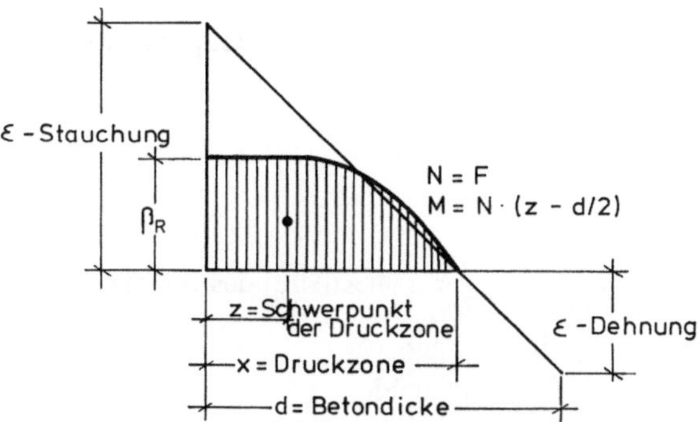

Abb. 17. Stauchung und Dehnung im Beton
Compressive strain und stress in concrete

6. Berechnung des Sicherheitsfaktors für die Normalkraft in der Spritzbetonschale

Beispiel

Schnittkräfte in der Ulme (aus Gebrauchslast). Die Schale wurde als elastisch gebetteter Ring berechnet.

1. *Normaler Spritzbeton, Schalenstärke 15 cm*

$$M = -0{,}77 \text{ Mp} \cdot \text{m/m}$$

$$N = -42{,}1 \text{ Mp} \cdot \text{m/m}$$

$$\text{Ausmitte } M/N = 0{,}0183$$

(Dies kommt näherungsweise hin für Faserbeton mit 10 cm Wandstärke.)

$$\sigma_N = N/0{,}15 = 28{,}1 \text{ kp/cm}^2$$

$$\sigma_M = M \cdot 6/0{,}15^2 = 20{,}5 \text{ kp/cm}^2$$

$$\varepsilon_N = \frac{1}{300\,000} \cdot \sigma_N = 960 \cdot 10^{-7}$$

$$\varepsilon_M = \frac{8^*}{300\,000} \cdot \sigma_M = \pm 5467 \cdot 10^{-7}$$

$$\varepsilon_N + \varepsilon_M = 6427 \cdot 10^{-7} = 0{,}6427\ ^0/_{00} \Rightarrow \text{Stauchung}$$

$$\varepsilon_N - \varepsilon_M = -4507 \cdot 10^{-7} = -0{,}4507\ ^0/_{00} \Rightarrow \text{Dehnung}$$

* Das Material wurde auf Biegung achtmal so weich angenommen, um die Aufweichung durch Zugriß und Plastifizierung auszugleichen.

Bruchzustand: Verformungen wachsen proportional.

Stauchung: 3,5 ‰ (nach DIN 1045).

Dehnung: $0{,}4507 \cdot 3{,}5/0{,}6427 = 2{,}45$ (zurückgerechnet aus dem Verhältnis)

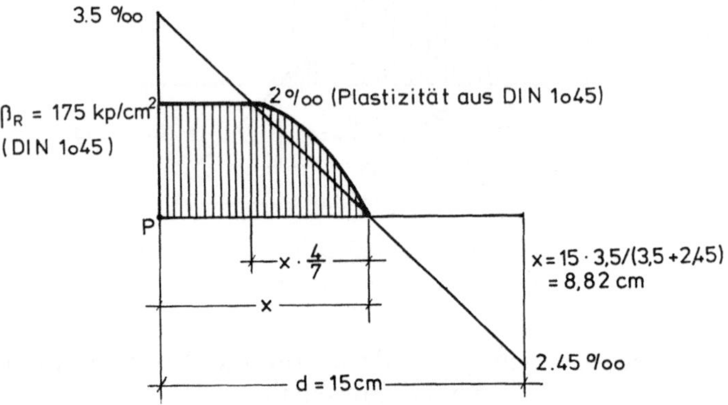

Abb. 18. Stauchung und Dehnung im Beton ohne Stahlfasern

Compressive strain und stress in concrete non-containing steel fibres (thickness of shotcrete d = 15 cm)

$$N = \beta_R\, x \left(1 - \frac{4}{7} \cdot \frac{1}{3}\right) = \beta_R\, x \cdot 0{,}8095 = 125 \text{ Mp}$$

Statisches Moment der Druckverteilung um P

$$M_s = \beta_R\, x^2 \left(\frac{1}{2} - \frac{4}{7} \cdot \frac{1}{3} \cdot \frac{6}{7}\right) = \beta_R\, x^2 \cdot 0{,}3367$$

Schwerpunktslage zu $P = M_s/N = x \cdot 0{,}3367/0{,}8095$

$= 8{,}82 \cdot 0{,}4159 = 3{,}668$ cm

Ausmitte: $1/2 \cdot 15 - 3{,}668 = 0{,}0383$

Sicherheit: $\eta = 125/42{,}1 = 2{,}97$

2. Faserspritzbeton $d = 15$ cm

Hier wird ein dreimal so großes Arbeitsvermögen angesetzt wie bei normalem Beton (eventuell etwas zu hoch angesetzt).

Druckfestigkeitserhöhung 15%, Zugfestigkeitserhöhung 0.

Bruchstauchung: $3{,}5 \cdot 3 = 10{,}5$ ‰.

Plastizitätsgrenze ist nach wie vor bei 2 ‰, β_R wird um 15% erhöht.

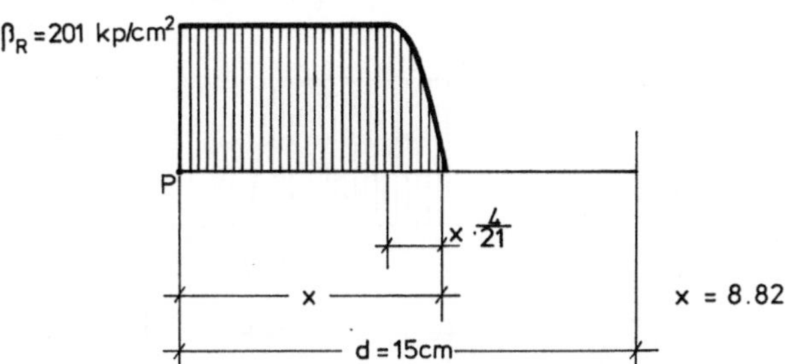

Abb. 19. Druckzone im Beton mit Stahlfasern
Zone of pressure in concrete containing steel fibres

$$N=\beta_R\,x\left(1-\frac{4}{21}\cdot\frac{1}{3}\right)=\beta_R\cdot x\cdot 0{,}9365=166\ \mathrm{Mp}$$

$$M_s=\beta_R\,x^2\left(\frac{1}{2}-\frac{4}{21}\cdot\frac{1}{3}\cdot\frac{20}{21}\right)=\beta_R\,x^2\cdot 0{,}4395$$

Schwerpunktslage zu P $M_s/N=x\cdot 0{,}4395/0{,}9365=4{,}14$ cm.

Ausmitte: $7{,}5-4{,}14=\underline{3{,}36\ \mathrm{cm}}$.

Sicherheitsfaktor: $\eta=166/42{,}1=3{,}94$.

Gesamtzuwachs des Sicherheitsfaktors: $166/125=1{,}33$.

Der Zuwachs des Sicherheitsfaktors, der allein durch das größere Arbeitsvermögen des Faserbetons bedingt ist, beträgt: $1{,}33/1{,}15=1{,}16$.

3. Faserbeton d = 10 cm

Es wird das gleiche Verhältnis zwischen Stauchung und Verkrümmung angesetzt.

Arbeitsvermögen 3fach, Druckfestigkeitserhöhung 15%.

Zugfestigkeitsvermögen wurde nicht angesetzt.

$$\varepsilon_N=960\cdot 10^{-7}$$

$$\varepsilon_M=5467\cdot 10^{-7}\cdot 0{,}1/0{,}15=3645\cdot 10^{-7}$$

$$\varepsilon_N+\varepsilon_M=4605\cdot 10^{-7}=0{,}4605\ ‰$$

$$\varepsilon_N-\varepsilon_M=-2685\cdot 10^{-7}=-0{,}2685\ ‰$$

Stauchung: 10,5 ‰

Dehnung: $0{,}2685 \cdot 10{,}5/0{,}4605 = 6{,}122$

$$x = 10 \cdot 10{,}5/(6{,}122 + 10{,}5) = 6{,}317 \text{ cm}$$

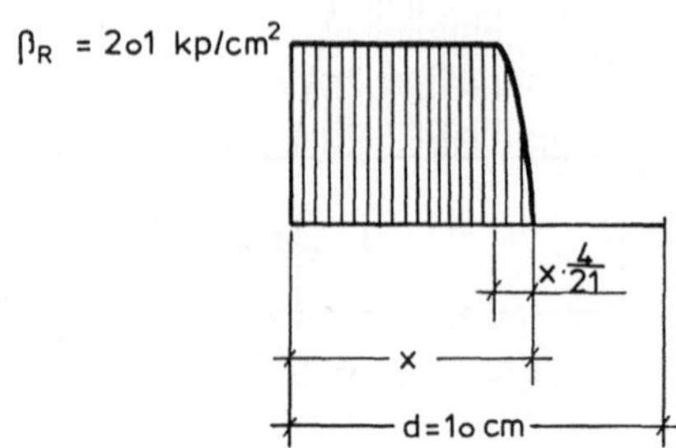

Abb. 20. Druckzone im Beton mit Stahlfasern
Zone of pressure in concrete containing steel fibres

$$N = \beta_R \cdot x = 0{,}9365 = 119 \text{ Mp}$$
$$M_s = \beta_R \cdot x^2 \cdot 0{,}4395$$

Schwerpunktslage zu P $M_s/N = x \cdot 0{,}4395/0{,}9365 = 2{,}96$ cm

Ausmitte: $5{,}0 - 2{,}96 = \underline{2{,}04 \text{ cm}}$ $\sim 1{,}83$

$\eta = 119/42{,}1 = \underline{2{,}83}$ $\sim$ unbewehrter Beton mit 15 cm Stärke

$$\eta = 2{,}97.$$

7. Anwendung im Tunnelbau

Ich möchte 4 Thesen voranstellen, die zugegeben sehr optimistisch sind:

1. Der moderne Tunnelbau geht von einer verformbaren vorübergehenden Sicherung aus. Eine dünnere Schale mit geringer Steifigkeit und mit der Fähigkeit elastische Gelenke zu bilden, kommt dieser Forderung besser nach. Stahlfaserbeton erfüllt diese Eigenschaften durch das größere Arbeitsvermögen besser als bisherige Ausbausysteme mit Spritzbeton.

2. Die Standzeit des Gebirges ist — wie der Begriff schon sagt — zeitabhängig und bestimmt somit das Verfahren, wie schnell eine vorübergehende Sicherung eingebracht werden kann. Faserspritzbeton hat nur einen Herstellungsgang. Gehen wir optimistisch vor, die Stahlbögen und die Matten können bei gebrächem und druckhaftem Gebirge entfallen. Bei Gebirgen mit geringer Standzeit könnte also in den Bauverfahren eine Weiterentwicklung eintreten; nämlich in der Größe und der Anordnung von Teilausbrüchen, oder zu noch mehr Mut zu Vollausbrüchen.

3. Der Stahlfaserbeton kann ein Beitrag zur einschaligen Bauweise sein. Die bessere Rissverteilung und das günstigere Schwindverhalten können zur besseren Wasserdichtigkeitseigenschaft führen. Dieser Nachweis ist noch

nicht geführt. Darüber laufen an der RUB Untersuchungen. Wir haben nicht die Probleme der Überdeckung, der genauen Lage der Bewehrung und des Verbundes, allerdings noch offene Fragen bezüglich der Korrosion.

4. Die erhöhte dynamische Beanspruchbarkeit kann für die Auskleidung von Druckstollen und Ölkavernen aufgrund der dort auftretenden hohen Wechselbeanspruchungen genutzt werden.

Weitere Anwendungsbeispiele sind:

1. Einschalige Konstruktion nach einer wandernden vorübergehenden Sicherung, z. B. nach einem Messerschild oder Schild mit einer Gleitschalung. Solche Verfahren wurden schon lange verfolgt, aber nicht hinreichend gelöst.

Bisher waren solche Verfahren wegen des Einbaues der Bewehrung nicht möglich. Nur mit sehr langen Nachlaufmessern und umzusetzender Schalung konnte ein Teil dieses Gedankens verwirklicht werden. Mit Faserbeton entfällt der Einbau der Bewehrung, die Schalung braucht nicht geöffnet zu werden. Es gibt hier noch eine Reihe anderer Probleme beim Gleiten, auf die ich hier nicht eingehen will. Doch zeigt ein anstehendes Projekt in Hamburg-Harburg, daß ein weiterer Schritt in dieser Richtung unternommen wird.

INNENSCHALUNG ALS UMSETZSCHALUNG

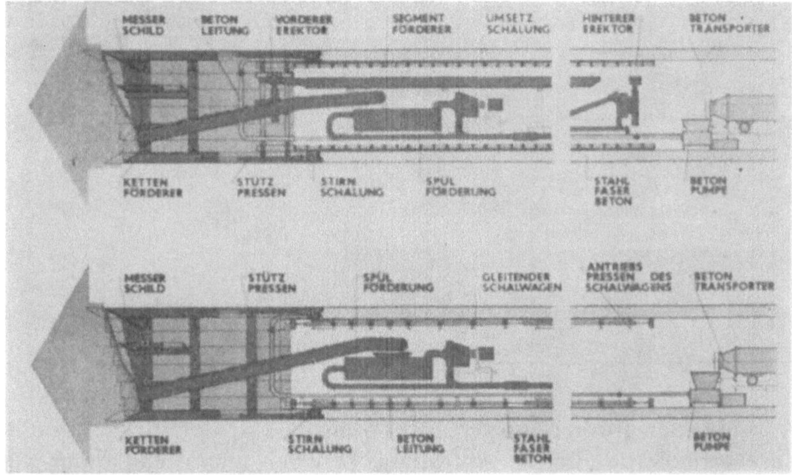

INNENSCHALUNG ALS GLEITSCHALUNG

Abb. 21. Sammler Hamburg-Harburg
Sewer

Daß die geplante Gleitschalung wie im Sondervorschlag vorgeschlagen nicht zum Einsatz kommt, liegt wohl daran, daß nicht alle Probleme in ein Projekt gepackt und dort gelöst werden können. Es handelt sich ja hierbei um einen Auftrag und nicht um ein Forschungsprojekt. So wird die Schalung nach kontinuierlichem Pumpen und Verschiebung der Stirnschalung umgesetzt.

2. Ein großer Anwendungsbereich kann sich im Bergbau ergeben. Der Bergbau ist bisher nicht als besonders betonfreundlich zu bezeichnen, und zwar liegt das an dem für den Bergbau ungeeigneten Bruchverhalten des Betons. Stahlfaserbeton ist durch das größere Arbeitsvermögen für den Bergbau ein interessanter Baustoff geworden.

Im Bergbau sind auch die höheren Kosten nicht von so großer Bedeutung, wenn dadurch erhebliche Vorteile gegenüber dem bisherigen Ausbau erreicht werden können. Größere Probleme stellen sich jedoch derzeit noch in der Förderung des Spritzbetons zum Einbauort.

3. Probeneinsatz im U-Bahnbau in Mülheim.

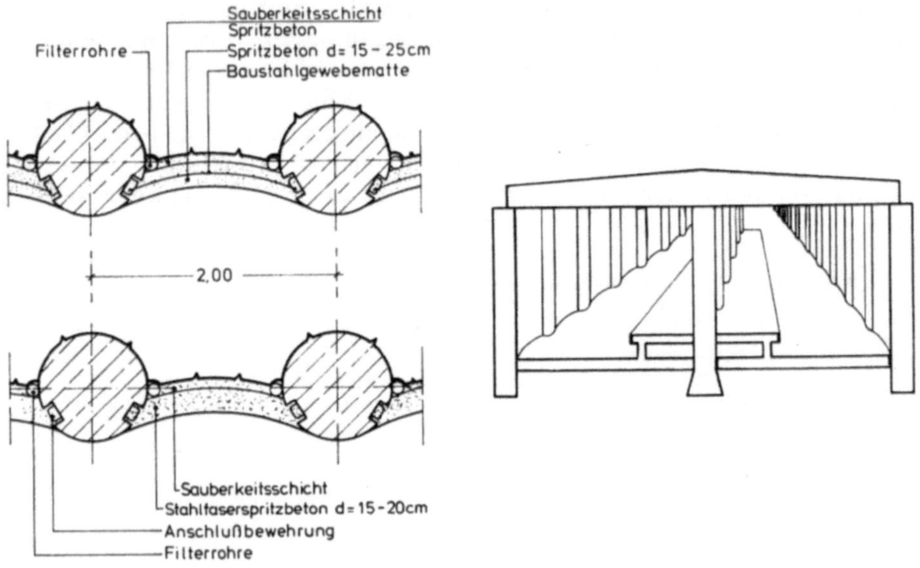

Abb. 22. Aufgelöste Bohrpfahlwand mit Spritzbetonkappen

Bored pipe wall in open order with shotcrete caps

4. Eines der größten ausgeführten Beispiele ist der Einsatz von 4500 m³ Stahlfaserspritzbeton für ein Auffangbecken bei einem Tanklager der Scan-Raffinerie, Brofjorden [4]. Wegen der Forderung nach möglichst rissefreiem Beton war ein Vergleich mit Stahlfaserspritzbeton schon im Jahre 1974 wirtschaftlicher.

Nach 2 Jahren wurden Kontrollen durchgeführt und es zeigte sich, daß man auch heute noch mit der damaligen Entscheidung voll einverstanden ist.

8. Weitere Forschungen an der RUB

Wir bearbeiten zur Zeit in Verbindung mit Stahlfaserbeton 3 Forschungsprogramme:

1. Frühfestigkeitsentwicklung von Stahlfaserspritzbeton. Mit der Bergbauforschung in Essen wurden auf dem Versuchsstand folgende Versuche gefahren.

— Spritzen von 17 m³ Faserspritzbeton.

— 10 Versuchsreihen mit verschiedenen Faserarten und Fasermengen.

Abb. 23. Spritzwand auf der Anlage Bergbauforschung Essen

Test wall on the plant of "Bergbauforschung Essen"

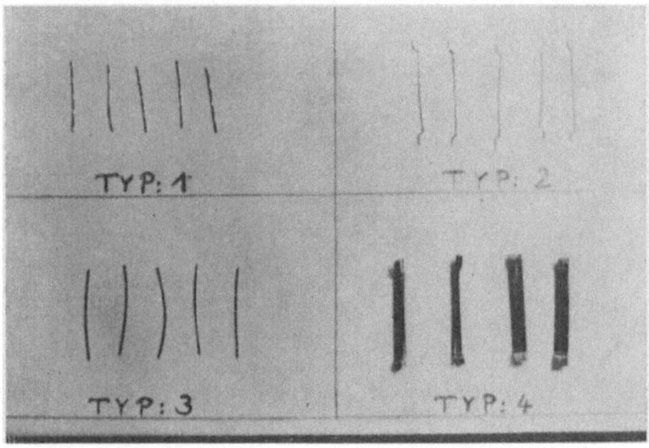

Abb. 24. Fasertypen für Versuchsprogramm

Type of fibres for test programme

2. Weitere Versuche mit Ausfallkörnung, um Aussagen über den Rückprall zu erhalten.

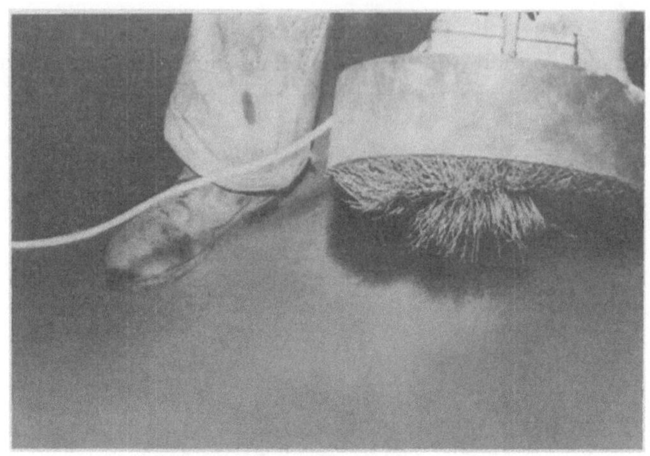

Abb. 25. Einsammeln der Stahlfasern (Rückprallanalyse)
Collecting of steel fibres (rebound analysis)

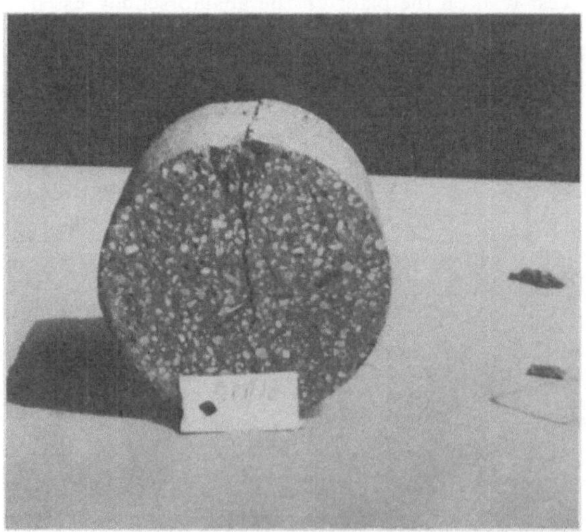

Abb. 26. Spaltzugversuch mit Fasern
Splitting-tensile test with fibres

3. Rückpralluntersuchungen mit deren Einflußfaktoren aus der Verfahrenstechnik und Wasserdichtigkeitsuntersuchungen.

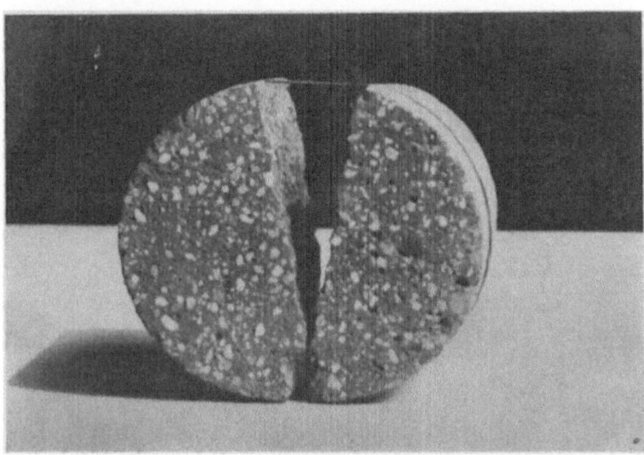

Abb. 27. Spaltzugversuch ohne Fasern

Splitting-tensile test without fibres

Das Forschungsprogramm wird durch das Land Nordrhein-Westfalen finanziert.

Der Versuchsstand ist an der Ruhr-Universität aufgebaut und wurde gerade in Betrieb genommen.

Ich hoffe, daß ich später noch einmal über die Erkenntnisse berichten kann.

Literatur

[1] Zerna, W., Schnütgen, B., Stiller, W.: 1. und 2. Bericht über die bautechnische Erprobung von Stahlfaserspritzbeton in Untertageanlagen der Bundeswehr 1975/1976.

[2] Schnütgen, B.: Das Festigkeitsverhalten von mit Stahlfasern bewehrtem Beton unter Zugbeanspruchung. Technisch-wissenschaftliche Mitteilungen der RUB, Mitteilung Nr. 75-8.

[3] Wischers, G.: Faserbewehrter Beton. Beton *3* (1974).

[4] Nordforsk's projektkommitte för FRC-material Cement-och betonginstitutet Stockholm, 1977.

[5] Maidl, B., Rapp, R.: Versuchsberichte über Stahlfaserspritzbeton bei der Bergbauforschung Essen. (Unveröffentlicht).

[6] Maidl, B., Rapp, R.: Versuchsbericht über Baustelleneinsatz von Stahlfaserspritzbeton auf der U-Bahnbaustelle Oberer Hingberg Mühlheim. (Unveröffentlicht).

[7] Lucke, W. N.: Versuchs- und Entwicklungsarbeiten mit Tunnelauskleidungen in den USA. Forschung und Praxis, Heft 15.

[8] Maidl, B., Rapp, R.: Spritzbeton im Tunnel- und Stollenbau. Bericht im Taschenbuch für Tunnelbau. Essen: Verlag Glückauf 1978.

[9] Maidl, B., Geißler, E.: Dimensionierung von Faserspritzbetonsicherungen für das U-Bahnbaulos 20 Frankfurt und dem Sammler Hamburg–Harburg. (Unveröffentlichte Sondervorschläge 1976).

Anschrift des Verfassers: Prof. Dr.-Ing. B. Maidl, Ordinarius am Institut für konstruktiven Ingenieurbau, Lehrstuhl für Bauverfahrenstechnik und Baubetrieb, Ruhr-Universität Bochum, Universitätsstraße 150, D-4630 Bochum, Bundesrepublik Deutschland.

Rock Mechanics, Suppl. 7, 249—265 (1978)

Rock Mechanics
Felsmechanik
Mécanique des Roches

Optimierung des Ausbaues in untertägigen Hohlräumen Beispiel: Ankerausbau

Von

S. Gałczyński und **J. Dudek**

Mit 8 Abbildungen

Zusammenfassung — Summary

Optimierung des Ausbaues in untertägigen Hohlräumen. Beispiel: Ankerausbau. Beim Ankerausbau kann man zwei Arbeitsgänge des Systems Anker und Gebirge unterscheiden. In dem ersten nimmt man an, daß der Ausbau eine steife Konstruktion bildet (steife Platte, steifes Gewölbe). In diesem Zusammenhang ist das System Anker – Gebirge mit einem Druck belastet, der aus dem primären Spannungszustand hervorgeht.

Die Autoren stellen hier eine neue Lösung vor, wobei ein nachgiebiges System betrachtet wird. Die verankerte Zone bildet eine Konstruktion, die sich verschieben kann. Die Verschiebung bewirkt eine Entspannung des Hangenden und der Druck auf den Ausbau fällt bis zum Wert des Auflockerungsdruckes.

Man hat Formeln ermittelt, die die Berechnung der Tragfähigkeit des Ankerausbaues ermöglichen. Am Ende wird der Einfluß der Bogenpfeilhöhe auf die Stabilität der Firste analysiert.

Optimization of Lining in Underground Excavations. Example: Roof Bolting System. The analysis of the performance characteristic of the roof bolting system is presented in the paper. The elements of such a system are the different types of bolts and the bolted rock. The load-bearing capacity of a separate bolt is a sufficiently recognized problem, but this problem for the arrangement of several bolts together with rock mass requires the separate optimum solution. Treating the bolts as the elements anchoring the stratified rock, suspending the immediate roof to the safe roof or the loose block to the pressure arch or, finally, as the elements reinforcing the rock mass, it is always considered that the roof bolting system is a rigid, statical arrangement. In such arrangement the bolts can be applied in the shallow excavations set in a strong rock. From the analysis of the formulas for calculating the support factors results that from the technical and economical point of view, the rigid arrangement is not advantageous for the deep excavations in a weak rock.

The term of susceptible roof bolting system is introduced. The system is created by the part of the bolted roof which works on the principle of a piston shifting along the surfaces of shearing of the overburden and bearing the minimal pressure of the completely decompressed rock mass. The equations to calculate the susceptible roof bolting system parameters as length and load of bolts are given. The bases of selection of the susceptible roof bolting system confirm the possibility and the technical and economical usefulness of its application in the difficult geotechnical conditions.

1. Einführung

Der Bau von Hohlräumen, wie Tunneln, Kavernen, Strecken und Stollen, spielt besonders im Bergbau, aber auch im unterirdischen Verkehrswegebau und sonstigen unterirdischen Objekten, eine sehr wichtige Rolle. Zumeist handelt es sich um Hohlräume für dauernden oder langfristigen Bestand, die im allgemeinen einen ausreichenden standfesten Ausbau erfordern. Nur unter geologisch besonders günstigen Bedingungen oder bei kurzfristiger Inanspruchnahme kann ein Ausbau entfallen bzw. sich auf örtliche Sicherungen beschränken.

Die Zahl der Strecken ohne Ausbau wird im polnischen Bergbau immer geringer. Es geht im folgenden um eine Methode zur Feststellung von Parametern, die eine optimale Ausnutzung des Hohlraumes ermöglichen. Die bisher festgestellten Kriterien [2] zur Optimierung der Dimensionen von Hohlräumen ohne Ausbau ermöglichen den Entwurf sehr wirtschaftlicher Lösungen.

2. Theorien über das Zusammenwirken von Gebirge und Anker

Das Problem wird jedoch kritisch, wenn der Hohlraum nicht stabil ist und ein Ausbau unbedingt erforderlich wird. Die Wahl des Ausbaues und seine Optimierung sind dann schwer zu lösen, da sie von vielen verschiedenen Faktoren abhängig sind. In dieser Arbeit soll versucht werden, auf der Basis des „Ankerausbaues" Lösungen zu formulieren, die zu einer wirtschaftlich günstigen Ausbauart führen.

Zur Zeit gibt es schon eine große Zahl verschiedener Ankerausbauarten, für die genügend sichere Verfahren bestehen, sie bei unterschiedlichen geologischen Bedingungen einzubauen. Es gibt auch bereits eine Reihe von Kriterien für die Festlegung der Ankerlänge und des Abstandes der Anker [4, 5, 6, 7]. Ähnlich wie bei den Strecken ohne Ausbau wird die beste Lösung dann erzielt, wenn eine maximale Ausnutzung der Gebirgs-Tragfähigkeit sowohl in der nicht gestörten als auch in der durch Abbauarbeiten gestörten Zone möglich ist. Auch die Wahl zusätzlicher Sicherungsmittel zur Erhöhung des Ausbauwiderstandes, etwa durch Baustahlgewebe, Spritzbeton usw., erfordert eine Erarbeitung von Optimierungskriterien. Vom Standpunkt des Zusammenspieles von Ankern und Gebirge kann man zwei Typen des Ankerausbaues unterscheiden:

1. Der Anker als individuelles Konstruktionselement mit einer bestimmten Tragfähigkeit der Ankerstange, des Kopfes und des Endstückes in Verbindung mit einer zusätzlichen Sicherung des Abbauraumes.
2. Anker als mittragende Elemente, die gemeinsam mit dem Gebirgsmassiv einen Ausbau um den Abbauraum herum bilden.

Stellt die Beurteilung der Festigkeit einer einzelnen Ankerstange fast kein technisches Problem dar und gibt es die Möglichkeit einer ständigen Kontrolle des Ankerzustandes durch Messung der Vorspannung, so erfordert die Bestimmung der Tragfähigkeit des Ausbaues als Einheit eine genauere Analyse der Mitwirkung des Gebirges. Dieses muß einmal als Belastung, ein anderes Mal als ein mit dem Anker wirkendes Element betrachtet werden. Die optimale Auswahl der Ausbauparameter bezieht sich auf die richtige Berechnung der Länge, des Abstandes und der Vorspannung der Anker sowie der Maßnahmen zur zusätzlichen Sicherung des Abbauraumes.

Im Lichte der bisher festgestellten Ankerarbeitstheorien kann man die Wirkung der Ankerung betrachten als [5]:

- Zusammenheften der geschichteten Firste,
- Anheften der Firstzone an die sich außerhalb der Reichweite der Gewölbewirkung befindlichen Gesteine,
- Anheften der Schichten einer schwachen Firste an das höher liegende feste, nicht aufgelockerte Gebirge,
- Armierung des Gebirgsbalkens, analog der Bewehrung eines Stahlbetonbalkens,
- Elemente, die den Spannungszustand im Gebirgsmassiv beeinflussen, indem sie vorwiegend Zugspannungen eliminieren,
- Verbesserung der Gebirgseigenschaften, z. B. durch Übernahme von Zugspannungen am Rande des Abbauraumes.

Aus vorstehender Aufzählung geht deutlich hervor, daß die Anker Haupttragelemente sind und der Ausbau, als Einheit, die Rolle einer tragenden Konstruktion erfüllt, die mehr belastet ist als jeder nachgiebige Ausbau.

3. Ankerausbau als steife tragende Konstruktion

Das Zusammenspannen der geschichteten Firste vergrößert deren Steifigkeit, wodurch Verschiebungen durch Biegung und durch Entspannung des Hangenden eingeschränkt werden. Außerdem wird durch diese Erhöhung der Steifigkeit die Firste als eingespannte Platte oder Balken wirksam, wobei über den Auflagern (Stößen) eine Konzentration der Spannungen — hauptsächlich der Schubspannungen — auftritt [8]. Das Anheften der aufgelockerten Firstpartien außerhalb der Zone der Gewölbewirkung oder das Anheften dieser Schichten an die höher liegenden festen Gesteine stellt, wenn auch nur zum Teil, den primären Spannungszustand im Hangenden wieder her.

Eine Armierung des „Gebirgsbalkens“ oder Verfestigung dieses Bereiches durch Ankerungen beugt der Zerstörung der Firste direkt über dem Abbauraum vor. Auch die Spannungsänderung um den Abbauraum herum kann man als Wiederherstellung des primären Spannungszustandes betrachten. In den folgenden Untersuchungen der Ankerwirkungsweise wird angenommen, daß keine Möglichkeit zur Entspannung des Gebirges in und hinter der verankerten Zone gegeben ist. So betrachtet stellt sich der Ausbau als steifes, statisches System dar.

3.1 Verankerte Firste als steife Platte

Wenn man den durch Anker abgetrennten Gebirgsteil oder einige zusammengeheftete Gebirgsschichten als eingespannte oder frei unterstützte Platte betrachtet, erhält man aus der Gleichung der Momente (Abb. 1) [8]:

$$\frac{4\,p_z\,b^2}{16} = \frac{l^2}{6} \cdot R_z, \tag{1}$$

wobei p_z = primärer Spannungszustand,
b = halbe Breite des Hohlraumes,
l = Länge der Anker,
R_z = Zugfestigkeit des Gebirges bei der Biegung.

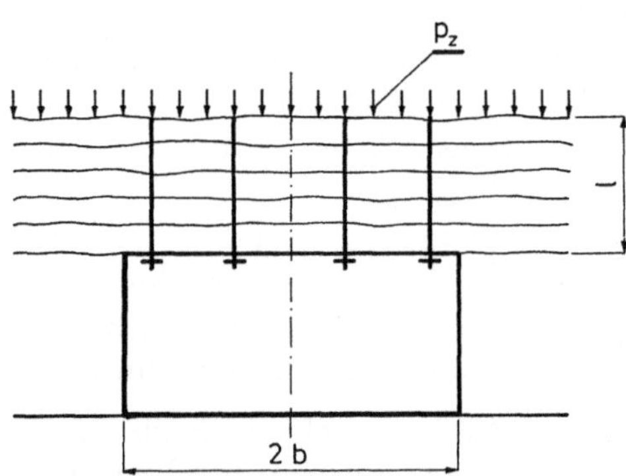

Abb. 1. Schema des Abbauraumes, der mit Ankern ausgebaut ist. Anker und Gebirge bilden eine steife Platte, die mit dem primären Druck belastet ist

Scheme of bolted excavation. Bolts and rock create the rigid plate loaded by primary pressure

Aus Gl. (1) geht hervor, daß sich die Längen der Anker, die das ganze Gewicht des Hangenden übertragen könnten, ergibt mit:

$$l = b\sqrt{\frac{3}{2}\,\frac{p_z}{R_z}} \tag{2}$$

Zur Eliminierung der horizontalen Verschiebung der Schichten oder einer Abscherung des Gebirges in den vertikalen Flächen werden die Anker vorgespannt. Diese Vorspannung überträgt sich auf das Gebirge und erzeugt bei Gebirgsbewegungen mit Gleitung auf den Flächen Reibungskräfte. Es sind dies Hauptkräfte, die den Widerstand der horizontalen Gleitung der zusammengehefteten Gebirgsschichten erzeugen. Beim Abscheren in einem homogenen oder geschichteten Gebirge wirkt außer der inneren Reibung auch dessen Kohäsion.

Vom technischen Standpunkt kann vereinfachend für die Schätzung der Scherfestigkeit angenommen werden, daß sie einem Widerstand entspricht, dessen Größe direkt proportional zur Ankervorspannung ist. Die Proportionalitätsfaktoren sind dabei die Koeffizienten der inneren Reibung des Gebirgsmassivs im Sinne der Protodjakonovschen Koeffizienten. Im geschichteten Gebirge berücksichtigen sie nur die reine Reibung, im homogenen Gebirge sind sie um den Wert der Kohäsion vergrößert.

Diese angenommene Vereinfachung gestattet Lösungen, die zuverlässiger sind als die mit gesonderter Berücksichtigung von Reibungskräften und Kohäsion. Es gibt eine sehr große natürliche Streuung der Kohäsion im Gebirgsmassiv und auch eine große Empfindlichkeit der Formeln, die diese Parameter berücksichtigen. Schon kleine Änderungen bewirken daher, daß sich für die berechneten Größen, z. B. den Gebirgsdruck, keine vergleichbaren Ergebnisse ergeben, obwohl die technischen Bedingungen sich nicht viel änderten.

Aus diesen Annahmen folgt, daß sich ein Gleichgewichtszustand der Gebirgsschichten in der horizontalen Richtung erst ergibt, wenn die Reibungskräfte zwischen den Schichten den Wert der Scherkräfte erreichen werden, d. h.:

$$n \cdot N \cdot f' = \frac{3\, p_z\, b^2}{2\, l}. \tag{3}$$

Aus dieser Gleichung kann man die erforderliche Vorspannung bei bekannter, aus Gl. (2) ermittelter Ankerlänge bestimmen:

$$N = \frac{\sqrt{6\, p_z \cdot R_z}}{2 \cdot n \cdot f'} \cdot b. \tag{4}$$

Ähnlich lautet die Gleichung des Gleichgewichtszustandes bei Abscheren eines homogenen Massivs oder von Gebirgsschichten:

$$\frac{n \cdot N}{2\, b} \cdot K \cdot l \cdot f = p_z \cdot b. \tag{5}$$

Die Größe der Vorspannung wird in diesem Falle:

$$N = \frac{2\sqrt{6\, p_z \cdot R_z}}{3\, n \cdot K \cdot f} \cdot b. \tag{6}$$

Hierbei sind

n — die auf 1 m Länge des Abbauraumes bezogene Anzahl der Anker,

f', f — entsprechender Koeffizient der Gebirgsreibung zwischen den Schichten ohne Kohäsion und in einer beliebigen Fläche mit Kohäsion,

K — Seitendruckkoeffizient.

Werden die Gln. (4) und (6) miteinander verglichen, stellt man fest, daß bei der Berechnung von

$$f' \geqslant \frac{3}{4} K \cdot f \tag{7}$$

die Gl. (6) benutzt werden muß.

Für Gesteine, die zur Ankerung verwendbar sind, soll der Koeffizient f mindestens die folgende Ungleichung erfüllen:

$$f \geqslant 3. \tag{7a}$$

Wenn man den Seitendruckkoeffizienten durch die Coulombsche Formel ausdrückt, welche die Abhängigkeit vom Winkel der inneren Reibung darstellt:

$$\varphi = \operatorname{arctg} f \quad \text{und} \quad K = \frac{1}{(\sqrt{f^2+1}+f)^2} \cong \frac{1}{4f^2} \tag{7b}$$

so erhält man:

$$f' \geqslant \frac{1}{16}.$$

Praktisch erfüllt jedes beliebige Gebirge die Bedingung (7) und muß nicht nach der Gl. (4) geprüft werden.

Wie Modellversuche beim Einbau von Ankern zeigen, die kürzer sind, als sich aus der Gl. (2) ergibt, entsteht eine Überschreitung der Zug-

Abb. 2. Modellversuche mit einer geschichteten Firste. Im Unterteil, in der Mitte und im Oberteil über den Stössen (Auflager) entstehen Risse. Erste Phase der Belastung

Model tests of stratified roof. In the bottom of a central part of excavation and over the sidewalls the developing cracking can be noted. First phase of loading

festigkeit bei Biegung des Gebirges in der Firste und ihr Bruch. Im Unterteil der Firste, in der Mitte und im Oberteil über den nicht nachgiebigen Auflagern entstehen Risse im Gebirge (Abb. 2). Dieses Modellschema wird auch durch Beobachtungen in Gruben bestätigt [1].

Der Zustand einer teilweise zerstörten Firste, die mit Ankern gesichert ist, bedeutet natürlich nicht, daß ein gefährlicher Zustand entstanden ist. Vielmehr hat das Tragvermögen der Konstruktion zugenommen und wird ein Teil der Belastung durch das Gebirge selbst übernommen. Das Gebirge ist jetzt ein Element, das zusammen mit dem Ausbau wirksam geworden ist.

3.2 Ausbau als steifes Gewölbe

Wenn in der verankerten Firste Risse entstehen, so bildet sich im Bereich dieser Zone ein Gewölbe, in dem nur Druck- und Scherspannungen übertragen werden. Die Form der Gewölbewirkung wird durch eine Parabel

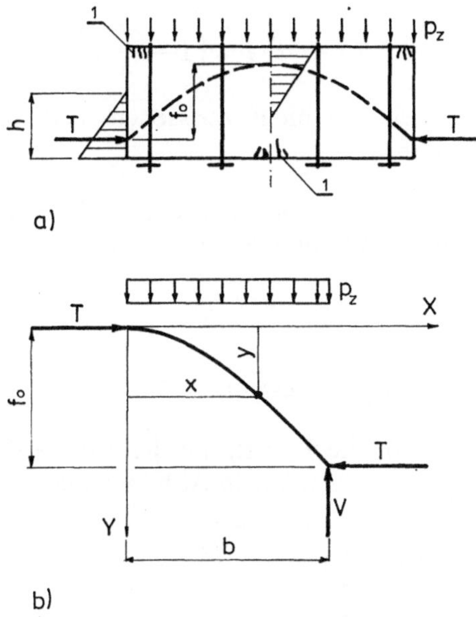

Abb. 3. a) Schema der Bildung des Gewölbes in der verankerten Zone, b) Parameter der Gewölbewirkung

a) Scheme of arch development in the bolted zone, b) Pressure arch parameters

zweiten Grades beschrieben, die sich wie für den Fall eines gleichmäßig verteilten Vertikaldruckes (Abb. 3) errechnet:

$$y = \frac{p_z \cdot x^2}{2\,T}. \tag{8}$$

Andererseits sollte der Seitenschub einer Gewölbewirkung, die keine Momente überträgt, die folgende Abhängigkeit erfüllen:

$$T = \frac{p_z \cdot b^2}{2 f_0} \tag{9a}$$

wobei T = Seitenschub der Gewölbewirkung,
f_0 = Pfeilhöhe des Druckgewölbes.

Man nimmt an, daß in den Kämpfern und im Bogenschluß nur Druckspannungen mit dreieckiger Verteilung auftreten. Dicke der Gewölbewirkung ist ein Teil der verankerten Platte, d. h.:

$$h = \alpha \cdot l. \tag{10}$$

Aus den geometrischen Abhängigkeiten geht hervor, daß die Pfeilhöhe des Druckgewölbes nach der folgenden Gleichung errechnet werden kann:

$$f_0 = l - 2 \cdot \frac{1}{3} \alpha \cdot l = \left(1 - \frac{2}{3} \alpha\right) l, \tag{11}$$

wobei α = Koeffizient, der den nicht zerissenen Teil an den verankerten Balken bestimmt.

Den Seitenschub kann man auch aus der Bedingung der Druckfestigkeit im Grenzzustand der Gewölbe bestimmen, d. h.:

$$T = \frac{\alpha \cdot l}{2} R_c, \tag{9b}$$

wobei R_c = Druckfestigkeit der Gesteine.

Diese Beziehung erlaubt die Ermittlung der Mindesthöhe des verankerten Balkens, d. h. der Länge der Anker in Abhängigkeit vom Koeffizienten α:

$$l = b \sqrt{\frac{p_z}{\left(1 - \frac{2}{3} \alpha\right) \alpha \cdot R_c}}. \tag{12}$$

Nach dem Differenzieren dieser Funktion nach dem unbekannten Koeffizienten α können wir seinen Extremwert errechnen als:

$$\frac{dl}{d\alpha} = -\frac{1}{2} b \sqrt{\frac{p_z}{R_c}} \left(\alpha - \frac{2}{3} \alpha^2\right)^{-\frac{3}{2}} \left(1 - \frac{4}{3} \alpha\right) = 0. \tag{13}$$

Weil $\alpha \neq 0$ und $1 - \frac{2}{3} \alpha \neq 0$, so erhält man:

$$\alpha = \frac{3}{4}. \tag{14}$$

Die zweite Ableitung für den festgestellten Wert des Koeffizienten α ist:

$$\frac{d^2 l}{d\alpha^2} = b\sqrt{\frac{p_z}{6 R_c}}. \tag{15}$$

Das Ergebnis ist größer als Null, was bedeutet, daß die Funktion (12) ihr Minimum erreicht hatte, d. h. für die angegebenen Bedingungen die kürzeste Ankerlänge ergibt:

$$l = 2b\sqrt{\frac{2 p_z}{3 R_c}}. \tag{16}$$

Aus dem Vergleich der Gln. (16) und (2) geht hervor, daß das System mit Gewölbewirkung nur dann günstiger wird, wenn die folgende Ungleichung erfüllt ist:

$$b\sqrt{\frac{3 p_z}{2 R_z}} > 2b\sqrt{\frac{2 p_z}{3 R_c}}. \tag{17a}$$

Dies wird nur möglich bei:

$$R_c > \frac{16}{9} R_z. \tag{17b}$$

Eine Ungleichung dieser Art wird praktisch bei allen ankerbaren Gebirgsarten erfüllt. Die Größe der Ankervorspannung kann man in diesem Falle

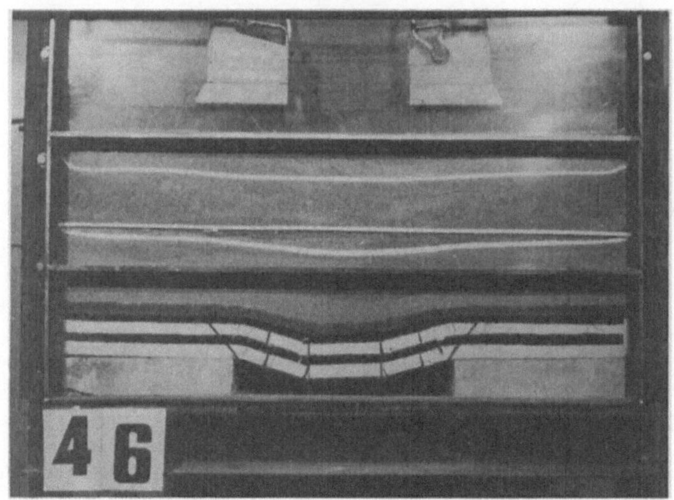

Abb. 4. Letzte Phase der Belastung der Firste bei Modellversuchen. Trotz zahlreichen Rissen ist die Firste nicht zerstört

Final phase of roof loading in model tests. The roof is not failed, although the numerous cracks are seen

aus der Gl. (5) berechnen. Unter der Annahme einer korrigierten Ankerlänge nach Gl. (16) erhält man:

$$N = \frac{\sqrt{6 p_z \cdot R_c}}{2 n K \cdot f} \cdot b. \tag{18}$$

Weitere Modellbeobachtungen, die durch den tatsächlichen Bruchverlauf in den Gruben bestätigt werden (Abb. 4), zeigen, daß die Tragfähigkeit der steifen Gewölbe möglicherweise überschritten wird. Bei unzureichender Tragfähigkeit des Gewölbes tritt Abscherung in den Ecken ein, die Konstruktion wird aber nicht zerstört. Man erkennt deutlich das intensive Zusammenwirken von Gebirge und Ausbau.

4. Nachgiebiger Ankerausbau

Der Abscherung der Firste folgt eine wesentliche Änderung in der Arbeit des Systems Ausbau + Gebirge. Die Verschiebung des verankerten Hangenden in Richtung des Abbauraumes verursacht deren Lockerung und Entspannung auf der ganzen Breite der Firste sogar im Bereich der Abscherungskeile, die in der Nähe der entlasteten Zone entstanden sind. Die Spannungskon-

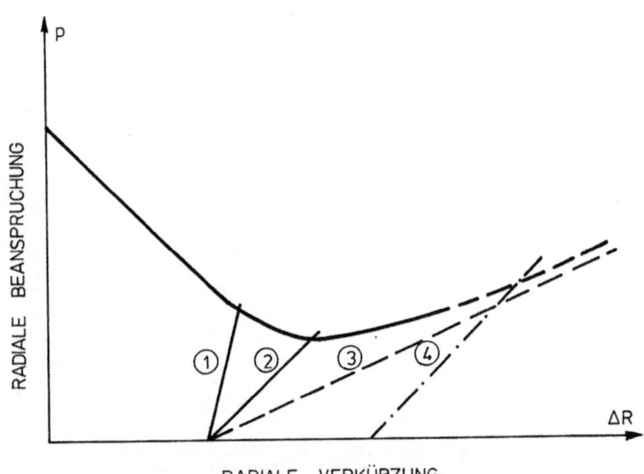

Abb. 5. Entspannung des Gebirges um den Abbauraum herum nach K. S. Lane, 1975
1 — ein zu steifer Ausbau; *2* — ein richtiger Ausbau; *3* — ein zu nachgiebiger Ausbau; *4* — zu spät ausgebauter Hohlraum

Rock mass decompression around the excavation acc. to K. S. Lane, 1975
1 — too stiff support; *2* — proper support, *3* — too flexible support; *4* — too delayed support

zentration verschiebt sich ins Massiv und verteilt sich auf die Auflager. Die Entspannung des Hangenden ist mit dem Abfall der Belastung, die am Anfang auf den verankerten Firstteil wirkte, verbunden. Unter günstigen Bedingungen kann der Zerstörungsprozeß der Firste aufgehalten werden.

Um zu einer Beurteilung des neu entstandenen Systems zu kommen, wird zuerst der Entspannungsvorgang, dann die Tragfähigkeit des Ankerausbaues analysiert.

4.1 Gebirgsdruck auf einen nachgiebigen Ausbau

Der verankerte Teil des Gebirges, der sich wie ein Kolben in der Abbauraum-Richtung verschiebt, ermöglicht eine Entspannung des Hangenden und Abfall des Primärdruckes bis zu einem kritischen Wert. Nach der Überschreitung dieses Wertes entsteht die Zerstörung der Firste und ein Zuwachs des Gebirgsdruckes (Abb. 5). Die Tragfähigkeit des Gebirges wird also am besten genützt, wenn der Druck den kleinsten Wert erreicht. Dieser Wert interessiert am meisten, weil er einer optimalen Belastung des Ausbaues

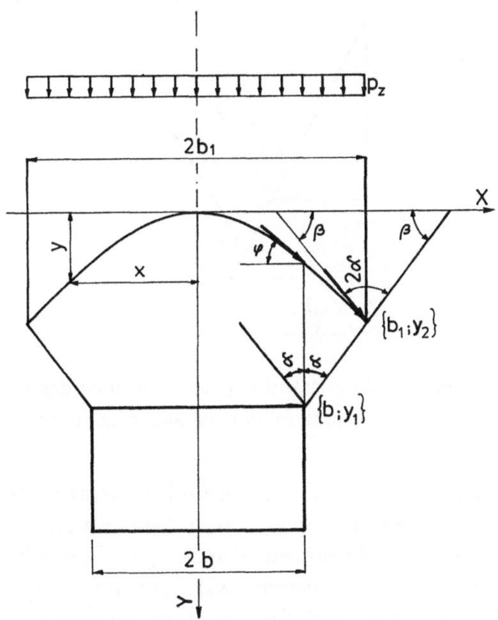

Abb. 6. Spannweite des Gewölbes im Gebirge
Span of pressure arch

entspricht. Wie schon bekannt ist, entsteht das Abscheren des Gebirges, das eine bestimmte Reibung hat, in zwei Richtungen, die von der Horizontalen um den Winkel

$$\pm \beta = \frac{\pi}{4} + \frac{\varphi}{2} \tag{19}$$

geneigt sind, wobei

φ = ein allgemeiner Reibungswinkel, der durch die Gl. (7b) erklärt ist.

Unter der Annahme, daß sich die vertikalen Scherkräfte über den Auflagern (Stößen) konzentrieren, kann man feststellen, daß die Scherfläche im Hangenden in Richtung des anstehenden Gebirges verläuft (Abb. 6). Bei Auflockerung des Gebirges wird im Hangenden ein Gewölbe entstehen.

Für das Gewölbe, das den vertikalen Gebirgsdruck überträgt, entspricht die Drucklinie der Gl. (8). Die Kämpfer dieses Gewölbes stützen sich auf die Gleitflächen der Abscherkeile. Für die Stabilität der Kämpfer auf diesen Flächen ist die Neigung der Resultierenden der Gewölbewirkung entscheidend. Sie soll solche allgemeinen Kräfte bewirken, die einen Gleichgewichtszustand des Systems sichern wie für einen Körper, der sich auf einer schiefen Ebene befindet. In diesem Falle wird sich die Resultierende des Gewölbes mit der Gleitlinie in einem Winkel 2α (Abb. 7) schneiden.

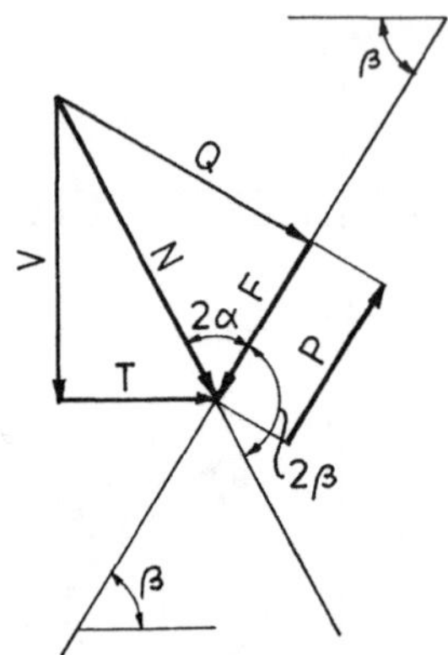

Abb. 7. Gleichgewichtszustand auf der Scherlinie des Gebirges
State of equilibrium in line of shearing in rock

Zweite Bedingung für den Gleichgewichtszustand des Druckgewölbes ist dessen Tragfähigkeit bei der gegebenen Spannweite des Abbauraumes. In diesem Bereich muß das Hangende durch das Gewölbe gestützt werden. Als Tragkraft des Gewölbes können wir die Reibungskraft im Gebirge betrachten, die mit dem Kämpferdruck im Gleichgewicht steht:

$$T \cdot f = p_z \cdot b. \tag{20}$$

Aus den Abhängigkeiten (8) und (20) kann man die Gleichung der Drucklinie berechnen:

$$y = \frac{x^2}{2b} \cdot f. \tag{8a}$$

Die Resultierende des Druckgewölbes wirkt tangential zur Drucklinie. Die Gleichung der Tangente der Drucklinie im Kämpferpunkt lautet:

$$y = \frac{b_1}{b} f (b_1 - x) + y_2. \tag{21}$$

Die Gleitlinie im Hangenden wird durch folgende Gleichung beschrieben:

$$y = -\operatorname{tg} \beta \, (b - x) + y_1. \tag{22}$$

Aus den geometrischen Abhängigkeiten kann man feststellen, daß beide Linien um den Winkel β zur Horizontalen geneigt sind (Abb. 6), d. h.:

$$tg\,\beta = \frac{b_1}{b} \cdot f. \tag{23}$$

Nach der Bestätigung der Abhängigkeit zwischen tg β und f und Einführung der Abkürzung:

$$\lambda = tg\,\beta = \sqrt{f^2+1} + f, \tag{24}$$

kann man aus der Gl. (23) die halbe Breite der Spannweite des Gewölbes bestimmen:

$$b_1 = \frac{b}{f} \cdot \lambda. \tag{25}$$

Die Belastung des Ausbaues wird vom Gebirge im Gewölbekern verursacht.

Aus einer Analyse des Gleichgewichtszustandes in den durch die Gl. (25) bezeichneten Punkten geht hervor, daß die Lage der Kämpferpunkte die Größe der druckentlasteten Zone in der Firste bestimmt. Für die Drucklinien, die sich unterhalb oder oberhalb der Grenzlinie befinden, ist das System unstabil, weil solche Linien der Gl. (8a) widersprechen. Im Falle einer Drucklinie, die sich oberhalb der berechneten Grenzlinie befindet, ist Gleichgewicht möglich, wenn die Gleitlinie vertikal verläuft, also die maximale Breite der druckentlasteten Zone begrenzt. In einem solchen System gibt es in der ganzen Höhe der druckentlasteten Zone oberhalb der Drucklinie einen Grenzgleichgewichtszustand und unterhalb der Drucklinie ein Abgleiten des Gewölbekerns in den Abbauraum. In diesem Falle muß der Gewölbekern durch Anker gehalten werden. Die Größe des Gewölbekerns bestimmt, unter Berücksichtigung der Reibung, die Belastung der Anker. Für einen infinitesimalen Streifen gilt die Gleichgewichtsbedingung

$$2\,b_1\,\gamma\,dz = 2\,b_1\,(p+dp) - 2\,b_1\,p + 2\,pKfdz. \tag{26}$$

Nach der Integration dieser Gleichung und bei Berücksichtigung der Belastung an der Oberfläche erhalten wir, analog zu Terzaghi, eine allgemeine Formel für die Größe des Gebirgsdruckes:

$$p_v = \frac{\gamma\,b\,\lambda}{K\,f^2}\left(1 - e^{-Kf^2\frac{H}{\lambda b}}\right) + q\,e^{-Kf^2\frac{H}{\lambda b}}. \tag{27a}$$

Beim Ankerausbau besitzt der Exponent nach unseren Untersuchungen folgenden Wert:

$$K \cdot f^2\,\frac{H}{\lambda\,b} \geqslant 4, \tag{28}$$

so daß die e-Funktion vernachlässigt werden kann. Der Gebirgsdruck kann dann durch folgende Formel bestimmt werden:

$$p_v = \frac{\gamma \cdot b \cdot \lambda}{K \cdot f^2}. \tag{27b}$$

Ist die Bedingung (7a) erfüllt, so kann angenähert gefolgert werden:

$$\lambda = 2f,$$

und man kann eine vereinfachte Formel für den Gebirgsdruck angeben:

$$p_v = \frac{2 \cdot \gamma \cdot b}{K \cdot f}. \tag{27c}$$

Nach der Ankerung haben wir es also mit zwei Gewölben zu tun: 1. mit einer Gewölbewirkung, welche die maximale Tragfähigkeit des Gebirges

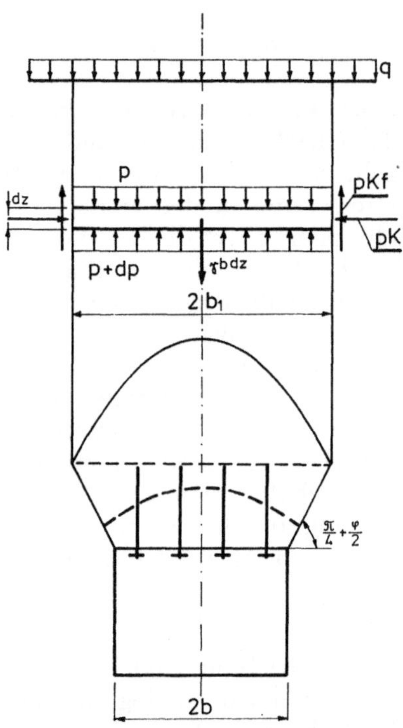

Abb. 8. Gleichgewichtszustand eines infinitesiemalen Streifens in der druckentspannten Zone
State of equilibrium of elementary segment in decompressed zone

angibt; 2. mit einem Stützgewölbe, das sich in der verankerten Zone befindet und dessen Tragfähigkeit von der Scherfestigkeit des Gebirges abhängt (Abb. 8).

4.2 Tragfähigkeit eines nachgiebigen Ankerausbaues

Durch Verankerung der Firste werden die Schichten verbunden und der Verformungsprozeß des Hangenden gestoppt. Diese Erscheinung tritt sehr deutlich dann auf, wenn die Anker vorgespannt sind. Die Vorspannung der Anker vergrößert die Vertikalspannungen in der verankerten Firste.

Die Vertikalspannungen entwickeln Horizontalspannungen, deren Wert sich errechnet nach:

$$p_0 = \frac{n \cdot N}{2b} \cdot K. \tag{29}$$

Diese Spannungen bewirken auf der Gleitfläche eine Reibung, die bei günstigen Bedingungen den vertikalen Gebirgsdruck des entspannten Hangenden ins Gleichgewicht bringt:

$$2\,p_0 \cdot l \cdot f = 2\,b p_v. \tag{30}$$

Aus dieser Abhängigkeit kann man die Grundparameter eines nachgiebigen Ankerausbaues, d. h. Länge, Abstand und Vorspannung der Anker, ermitteln. Unter Benützung der vereinfachten Gl. (27c) für den Gebirgsdruck erhält man:

$$n \cdot N \cdot l = \frac{4 \cdot \gamma \cdot b^3}{K^2 \cdot f^2}. \tag{31}$$

Werden die Anker nicht vorgespannt, so wirkt in der verankerten Zone ein Gebirgsdruck p_v, dessen Resultierende durch folgende Formel beschrieben wird:

$$n \cdot N = 2\,b \cdot \frac{2 \cdot \gamma \cdot b}{K \cdot f}, \tag{32a}$$

was bedeutet, daß die Anker die ganze druckentlastete Zone im Gewölbekern anheften sollen. Ihre Länge muß also sein:

$$l = \frac{b}{K \cdot f} = l_s. \tag{33a}$$

Dies ist genau die Höhe der druckentlasteten Zone, die sich nach der Formel von Terzaghi ergibt. Im Falle $K = 1$ erhalten wir die Pfeilhöhe des Gewölbes nach der Hypothese von Protodiakonov.

Wurden die Anker vor der Entspannung des Hangenden ins Massiv eingebracht, gilt:

$$n \cdot N = 2\,b p_z, \tag{32b}$$

bzw.

$$l = \frac{b \cdot p_v}{K \cdot f p_z} = l_s \frac{p_v}{p_z}. \tag{33b}$$

Die Länge der Anker ist somit direkt proportional zum Wert p_v, d. h. proportional zur Größe der Entspannung des Gebirges über der verankerten Zone. Als l_s haben wir die Länge der Anker beim steifen Ankerausbau bezeichnet. So wie sich der Gebirgsdruck p_v im entspannten Hangenden verkleinert, kann man im gleichen Maße die Anker bei einer ständigen Vorspannung verkürzen. Die optimale Länge erreichen die Anker bei voller Entspannung, die aus der Gl. (31) errechnet werden kann.

4.3 Optimierung der Querschnittsform des Hohlraumes

Für einen Hohlraum mit gewölbter Firste wurde schon anfangs der Seitenschub beachtet, der dem Abscheren des Hangenden einen zusätzlichen Widerstand leistet. Die Gl. (30) wird bei gewölbter Firste folgende Form annehmen:

$$(2\,p_0 \cdot l + T) \cdot f = 2\,b\,p_v. \tag{30a}$$

Ist die Pfeilhöhe f_1 bekannt und wird angenommen, daß das Gewölbe keine Biegemomente überträgt, kann man den Seitenschub nach folgender Formel bestimmen:

$$T = \frac{p_v \cdot b^2}{2\,f_1}. \tag{9c}$$

Führt man diese Abhängigkeit in die Gl. (30a) ein, so kann man schreiben:

$$n \cdot \mathrm{N} \cdot l = \frac{4\,\gamma \cdot b^3}{K^2 \cdot f^2}\left(1 - \frac{b}{4\,f_1}\right). \tag{31a}$$

Aus dem Vergleich der Gln. (31) und (31a) ist leicht festzustellen, daß folgende Abhängigkeit gilt:

$$0 < \left(1 - \frac{b}{4\,f_1}\right) \leqslant 1. \tag{40}$$

Aus ihr geht hervor, daß Gewölbeformen, welche die Bedingung

$$f_1 > \frac{b}{4} \tag{41}$$

erfüllen, verankert werden müssen, weil es eine Möglichkeit der Abscherung des Hangenden in den Kämpfern gibt. Andere Gewölbe mit kleineren Pfeilhöhen sollen als Flachfirste betrachtet werden. Eine ähnliche Analyse der Optimierung der Hohlraumform kann man für die Sohle und die Stöße durchführen. Weitere Forschungen im Sinne der Optimierung des Ankerausbaues müßten sich auf die Größe der Sicherheitskoeffizienten des Ausbaues konzentrieren sowie auf die Notwendigkeit einer zusätzlichen Sicherung des Abbauraumes mit Drahtverzug, Torkret oder Kappen usw. Eine optimale Auswahl des Sicherheitskoeffizienten und einer zusätzlichen Sicherung muß an die Art der verankerten Firste angepaßt werden.

5. Zusätzliche Sicherung der Hohlräume

Die Bruchgefahr ist von der Gebirgsart abhängig, besonders von der Streuung ihrer Parameter. Je größer diese Streuung, umso größer ist die Möglichkeit einer Abweichung von den Mittelwerten, für welche der Ausbau dimensioniert wurde. Weiters ist die Zuverlässigkeit des Ausbaues der Hohlraumfunktion anzupassen. Das heißt, daß Strecken und Stollen, die eine wichtige Rolle spielen, einen größeren Sicherheitsgrad besitzen müssen als Strecken von geringerer Bedeutung. Ihre Zerstörung oder ihr teilweiser

Verbruch verursachen keine großen technischen Schäden oder Störungen im Arbeitsablauf des Grubenbetriebes.

Verschiedene Ansprüche an die Sicherung der Hohlräume und die Einschätzung der Gebirgsparameter erfordern eine umfangreiche Analyse. Auf der dargelegten Basis könnte man optimale Parameter für die zusätzliche Sicherung der Hohlräume festlegen. Solche Analysen wurden mit Hilfe statistischer Methoden durchgeführt. Ihre Ergebnisse kann man als Richtlinie für die Projektierung zusätzlicher Sicherungen verwenden. Es wurden drei Klassifikationstabellen und ein Nomogramm zur Auswahl der Parameter aus diesen Tabellen erarbeitet [3].

Die Gebirgsklassifikation unterscheidet neun Gebirgsarten und berücksichtigt als Parameter Schichtung, Klüftung, Wassergehalt der Gesteine u. a. Die Hohlraumklassifikation wurde auf der Basis möglicher Brucherscheinungen und der entstehenden Gefahren erarbeitet. Es wurden vier Hohlraumtypen unterschieden, deren Bedeutung vom technischen Standpunkt unterschiedlich ist. Als Ergebnis wurde eine Klassifikation ermittelt, nach der diese Hohlraumtypen entsprechend den technischen Anforderungen gesichert werden müssen, um alle gewünschten geotechnischen Bedingungen voll zu erfüllen. Es wurden speziell sechs Klassen festgelegt, in die sich alle Hohlräume einordnen lassen.

Literatur

[1] Bachacou, J., Dudek, J., Raffoux, J.-F.: Contrôle de la stabilité des voies boulonnées. Symposium International de la ISRM "Protection contre l'éboulement des roches". Katowice 1973.

[2] Döring, T.: Geomechanische Probleme der optimalen Dimensionierung unterirdischer Hohlräume. Freiberger Forschungshefte A 545 (1975).

[3] Gałczyński, S., Dudek, J.: Prinzipien zur Klassifikation geankerter Hohlraumfirsten. Neue Bergbautechnik Nr. 2, 1977.

[4] Hugon, A., Costes, A.: Le boulonnage des roches en souterrain. Eyrolles 1959.

[5] Mohr, F.: Gebirgsmechanik. Goslar: H. Hübener Verlag 1963.

[6] Podgórski, K., Podgórski, W.: Obudowa kotwiowa wyrobisk górniczych, Katowice, wyd. Śląsk 1969.

[7] Raffoux, J.-F., Sinou, P., Tincelin, E.: Le boulonnage des voies et des galeries minières. Revue de l'Industrie Minérale Nr. 12 (1970).

[8] Timoshenko, S.: Résistance des matériaux. Paris: Dunod 1968.

Anschrift der Verfasser: Doz. Dr.-Ing. habil. Stefan Gałczyński, Dr.-Ing. Jerzy Dudek, Technische Hochschule Wrocław, Institut für Geotechnik, ul. Wybrzeże Wyspiańskiego 27, 50-370 Wrocław, Polen.

Verbindung [illegible] Schalldämmwerte [illegible] Mauerwerks [illegible].

Verschiedene [illegible] der Lochsteine [illegible] erfordert [illegible] Analyse [illegible] für die [illegible] Schalldämmung der [illegible] als [illegible] für die Projektierung [illegible] verwendet [illegible] als [illegible] Klassifikation [illegible] dieser [illegible] erarbeitet [6].

Die [illegible] Hohlsteine [illegible] [illegible] Rechenmodell [illegible].

Literatur

[1] [illegible]

[2] [illegible]

[3] [illegible]

[4] [illegible]

[5] [illegible]

[6] [illegible]

[7] [illegible] Revue d'Acoustique [illegible]

[8] [illegible]

Anschrift der Verfasser: [illegible] Technische Universität [illegible] Wybrzeże Wyspiańskiego 27, 50-370 Wrocław, [illegible]

Rock Mechanics, Suppl. 7, 267—269 (1978)

Rock Mechanics
Felsmechanik
Mécanique des Roches

Schlußwort

Von

G. Horninger

Meine Herren Vorsitzenden, meine sehr geehrten Herren!

Der Techniker sollte dem Geologen nicht Gelegenheit geben, das letzte Wort zu führen. Tut er es doch, dann darf er sich nicht wundern, wenn letzterer über Adam und Eva zurückgreift und von der Eiszeit redet. Besser gesagt von der Zwischeneiszeit, die nun bei uns bereits wieder gute 10000 Jahre dauert. Die Zwischeneiszeit ist nun dank Professor Müller seit gestern auch bei den Felsmechanikern ausgebrochen. Hoffentlich dauert es recht lange bis zur nächsten großen Eiszeit.

Herr Prof. Müller hat also dafür gesorgt, daß bei diesem Kolloquium zuerst einmal drei Berieselungen mit Geologie auf die Ingenieure niedergegangen sind. Er hat sich aber dann doch darauf besonnen, daß er eigentlich Ingenieur ist und in seinem ausgezeichneten Gespür für das, was der Ingenieur gerne hört, hat er sofort wieder auf einen Kälteeinbruch zurückgesteckt. Er hat da u. a. im Anschluß an eine Feststellung von Herrn Dr. Kurzmann vorgeschlagen, man möge doch die Futterkosten für die Geologen dadurch senken, daß man ihnen den Lohn nach den Enderfolgen zumißt. Wobei aber nicht zu übersehen ist, daß der Enderfolg schließlich auch noch durch das mit beeinflußt wird, was der Bauingenieur aus den Goldstücken, die ihm der Geologe hingelegt hat, gemacht hat; wie er z. B. beim Tunnelvortrieb vorgegangen ist, was alles er dort in Tagschichten — vielleicht auch in Nachtschichten, falls am Tag zu viel aufgepaßt worden ist — getan hat. Das sollte also nicht übersehen werden. Und so möchte ich etwa empfehlen, daß Herr Dr. Kurzmann seine gut gemeinten Vorschläge für die Beurteilung dessen, was an der Vorarbeit positiv und was negativ zu bewerten sei, um zwei Parameter erweitert: er sprach von den präliminierten Kosten und setzte diese einfach in Beziehung zu dem, was das Ganze am Ende gekostet hat. Die beiden Parameter, die noch zuzufügen wären, sind

a) die naturgegebenen größeren oder kleineren Schwierigkeiten, die in einem gegebenen Fall die Voraussage mit Unbestimmtheiten belasten. Es gibt z. B. einerseits über alle Zweifel erhabene Wetterlagen, bei denen man wirklich für die kommende Woche, auch ohne Meteorologe zu sein, voraussagen kann, daß sich nichts ändern wird, und anderseits Wetterlagen, bei denen eben irgendwelche nicht vorausbestimmbare Einflußgrößen die beste Voraussage umhauen können. So ähnlich geht es eben auch den Geologen. Wissen wir, welche Einflußgrößen im gegebenen Fall wirksam werden? Das wäre das eine;

b) als wichtiger Parameter, den Herr Kurzmann als Ingenieur wohl gerne übersehen hat anzuführen, die Frage, welche Geldmittel man denn überhaupt zuerst dem Geologen freigegeben hat, damit er alle jene Untersuchungen machen kann, die im gegebenen Falle notwendig wären, auf daß der Ingenieur dann auf ausreichend fundierte geologische Untersuchungen aufbauen kann. Das läßt sich nicht schematisch in Prozenten oder besser in Promille der Bausumme festlegen, sondern müßte auf das bezogen werden, was im gegebenen Fall eben notwendig ist. Auch für ein unbedeutendes, kleines Werk kann es unter schwierigen Verhältnissen u. U. notwendig sein, von vornherein schon etwas mehr als 2 bis 3‰ der Bausumme für die geologischen Vorarbeiten bereitzulegen.

Nun, Herr Prof. Weiss hat sich ja ohnedies mit Donnerstimme von dieser Klagemauer aus, die er sehr schnell in ein Anklageforum umzufunktionieren verstand, gegen die Auffassungen der Ingenieure gewandt, besser gesagt, zur Wehr gesetzt. Recht hat er gehabt.

Die gestern und heute von den Technikern vorgetragenen Themen waren ausgezeichnet gemengt. Zur Hälfte etwa betrafen sie die Theorie, an der den Geologen nur immer wieder der berühmte „Fels" störte: ein unbestimmtes Etwas mit gut gemeinten Nicht-Eigenschaften oder nur mit solchen Eigenschaften, die die finiten Elemente nicht stören. Ob abgesprochen oder nicht, wurde fast nur oder zumindest überwiegend über Anker und Ankerungen gesprochen. Es gäbe doch auch andere Stoffe, die zum Thema „Ausbau und Ausbauvorgang" gepaßt hätten. Zunächst schien es mir, als ob trotz der vielen Bergbautreibenden, die hier unter uns geweilt haben, das Verhältnis von Bergbau zu Tunnelbau etwas zu wenig berührt worden wäre. Es ist eigentlich gestern nur in einem einzigen Vortrag angeschnitten worden. Heute aber hat Herr Zischinsky sehr verdienstvoll gerade diese sachlichen Unterschiede, deren Feststellung viel zur Klärung beitragen kann, herausgearbeitet.

Zur anderen Hälfte der Vorträge, die der Praxis des Stollen- und Schachtbaues gewidmet waren, mögen Sie verzeihen, wenn ich wieder als Geologe — und Philosoph — sehr subjektiv bloß das herausgreife, was mir persönlich besonders imponiert hat. Es war dies zunächst die neue Methode der TIWAG, einen Schacht nach einer Art Wickelgamaschen-System hinunter auf- oder, wenn man will, abzuspulen; ohne Zweifel eine ausgezeichnete Lösung. Nicht minder interessant fand ich den Vortrag über den Pfaffensteiner Tunnel und den Arlberg-Vortrag.

Herr Reinhardt war heute Nachmittag der einzige Vortragende, der darauf hingewiesen hat, daß das, was beim Kolloquium in bezug auf Ingenieurbaukunst und in bezug auf Zusammenarbeit zwischen Ingenieuren und Geologen vorgebracht worden war, eigentlich zwei Monate später, kurz vor Weihnachten hätte kommen sollen: da war der ganze Christbaum vollgehängt mit dem, was man sich nur an Schönem über Zusammenarbeit wünschen möchte. Herr Reinhardt hat darauf hingewiesen, besser gesagt, hineingestochert, daß es in vielen, vielen Fällen der Praxis eben ganz anders ausschaut. Da ist im vorbereitenden Stadium für das Vorgutachten höchstens das Geld für eine oder gar für zwei Bohrungen da. Wenn dann aus dem Vorgutachten ein Hauptgutachten werden soll, dann bleibt meistens keine Zeit mehr für Erkundungsaufschlüsse. Dann muß letzten Endes der Geologe erst wieder aufgrund der unzureichenden Untersuchungen, die für das Vorgutachten reichen hatten müssen und die natürlich mit geringem Kostenaufwand zustande gekommen sind, alles das orakeln, was der Bauingenieur vom „Haupt"-Gutachten zu erfahren wünscht. Wenn unter solchen Voraussetzungen manches danebengeht, dann soll sich der Bauingenieur nicht wundern.

Darf ich vielleicht noch meine eigene Meinung zu einem Thema, das ein bißchen Funken gegeben hat, beitragen. Da war die Rede vom „Wollen" und „Können". Das Thema wurde heute Vormittag angeschnitten. Gestern gingen zwei Diskussionsbeiträge auf die Frage los, wieviel Firstensetzung man denn in einem Tunnel zulassen dürfe oder könne. Es wurde dabei auf den halben Meter im Arlberg-Tunnel und auch irgendwie auf den ganzen Meter im Schartnerkogel-Tunnel angespielt. Nun, von „zulassen" war da ohnedies nicht mehr die Rede. Ich habe das Gefühl, das war so, wie es uns mit den Terroristen geht: zulassen will sie keiner, sie sind aber leider da. Man muß dann nur schauen, wie man mit ihnen fertig wird. Da könnte man nur wieder an unseren seligen Kaiser Ferdinand denken, der in Wien 1848 die Revolution bestimmt nicht zulassen wollte. Wie man ihm aber gesagt hat, daß die Wiener nun doch revoltieren, hat er die erstaunte Frage gestellt: „Ja derf'ns denn des?" (Ja, dürfen sie denn das?).

Ein großer Trost für uns alle war, daß uns beim Vortrag Prof. Müllers zur Neuen Österreichischen Tunnelbauweise die Glaubensspaltung erspart geblieben ist; und zwar dank Müller. Zuerst hat er ein Feuerchen angezündet und dann hat er es selbst wieder in blendender Manier gedämpft. Wir werden an ihm keinen Muhammed Ali der Felsmechanik bekommen. Ich sage nur: Gott sei Dank!

Anschrift des Verfassers: Prof. Dr. Georg Horninger, Institut für Geologie, Technische Universität Wien, Karlsplatz 13, A-1040 Wien, Österreich.

Geomechanik gebirgsbildender Vorgänge und deren Auswirkungen auf Felsbauten ober und unter Tage
Geomechanics of Orogenetic Events and Their Effects on the Construction of Rock Structures on Subsurface and Underground

Vorträge des Hans-Cloos-Kolloquiums (25. Geomechanik-Kolloquium) der Österreichischen Gesellschaft für Geomechanik. Contributions to the Hans-Cloos-Colloquium (25th Geomechanical Colloquium) of the Austrian Society for Geomechanics. Salzburg, 14. und 15. Oktober 1976

Herausgeber: Österreichische Gesellschaft für Geomechanik, Salzburg

(Rock Mechanics – Felsmechanik – Mécanique des Roches / Supplementum 6)
105 Abbildungen. IV, 194 Seiten. 1978.
Geheftet S 660,—, DM 95,—
Vorzugspreis für Abonnenten von „Rock Mechanics – Felsmechanik – Mécanique des Roches“:
Geheftet S 561,—, DM 81,—

ISBN 3-211-81435-3

Inhaltsverzeichnis: L. Müller-Salzburg: Entwicklungstendenzen in der Geomechanik. — T. E. Gattinger: Aktuelle Krustenbewegungen in den Alpen und ihre Bedeutung für das Baugeschehen. — J. A. Franklin and O. Hungr: Rock Stresses in Canada. Their Relevance to Engineering Projects. — G. Horninger: Berührungspunkte und Reibungsflächen zwischen Geologie und Felsmechanik. — A. E. Scheidegger: Comparative Aspects of the Geotectonic Stress Field. — R. Richter: Über den plastischen Gesteinszustand. — G. Feder: Versuchsergebnisse und analytische Ansätze zum Scherbruchmechanismus im Bereich tiefliegender Tunnel. — G. Seeber: Zur Tunnelberechnung in druckhaftem Gebirge. — D. Kirschke: Die statische Berechnung als Bindeglied zwischen Ingenieurgeologie und Felsbaupraxis. — A. M. Heltzen: Betrachtungen über Gebirgsdruck und dessen Auswirkung auf den Sprengvortrieb von Tunneln und größeren Hohlräumen in Fels. — P. Göbl: Bau- und Vortriebsweise im Gebirge mit tektonischen Restspannungen. — H. Boldt: 6000 m Vollschnittauffahrung im linksrheinischen Steinkohlenbergbau. — P. Bauernfeind, F. Müller und L. Müller-Salzburg: Tunnelbau unter historischen Gebäuden in Nürnberg. — G. Laue, L. Müller-Salzburg und M. Will: Die bergmännische Auffahrung von U-Bahnhöfen unter geringer Überdeckung. — U. Zischinsky: Geologische und betriebliche Einflüsse auf die Nachbrüche in einer vollmechanisch aufgefahrenen Strecke.

Springer-Verlag Wien New York

Neue Erkenntnisse im Hohlraumbau — Fundierungen im Fels
Latest Findings in the Construction of Underground Excavations — Rock Foundations

Vorträge des 24. Geomechanik-Kolloquiums der Österreichischen Gesellschaft für Geomechanik. Contributions to the 24th Geomechanical Colloquium of the Austrian Society for Geomechanics.
Salzburg, 2. und 3. Oktober 1975.

Herausgeber: Österreichische Gesellschaft für Geomechanik, Salzburg

(Rock Mechanics – Felsmechanik – Mécanique des Roches / Supplementum 5)
165 Abbildungen im Text und auf einer Ausschlagtafel.
IV, 262 Seiten. 1976.
Geheftet S 880,—, DM 128,—
Vorzugspreis für Abonnenten von „Rock Mechanics – Felsmechanik – Mécanique des Roches":
Geheftet S 748,—, DM 108,80

ISBN 3-211-81384-5

Inhaltsübersicht: G. Horninger: Geologische Erfahrungen vom Bau der Kavernengaragen Mönchsberg-Nord, Salzburg. — U. Koerner, J. Nečas: Geologische Erkundung, felsmechanische Messungen und statische Bemessung beim Bau der Trinkwasserkaverne Lörrach (Rheintalgrabenrand). — H. Krause: Geologische Erfahrungen beim Einsatz von Tunnelvortriebsmaschinen in Baden-Württemberg. — K. W. John: Felsgründungen von großen Talsperren. Probleme – Lösungen. — H.-U. Werner: Gründungsprobleme beim Bau von Seilschwebebahnen. — H. Pöchhacker: Wirtschaftliche Aspekte des Hohlraumbaues in druckhaftem Gebirge. — D. Prader: Heitersbergtunnel der Schweizerischen Bundesbahnen. Vortrieb mit 10,67 m Durchmesser. Beschreibung und grundsätzliche Überlegungen. — E. H. Weiss: Die baugeologische Prognose für den Schnellstraßentunnel durch den Arlberg, Tirol – Vorarlberg. — M. John: Die geotechnischen Messungen im Arlbergtunnel und deren Auswirkungen auf das Baugeschehen. — B. Sharma: Model Tests for Slope Tunnels in Jointed Rocks. — A. Döllerl: Geotechnische Probleme und ihre bauliche Bewältigung beim U-Bahnbau in Wien. — H. Krimmer: Erfahrungen bei der Weiterentwicklung der Spritzbetonbauweise im Frankfurter U-Bahnbau. — G. Greschik: Geotechnische Probleme beim Bau der U-Bahn in Budapest. — A. Golta: Schwellvorgänge im Planum schweizerischer Bahntunnels. — K. Kuhnhenn, G. Spaun: Schäden an einer Kaverne und Druckrohrleitung durch Felsgleitungen.

Springer-Verlag Wien New York